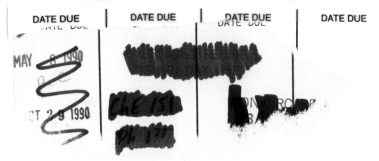

PHYSICAL FLUID DYNAMICS

PHYSICAL FLUID DYNAMICS

Second Edition

D. J. Tritton
Department of Physics, University of Newcastle upon Tyne

CLARENDON PRESS • OXFORD
1988

Oxford University Press, Walton Street, Oxford OX2 6DP

Oxford New York Toronto
Delhi Bombay Calcutta Madras Karachi
Petaling Jaya Singapore Hong Kong Tokyo
Nairobi Dar es Salaam Cape Town
Melbourne Auckland
and associated companies in
Berlin Ibadan

Oxford is a trade mark of Oxford University Press

Published in the United States
by Oxford University Press, New York

First edition published by
Van Nostand Reinhold (UK) 1977
Second edition 1988

British Library Cataloguing-in-Publication Data
Tritton, D. J.
Physical fluid dynamics—2nd ed.
1. Fluid dynamics
I. Title
532'.05   QC151
ISBN 0–19–854489–8
ISBN 0–19–854493–6 Pbk

Library of Congress Cataloging-in-Publication Data
Tritton, D. J.
Physical fluid dynamics/D.J. Tritton.—2nd ed.
p.   cm.
Bibliography: p.   Includes index
1. Fluid dynamics.   I. Title.
QC151.T74   1988   532'.05—dc19   87–34162
ISBN 0–19–854489–8
ISBN 0–19–854493–6 (pbk)

Typeset by The Universities Press (Belfast) Ltd
Printed in Great Britain
at The University Printing House, Oxford
by David Stanford
Printer to the University

PREFACE TO THE SECOND EDITION

I have been gratified by the comment, received from many people, that the first edition was a valuable addition to the range of textbooks on fluid dynamics—mainly because of its distinctive approach. This second edition endeavours to retain that approach, whilst introducing topics that have assumed greater importance since the first edition was prepared.

The principal changes and additions are as follows. The discussion of separation is extended—and given a chapter of its own (Chapter 13)—to include a fuller explanation and to cover all cases, not just boundary layer separation. The treatment of instabilities has been substantially reorganized; the phenomenological description and the explanation of linear stability theory are now more intermingled in a single chapter (Chapter 17). Bénard convection is no longer used as an example in the main discussion of instabilities, but is given a brief introduction in Chapter 4 and a full treatment in Chapter 22; the treatment in the first edition was rather fragmented, and there have been substantial advances in our understanding since that material was written. The discussion is extended in Chapter 23 to cover double diffusive convection. Partly as a consequence of these changes, the material on other convection topics in Chapters 4 and 14 is rearranged—but covers much the same ground as before. In the discussion of the structure of turbulent motion (Chapter 21) the parts concerned with large-scale motions (coherent structures) have been completely rewritten. Chapter 24 discusses modern ideas about 'chaos'—not, strictly speaking a fluid dynamical topic, but certainly one that students of fluid dynamics need to know about.

Outside the main text, the principal change is the addition of hints and answers for the problems. I have mixed feelings about the value of these. The temptation for a student to look at the hint too soon can be irresistable. However, this addition has been urged on me by a number of readers and on balance I now think that they are right.

There are quite a lot of minor changes. I have gone through the whole text, considering whether clarification or updating was needed—and bearing in mind the many helpful comments received from readers. Illustrations have been replaced where more appropriate ones have become available. But much remains unchanged; I hope that the right balance has been struck between the dynamics of a new edition and continuity with the old.

So many people have helped in one way or another that to list them all is impossible and to list some may be arbitrary. However, I have been particularly helped by expert advice on some of the sections in Chapter 26 from: Professor D. Etling of the University of Hannover (Section 26.2); Dr H. Tsoar of the Ben Gurion University of the Negev (26.4); Dr P. A. Davies of the University of Dundee (26.7); Mr R. G. Gawthorpe of the Railway Technical Centre, Derby (26.10); Dr A. D. W. Jones of B.P. Research Centre, Sunbury (26.11); Dr M. E. J. Holwill of King's College London (26.12); and Dr R. P. Clark of the M.R.C. Clinical Research Centre, Harrow (26.13).

I hope that the people who have helped with other parts of the book will accept a blanket, but very sincere, thank you. Many people have provided originals of photographs and diagrams, for which I am most grateful. The originators of all photographs and sets of data are indicated in the captions or by the references. Acknowledgements to bodies who have given permission for the reproduction of copyright material are listed separately.

In the preparation of the book I have been much helped by the technical skills of Mrs D. Cooper, who draughted the new diagrams, Mr J. Hunter and Mr G. Robb and the secretarial skills of Mrs L. Whiteford.

Finally, my thanks go to the staff of the Oxford University Press for their friendly and efficient collaboration.

Newcastle upon Tyne   D. J. T.
August 1987

FROM THE PREFACE TO THE FIRST EDITION†

To classify a book as 'experimental' rather than 'theoretical' or as 'pure' rather than 'applied' is liable to imply unreal distinctions. Nevertheless, some classification is necessary to tell potential readers whether the book is for them. In this spirit, this book may be said to treat fluid dynamics as a branch of physics, rather than as a branch of applied mathematics or of engineering. I have often heard expressions of the need for such a book, and certainly I have felt it in my own teaching.

I have written it primarily for students of physics and of physics-based applied science, although I hope others may find it useful. The book differs from existing 'fundamental' books in placing much greater emphasis on what we know through laboratory experiments and their physical interpretation and less on the mathematical formalism. It differs from existing 'applied' books in that the choice of topics has been made for the insight they give into the behaviour of fluids in motion rather than for their practical importance. There are differences also from many existing books on fluid dynamics in the branches treated, reflecting to some extent shifts of interest in recent years. In particular, geophysical and astrophysical applications have prompted important fundamental developments in topics such as convection, stratified flow, and the dynamics of rotating fluids. These developments have hitherto been reflected in the contents of textbooks only to a limited extent.

Much of the book is based on lectures I have given to final year physics students at the University of Newcastle upon Tyne, though I have substantially expanded parts of the material. I have also been influenced by teaching fluid dynamics both at a more elementary level and also to postgraduate students of geophysics and planetary physics. I have tried to learn which approaches to various topics students find most informative, and I hope this experience has led to improvements in what I have written.

I have had the final year physics students particularly in mind when deciding what background knowledge to assume of the reader. The mathematical methods used should all be familiar to such students (although substantial parts of the book would be intelligible to a reader at

† In this shortened version, the chapter and section numbers refer to the present edition.

an earlier mathematical stage). Also, it is not really paradoxical that a book aimed at physics students should contain less explanation of basic physics (such as related thermodynamics) than some other fluid dynamics texts; knowledge of this can reasonably be assumed.

Nevertheless, I hope that the book will be of value to a variety of other types of reader and I have tried to extend its usefulness wherever this has been possible without distorting the primary aim. Workers in widely various types of applied science need an understanding of the phenomena of fluid motion. I have in mind particularly, but not solely, geo- and planetary physicists (meteorologists and oceanographers as well as students of planetary interiors), of whose needs I probably have the most immediate appreciation. Where fluid dynamical topics are taught as part of a geophysics course, I believe that this should be a suitable textbook. I believe also that postgraduate students and other research workers, faced with a project involving fluid dynamics, should find this book of value. It is not uncommon for otherwise interesting geophysical papers to be marred by the employment of fluid dynamical concepts in a way that shows serious misunderstanding of their significance. But these misunderstandings are not surprising in the absence of convenient sources of information. Such readers should not imagine that there are easy short-cuts to an understanding of the phenomena of fluid motion, but I trust that this book may make the long route a little less arduous.

The choice of topics in a book of this sort must be controversial. The size of the subject necessitates many arbitrary omissions. The reader wishing to discover just what is included and what is omitted may find it useful to refer not only to the contents list but also to Section 1.2. The major limitation is the restriction to incompressible flow. This will, I know, make the book inadequate for some courses. Compressible flow is, however, such a large topic that it really requires a book of its own; just a chapter would perhaps do it more of an injustice than total omission. Even within the limitation to incompressible flow, I am much aware of omissions that some readers will regret. What I hope I have achieved is a text giving students sufficient knowledge of the basic concepts of fluid dynamics and sufficient insight into the consequences of these concepts that they will be able to use other (probably more advanced) sources to obtain information that is omitted here.

The systematic development of the subject occupies Chapters 5 to 24. After the introduction in Chapter 1, Chapters 2 to 4 treat three particular topics in a descriptive (thought hopefully not too superficial) way. These topics have been placed ahead of the systematic treatment for two reasons. Primarily they are intended to give the reader some understanding of the type of phenomena with which one is concerned; it is a long haul through the basic concepts of Chapters 5 to 8 if one does not know what it is all for. Additionally, I hope that Chapters 2 to 4 may make the book more accessible to students at an earlier stage of their studies.

Chapter 25 considers experimental methods; it is convenient to discuss these all together rather than alongside the experimental results. Chapter 26 illustrates applications of fluid dynamics—geophysical, biophysical, technological, and so forth. Its nature is explained more fully in Section 26.1. In a sense, this chapter is included for 'entertainment', but some knowledge of its range of applications should add to the reader's appreciation of the way in which fluid dynamics has developed.

Much of the illustrative material throughout the book has been drawn from research papers. It is a feature of fluid dynamics—probably more than of most branches of physics—that the details of rather simply specified topics are complex and still imperfectly understood. (Others must have shared my experience of having difficulty in convincing a postgraduate student that the topic proposed for him had not been fully elucidated long ago.) It is thus often appropriate to use even for introductory purposes (e.g. Chapters 2 to 4) topics that are still the subject of research. This feature of the subject also makes it easier for a book to serve both as an undergraduate text and as a source of information for research workers.

The figure captions often contain details that students will wish to ignore; these are intended for the more advanced reader who wants to know the particular conditions to which the data refer. However, I hope that these details may sometimes convey to students something of the 'flavour' of experimental fluid dynamics.

The book is more fully referenced than most undergraduate texts. Some of the references indicate sources of material—of illustrations or of ideas. (Sometimes I have thought a reference for a specific point appropriate, because the simplification justifiable for the newcomer to the subject may lead the more experienced reader to ask, 'Just what does he mean by that?') Other references have been included for the reader who is using the book as an information source and wishes to follow up a topic in detail. No attempt at completeness has been made; that would involve far too many references. I have tried to give an appropriate point of entry into the literature of a topic, more often a recent review or significant paper than the pioneering research work. In general, the role of a reference should be apparent from the context. In addition to specific references, there is a bibliography of related books (and films).

The primary purpose of the book being pedagogical, I have compiled a selection of problems for the student to work, although the emphasis on experimentally based information means that not all topics lend themselves readily to problem working. Some of the problems involve bringing together ideas from different chapters, so they are all collected together at the end of the book. The problems are very variable in difficulty and are intended more for selection by the teacher than for use by students on their own. A few of the problems amount to little more than the substitution of numbers into equations; the quantities have been chosen to be physically realistic and it is hoped that students will attempt to visualize the situation rather than just make the substitution mechanically. At the opposite end of the spectrum, a few of the problems will be solvable only by students with knowledge of mathematical methods not used in the text. Some of the problems are based on examination questions set to physics students at the University of Newcastle upon Tyne; I am grateful to the University for permission to include these.

Section 6.2 required specific assistance, Professor S. C. R. Dennis provided numerical details of the velocity field. He kindly recomputed these more accurately and in a more suitable form than in Ref. [140]; these computations were carried out in the Data Handling Division at CERN, Geneva. The subsequent computations leading to Figs. 6.3–6.6 were made in the Computing Laboratory of the University of Newcastle upon Tyne by Mrs S. Hofmann, for whose interest and skill I am most grateful.

ACKNOWLEDGEMENTS

I am grateful to the following bodies for permission to reproduce figures of which they hold the copyright:

Academic Press (Fig. 18.3)
American Geophysical Union (Fig. 23.4)
American Institute of Aeronautics and Astronautics (Fig. 21.1)
American Institute of Physics (Figs. 4.10, 16.9)
American Society of Mechanical Engineers (Fig. 15.1)
Annual Reviews Inc. (Fig. 15.3)
Artemis Press (Fig. 26.6)
Barth, J. A. Verlag (Fig. 18.6)
Basil Blackwell Ltd. (Fig. 26.3)
Cambridge University Press (Figs. 4.3, 4.4, 4.5, 4.9, 15.4, 15.7, 17.12, 18.1, 18.4, 18.5, 21.10, 21.18, 21.19, 21.22, 22.3, 22.4, 22.5, 23.2, 23.3, 23.7, 23.8, 24.10)
Central Electricity Generating Board (Figs. 26.12, 26.13)
Chapman and Hall Ltd. (Fig. 26.18)
Company of Biologists Ltd. (Fig. 26.20)
Her Majesty's Stationery Office, by permission of the Controller (Fig. 19.1)
Hydraulics Research Ltd. (Figs. 26.8 and 26.9)
Indian Association for the Cultivation of Science (Fig. 22.13)
Les Éditions de Physique (Fig. 24.9)
Macmillan Magazines Ltd. (Figs. 23.5, 26.2)
New York Academy of Sciences (Fig. 24.8)
Pergamon Journals Ltd. (Fig. 3.12)
Physical Society of Japan (Figs. 3.4, 12.1, 12.6, 12.7, 17.20)
Royal Meteorological Society (Figs. 16.14, 16.17)
Royal Society (Fig. 17.11)
Sedimentology (Fig. 26.5)
Springer Verlag (Figs. 2.10(a), (b), 3.7, 14.6, 22.6, 22.7, 22.8)
VDI Verlag (Figs. 14.2, 14.5, 22.14)

Those silent waters weave for him
A fluctuant mutable world and dim,
Where wavering masses bulge and gape
Mysterious, and shape to shape
Dies momently through whorl and hollow,
And form and line and solid follow
Solid and line and form to dream
Fantastic down the eternal stream;
An obscure world, a shifting world,
Bulbous, or pulled to thin, or curled,
Or serpentine, or driving arrows,
Or serene slidings, or March narrows.

From 'The Fish' by Rupert Brooke

CONTENTS

1

INTRODUCTION

1.1 Preamble

We know from everyday observation that liquids and gases in motion behave in very varied and often complicated ways. When one observes them in the controlled conditions of a laboratory, one finds that the variety and complexity of flow patterns arise even if the arrangement is quite simple. Fluid dynamics is the study of these phenomena. Its aims are to know what will happen in a given arrangement and to understand why.

One believes that the basic physical laws governing the behaviour of fluids in motion are the usual well-known laws of mechanics—essentially conservation of mass and Newton's laws of motion, together with, for some problems, the laws of thermodynamics. The complexities arise in the consequences of these laws. The way in which observed flow patterns do derive from the governing laws is often by no means apparent. A large theoretical and conceptual structure, built, on the one hand, on the basic laws and, on the other hand, on experimental observation, is needed to make the connection.

In most investigations in fluid mechanics, the physical properties of the fluid, its density, viscosity, compressibility, etc., are supposed known. The study of such properties and the explanation in terms of molecular structure of their values for different fluids do, of course, constitute another important branch of physics. However, the overlap between the two is, in most cases, slight. This is essentially because most flows can be described and understood entirely in macroscopic terms without reference to the molecular structure of the fluid.

Thus, for the most part, fluid dynamical problems are concerned with the behaviour, subject to known laws, of a fluid of specified properties in a specified configuration. One might for example want to know what happens when oil flows through a pipe as the result of a pressure difference between its ends; or when a column of hot air rises above a heat source; or when a solid sphere is moved through a tank of water—the whole of which might be placed on a rotating table. Ideally one would like to be able to solve such problems through an appropriate mathematical formulation of the governing laws; the role of experiment would then be to check that solutions so obtained do correspond to reality. In fact the mathematical difficulties are such that a formal theory often has to be supplemented or replaced by experimental observations and measurements. Even in cases where a fairly full mathematical description of a flow has been obtained, this has often been possible only after experiments have indicated the type of theory needed. The subject involves an interplay between theory and experiment. The proportion each contributes to our understanding of flow behaviour varies greatly from topic to topic. This book is biased towards some of those topics where experimental work has been particularly important.

On the other hand, the book is primarily concerned with 'pure' fluid mechanics rather than with applications. It attempts to develop an understanding of the phenomena of fluid flow by considering simple configurations—simple, that is, in their imposed conditions—rather than more complicated ones that might be of importance in particular applications. This is not to deny the influence of applications on the development of the subject. Fluid dynamics has numerous and important applications to engineering, to geo- and astrophysics, and to biophysics; if this is not evident a glance forward to Chapter 26 will make the point. The topics chosen for investigation, even in fundamental studies, owe much to the applications currently considered significant. For example, it is doubtful whether the intriguing phenomena that arise when the whole body of the fluid is rotating would have received much attention but for the importance of such effects in atmospheric and oceanic motions.

1.2 Scope of book

Fluid dynamics has many facets. It is necessary in any book to make some restrictions to the range of topics considered. The principal restrictions in this book are the following six.

1. The laws of classical mechanics apply throughout. Since other restrictions will limit the flows considered to low speeds, the exclusion of relativistic effects is not significant. The exclusion of quantum mechanical effects just means that we are not dealing with liquid helium.

2. The length scale of the flow is always taken to be large compared with the molecular mean-free-path, so that the fluid can be treated as a continuum. More precise meaning will be given to this statement in Section 5.2. It excludes the flow of gases at very low pressures (rarefied gases) from our considerations.

3. We consider only incompressible flow; that is flow in which the pressure variations do not produce any significant density variations. In isothermal flows this means that the density is a constant; in other flows that it is a function of the temperature alone.

There are two ways in which fulfilment of this condition can come about. The fluid may have such a small compressibility (such a large bulk modulus) that, even if large pressure variations are present, they produce only slight density variations. Or the pressure variations may be sufficiently small that, even if the compressibility is not so small, the density variations are small. Liquid flows can usually be treated as incompressible for the former reason. More surprisingly, gas flows can often be similarly treated for the latter reason—whenever, as we shall see in Section 5.8, the flow speed is everywhere low compared with the speed of sound. Thus, although this restriction excludes the many interesting phenomena of the high-speed flow of gases (compressible flow), it still retains a wide range of important situations in the dynamics of gases as well as of liquids.

4. We consider only Newtonian fluids. This is a statement about the physical properties that affect the stresses developed within a fluid as a result of its motion and thus enter the dynamics of a flow. To see what is meant by a Newtonian fluid we consider a simple configuration, Fig. 1.1; the relationship of this to a general flow configuration, and thus a more rigorous definition of a Newtonian fluid, will be considered in Section 5.6. In Fig. 1.1, all the fluid is moving in the same direction but with a speed that varies in a perpendicular direction; i.e. the only non-zero component of the velocity is the x-component, u, and this is a function, $u(y)$, of the coordinate y. Across any arbitrary plane, perpendicular to y, within the fluid, a stress will act. As drawn in Fig. 1.1, the faster fluid above the plane AB will drag the fluid below forward, and the slower fluid below will drag the fluid above back. Equal and opposite forces will thus act on the fluid above and below as shown (although the arrows are drawn on the sides of the fluid *on* which they act, the lines of action of

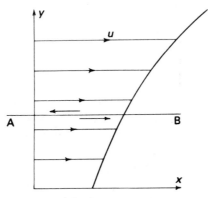

FIG. 1.1 Schematic diagram of viscous stress generated by simple velocity variation, $u(y)$. Short arrows represent forces acting in plane of AB, drawn on side of fluid on which force acts.

both forces are actually in the dividing plane AB). The generation of this internal stress is known as viscous action. In a Newtonian fluid the stress is directly proportional to the velocity gradient; if τ is the force per unit area,

$$\tau = \mu \frac{\partial u}{\partial y} \tag{1.1}$$

where μ is the coefficient of viscosity of the fluid, often called just the viscosity. Another way of stating the definition of a Newtonian fluid is to say that it is a fluid of constant viscosity. Constant here means that it does not depend on the velocity field; the viscosity of a Newtonian fluid may (and usually does) vary with temperature or it may (but usually does not) vary with pressure.

All gases and liquids with small molecules interacting in a simple way are observed to be very closely Newtonian. Non-Newtonian behaviour can arise in liquids with long molecules (polymers), in solutions of polymers, in liquids where the molecules tend to gather in more organized structures than usual, and in suspensions and emulsions (the former name being used when solid material is dispersed through a liquid and the latter when drops of one liquid are dispersed through another). Thus all the common fluids one is likely to use in the laboratory are Newtonian, as are many of the fluids occurring in important applications. However, by restricting ourselves to Newtonian fluids, we are excluding various fluids of importance in biophysics (e.g. blood flowing through small capillaries), industrial applications (e.g. many paints), and chemical engineering.

5. We shall not (with two exceptions, Sections 4.5 and 17.2) consider flows with free surfaces. Topics such as waves on a water surface, flow in open channels, and the dynamics of bubbles and drops are thus outside the scope of the book.

6. We shall not consider any problems in which electromagnetic effects are important, either purely electrohydrodynamic problems (such as the behaviour of a liquid of variable dielectric constant in an electric field) or magnetohydrodynamic problems (such as the flow of an electrically conducting fluid in a magnetic field).

All the above restrictions mean, in varying degrees, that important and interesting topics are being omitted. But the range of phenomena that remains within our scope is wide and varied. We shall be dealing primarily with flows induced by imposed pressure gradients, flows arising from the relative motion of boundaries or of one boundary and ambient fluid, convection—that is, flows induced by or associated with temperature variations—and flows strongly affected by rotation of the whole system or by density stratification. This may not at first sight seem a very broad spectrum of topics, but each has many facets. A major reason for this is the frequent occurrence of instabilities, leading to the breakdown of one type of flow into another. The next three chapters will illustrate the variety and complexity of the phenomena that occur, and illustrate in particular that the complexity can arise even for very simple imposed conditions. We may also remark here that the examples of applications of fluid dynamics in Chapter 26, drawn from widely assorted branches of applied science, have all been chosen as cases that can be understood, at least in part, within the limitations of this book.

Throughout most of this book, except where specific experimental arrangements are concerned, we shall talk about fluids, without making any distinction between liquids and gases. This is because the range of situations within the above limitations is just the range in which liquids and gases exhibit the same phenomena (provided the comparison is made in the correct quantitative way—see Chapter 7); circumstances in which any one of the limitations, except (6), did not apply would normally refer specifically either to a liquid or to a gas. Points (3) and (4) are particularly important. The facts that in many flows gases behave as if they were incompressible and that the same law of viscous behaviour applies to gases and many liquids are central to the development of a common dynamical description for the two phases.

1.3 Notation and definitions

A list of symbols used in this book is given on pages 462–8. This is intended as an aid when a symbol reappears in the text some time after it has been first introduced.

Here we may just note the symbols for the basic fluid dynamical quantities appearing throughout the book. Fluid velocity is denoted by u (with Cartesian components (u, v, w) in directions (x, y, z)), the pressure by p, the temperature by T, fluid density by ρ, the viscosity (already introduced in Section 1.2) by μ, the kinematic viscosity (μ/ρ) by ν, and time by t.

When Cartesian coordinates are used, their choice is made on the following basis. If there is one predominant flow direction, this is the x-direction. For two-dimensional flow (see below) the coordinates are x and y. When gravitational forces are significant, the z-direction is vertically upwards; this choice overrides the two preceding sentences when they would imply a different choice.

Many of the dynamical and physical quantities appearing in fluid dynamics are standard quantities with accepted definitions familiar to a physicist. It has not been thought necessary to define such quantities in the text, but their dimensions are quoted in the list of symbols when this provides a useful reminder of the definition.

It is conventional to use a double letter symbol for most non-dimensional quantities (e.g. Re for Reynolds number, Ra for Rayleigh number). Such symbols will be printed non-italicized to distinguish them from products of two quantities.

Two terms that will be used frequently require definition. Both refer to particular classes of flow that are often considered because of their relative simplicity.

Firstly, a steady flow is one which does not change with time. An observer looking at such a flow at two different instants will see exactly the same flow pattern, although the fluid at each position in this pattern will be different at the two instants. Mathematically (using the Eulerian system to be introduced in Section 5.3), steady flow may be expressed

$$\partial/\partial t = 0 \tag{1.2}$$

where the derivative operates on any parameter associated with the flow. Steady flow can occur only if all the imposed conditions are constant in time. This means that a flow is steady only in the appropriate frame of reference (flow past a fixed obstacle may be steady, but the same situation seen as the obstacle moving through the fluid is not steady, even though the two cases are dynamically equivalent—see Section 3.1). However, one would normally choose that frame for study of the flow. A flow that is changing with time is, of course, called unsteady. An intrinsically unsteady flow is one that is not steady in any frame of reference. Such a flow must occur if there is no frame in which the imposed conditions remain fixed. We shall be seeing that intrinsically unsteady flow also sometimes arises spontaneously even when the imposed conditions are steady.

Secondly, a two-dimensional flow is one in which the motion is confined to parallel planes (the velocity component in the perpendicular direction is zero everywhere) and the flow pattern in every such plane is the same. Formally,

$$w = 0, \quad \partial/\partial z = 0. \tag{1.3}$$

Such a motion may occur in an effectively two-dimensional geometry with the ends in the third direction so distant that they have negligible effect on the flow in the region of interest.

The significance of these concepts will become clearer through specific examples in the following chapters.

2

PIPE AND CHANNEL FLOW

2.1 Introduction

In this and the next two chapters, we take three geometrically simple flow configurations and have a look at the principal flow phenomena. These will provide a more specific introduction than the last chapter to the character of fluid dynamics. We consider these examples now, before starting on the formal development of the subject in Chapter 5; we can then approach the setting up of the equations of motion with an idea of the types of phenomena that one hopes to understand through these equations. Although these chapters are primarily phenomenological, the present chapter will also be used to introduce some simple theoretical ideas.

The first topic is viscous incompressible flow through pipes and channels. Consider a long straight pipe or tube of uniform circular cross-section. One end of this is supplied by a reservoir of fluid maintained at a constant pressure, higher than the constant pressure at the other end. A simple arrangement for doing this in principle, using a liquid as the working fluid, is shown in Fig. 2.1; practical arrangements for investigating the phenomena to be described require some refinement of this arrangement. Fluid is pushed through the pipe from the high pressure end to the low. We suppose that the gravitational force on the fluid is irrelevant, either because the pipe is horizontal or because this force is small compared with the forces associated with the pressure differences. Although there are other experiments that one can do with pipes, this configuration is usually known as pipe flow.

Channel flow is the two-dimensional counterpart of pipe flow. Flow is supposed to occur between two parallel planes close together. The pressure difference is maintained between two opposite sides of the gap. The other two sides must be walled but are supposed to be so far away from the working region that they have no effect there. This is obviously a more difficult arrangement to set up experimentally, and the description of observations will be given in the context of pipe flow. However, a simple piece of theory can be developed about one possible flow pattern, and it is convenient to consider this first for channel flow.

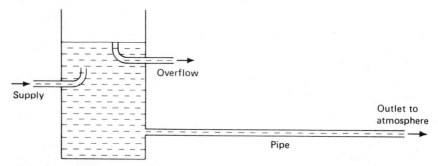

FIG. 2.1 Simple pipe flow: experimental arrangement. The pipe length is reduced in scale.

2.2 Laminar flow theory: channel

Figure 2.2 shows the notation for this theory. The channel width is taken as $2a$ and its length as l; we are supposing that l is much larger compared with a than one can show in the diagram. Pressures p_1 and p_2 $(p_1 > p_2)$ are maintained at the ends of the channel. Coordinates are chosen as shown with the zero of y on the mid-plane of the channel.

It is observed (see Section 5.7) that the fluid immediately next to the walls remains at rest—a fact known as the no-slip condition. The speed, u, with which the fluid moves in the x-direction must thus be a function of y—zero at $y = \pm a$, non-zero elsewhere. This distribution $u(y)$ is known as the velocity profile.

The remoteness of the other walls, in the z-direction, is taken to imply that the flow is two-dimensional (see Section 1.3). We can thus consider all processes to be occurring in the plane of Fig. 2.2.

For the present theory we also suppose that the velocity profile is the same at all distances down the channel; that is at all x. Since the density ρ is taken to be constant as discussed in Section 1.2(3), this is obviously one way of satisfying the requirement that the flow should conserve mass. The same velocity profile transports the same amount of mass per unit time past every station.

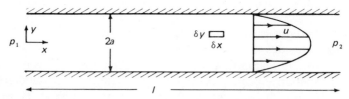

FIG. 2.2 Definition diagram for channel flow. Width of channel is shown exaggerated with respect to length.

The pressure varies with x, obviously, but is constant across the pipe at each x. We shall see below that a pressure gradient in a given direction generates a force in that direction; there is nothing to balance such a force in the y- or z-direction.

We consider now the forces acting on a small element of fluid of sides δx and δy as shown in Fig. 2.2 and side δz in the third direction. There are two processes giving rise to such forces—the action of viscosity as described in Section 1.2(4), and the pressure.

The expression for the viscous force illustrates an important general point that, although the viscous stress depends on the first spatial derivative of the velocity, the viscous force on a fluid element depends on the second derivative. The net force on the element is the small difference of the viscous stresses on either side of it. Figure 2.3 extends Fig. 1.1 to show this. Per unit area perpendicular to the y-direction, forces $\mu(\partial u/\partial y)_{y+\delta y}$ and $-\mu(\partial u/\partial y)_y$ act in the x-direction on the region between planes AB and CD. The net force on our element is

$$\left[\mu\left(\frac{\partial u}{\partial y}\right)_{y+\delta y} - \mu\left(\frac{\partial u}{\partial y}\right)_y\right]\delta x\,\delta z = \frac{\partial}{\partial y}\left(\mu\frac{\partial u}{\partial y}\right)\delta y\,\delta x\,\delta z$$

(when δy is small enough)

$$= \mu\frac{\partial^2 u}{\partial y^2}\,\delta x\,\delta y\,\delta z \text{ (when } \mu \text{ is constant).} \quad (2.1)$$

The viscous force per unit volume is $\mu\,\partial^2 u/\partial y^2$. In the present case, for which u is independent of x and z, this may be written $\mu\,d^2 u/dy^2$. From the general shape of the velocity profile in Fig. 2.2, or from the physical expectation that viscous action will oppose the flow, we anticipate that $d^2 u/dy^2$ will be negative.

The pressure decreases as one goes downstream; there will be slightly different pressure forces acting on the two ends of the element. Since

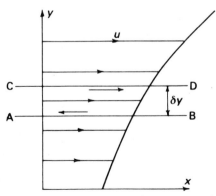

FIG. 2.3 Extension of Fig. 1.1 to show viscous stresses acting on fluid element.

pressure is force per unit area, that at the upstream end acting in the downstream direction on the element is $p_x \, \delta y \, \delta z$ (p_x denoting the value of p at x) and that at the downstream end acting in the upstream direction is

$$p_{x+\delta x} \, \delta y \, \delta z = \left(p_x + \frac{\partial p}{\partial x} \, \delta x \right) \delta y \, \delta z \qquad (2.2)$$

(for small enough δx). The net force in the downstream direction is

$$-\frac{\partial p}{\partial x} \, \delta x \, \delta y \, \delta z \qquad (2.3)$$

or $-\partial p / \partial x$ per unit volume. This will be positive.

Again the partial derivative (which has been written because we shall later be looking at this matter in a more general context) may be replaced by a total derivative, because p varies only with x. Further, the assumption of an unchanging velocity profile makes the dynamical processes the same at all stations downstream; the pressure force per unit volume—i.e. the pressure gradient—must be independent of x. Hence,

$$-\frac{\partial p}{\partial x} = -\frac{dp}{dx} = \frac{p_1 - p_2}{l} = G, \text{ say.} \qquad (2.4)$$

The momentum of the element $\delta x \, \delta y \, \delta z$ is not changing; each fluid particle travels downstream at a constant distance from the centre of the channel and so with a constant speed. Hence, the total force acting must be zero:

$$\mu \frac{\partial^2 u}{\partial y^2} \, \delta x \, \delta y \, \delta z - \frac{\partial p}{\partial x} \, \delta x \, \delta y \, \delta z = 0, \qquad (2.5)$$

that is

$$\mu \frac{d^2 u}{dy^2} = -G. \qquad (2.6)$$

With the boundary conditions

$$u = 0 \quad \text{at} \quad y = \pm a \qquad (2.7)$$

this integrates to give

$$u = \frac{G}{2\mu} (a^2 - y^2). \qquad (2.8)$$

We have ascertained that the velocity profile is a parabola.

The mass of fluid passing through the channel per unit time and per unit length in the z-direction is

$$\int_{-a}^{a} \rho u \, \mathrm{d}y = 2G\rho a^3/3\mu. \tag{2.9}$$

2.3 Laminar flow theory: pipe

The corresponding flow in a pipe is usually known as Poiseuille flow (or sometimes, in the interests of historical accuracy, Hagen–Poiseuille flow). This case is marginally more complicated because of the cylindrical geometry. Figure 2.4 shows a cross-section of the pipe; the flow direction (the x-axis) is normal to the page. The velocity profile now represents the speed as a function of radius, $u(r)$, and we again consider the case when this is independent of x. We consider an element of fluid as shaded in Fig. 2.4 and having length δx in the flow direction. The viscous forces on the two faces of this now differ slightly not only because the velocity gradients differ but also because the two faces have different areas. The force on one face is

$$\mu(\partial u/\partial r)r \, \delta\phi \, \delta x \tag{2.10}$$

and, by an argument parallel to that in Section 2.2, the net viscous force on the element is

$$\mu \frac{\partial}{\partial r}\left(r \frac{\partial u}{\partial r}\right) \delta r \, \delta x \, \delta\phi. \tag{2.11}$$

The pressure force on one end of the element is

$$pr \, \delta\phi \, \delta r \tag{2.12}$$

and the net pressure force

$$-(\partial p/\partial x)r \, \delta x \, \delta\phi \, \delta r. \tag{2.13}$$

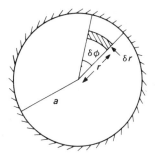

FIG. 2.4 Definition sketch for pipe flow.

Arguing in the same way as for channel flow we get

$$\mu \frac{\mathrm{d}}{\mathrm{d}r}\left(r\frac{\mathrm{d}u}{\mathrm{d}r}\right) = -Gr. \qquad (2.14)$$

Integration gives

$$u = -\frac{Gr^2}{4\mu} + A\ln r + B. \qquad (2.15)$$

A must be zero for the velocity not to become infinite on the axis and B can be evaluated from the fact that

$$u = 0 \quad \text{at} \quad r = a \qquad (2.16)$$

giving

$$u = \frac{G}{4\mu}(a^2 - r^2). \qquad (2.17)$$

The velocity profile is a paraboloid with a maximum speed

$$u_{\max} = \frac{Ga^2}{4\mu}. \qquad (2.18)$$

The mass per unit time, or mass flux, passing through the pipe is

$$\int_0^a \rho 2\pi r u \, \mathrm{d}r = \frac{\pi\rho Ga^4}{8\mu} = \frac{\pi\rho(p_1 - p_2)a^4}{8\mu l}. \qquad (2.19)$$

This is a quantity of some importance as it can readily be measured. Agreement with observation (under circumstances to be delimited below) provides an important check of the validity of underlying hypotheses, such as the no-slip condition and the applicability of continuum mechanics (see Chapter 5). Alternatively, an unknown viscosity can be determined from the rate of flow of the fluid through a tube under a known pressure gradient. This is the principle of one important type of viscometer.

An average speed can be defined as the mass flux divided by the density and cross-sectional area

$$u_{\mathrm{av}} = \frac{Ga^2}{8\mu}. \qquad (2.20)$$

2.4 The Reynolds number

We have determined above one theoretically possible flow behaviour. It is not the only possibility. Sometimes the actual flow behaviour corresponds to the theory, sometimes not. In order to specify the circumstances in which the different types of flow occur, we need to introduce the concept of the Reynolds number.

There are several types of variable associated with the pipe flow configuration: the dimensions of the pipe, the rate of flow, the physical properties of the fluid. For present purposes, we will suppose that the situation is fully specified if we know:

$d(=2a)$, the pipe diameter

u_{av}, the average flow speed

ρ, the fluid density

μ, the viscosity.

Two omissions from this list require some comment. The pipe length is not included on the supposition that, provided the pipe is long enough, the type of flow is determined before the downstream end can have any influence. The pressure gradient is not included because it cannot be varied independently of the above parameters. To produce the same average flow-speed of the same fluid through the same pipe will require the same imposed pressure gradient—whether or not the Poiseuille flow relationship applies. Hence, the pressure gradient need not be included in the specification of the situation. It is arguable that one should include the pressure gradient and omit the average speed, since the former is the controlled variable in most experiments. In Section 18.3 we shall see that there is one case in which this would certainly be the better procedure. However, this has not been conventional practice, and to adopt it would make for confusion. We notice that u_{av} relates directly to the total rate of flow through the pipe and must be the same at every station along the length—unlike u_{max} and other speeds one might define that depend on the detailed velocity profile.

We are now concerned with the question: what type of flow occurs for given values of d, u_{av}, ρ, and μ? But these four parameters are dimensional quantities, whereas the concept of a 'type of flow' does not have dimensions associated with it. Just as it is meaningless to write down an equation, $A = B$, with A and B of different dimensions, so it is meaningless to associate a type of flow with certain values of any dimensional quantity. For example, one would not expect a particular type of flow to occur over the same range of u_{av} for pipes of different diameter or for different fluids. The values of u_{av} specifying the range must be expressible (in principle, whether or not in practice) in terms of the things that determine it. There must be some (known or unknown) expression for it, and this expression must bring in other dimensional quantities in order to be dimensionally satisfactory. Thus we look for the factors that determine the type of flow in terms of dimensionless, not

dimensional, parameters. Such dimensionless parameters must be provided as combinations of the specifying dimensional parameters. (We
have arrived by a plausibility argument at a conclusion that we shall be
examining more systematically in Chapter 7.)

There is one dimensionless combination of the four parameters for the
present configuration:

$$\text{Re} = \frac{\rho u_{av} d}{\mu}. \tag{2.21}$$

This is known as the Reynolds number of pipe flow.

It is one example of the general definition of the Reynolds number as

$$\text{Re} = \frac{\rho UL}{\mu} = \frac{UL}{\nu} \tag{2.22}$$

where U and L are velocity and length scales; that is, typical measures of
how fast the fluid is moving and the size of the system. As the book
proceeds, we shall be seeing the relevance of forms of Reynolds number
to a variety of situations.

Hence, it is appropriate to discuss the different types of flow in a pipe
in terms of different ranges of Reynolds number.

2.5 The entry length

When the Reynolds number is less than about 30, the Poiseuille flow
theory always provides an accurate description of the flow. In fact eqn
(2.19) was first established empirically by Hagen and Poiseuille and the
theory was given later by Stokes—a somewhat surprising historical fact if
one knows how easy it is to fail to verify the theory as a result of having
the Reynolds number too high!

At higher Reynolds numbers, the Poiseuille flow theory applies only
after some distance down the pipe. The fluid is unlikely to enter the pipe
with the appropriate parabolic velocity profile. Consequently, there is an
entry length in which the flow is tending towards the parabolic profile. At
low Reynolds number, this is so short that it can be ignored. But it is
found both experimentally and theoretically that, as the Reynolds
number is increased, this is no longer true. The details of the entry length
depend, of course, on the actual velocity profile at entry, which in turn
depends on the detailed geometry of the reservoir and its connection to
the pipe. However, an important case is that in which the fluid enters
with a uniform speed over the whole cross-section. Because of the no-slip
condition, the fluid next to the wall must immediately be slowed down.
This retardation spreads inward, whilst fluid at the centre must move

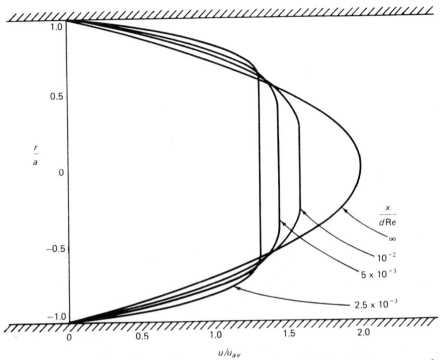

FIG. 2.5 Laminar velocity profiles in pipe entry length. Based on average of various experimental and theoretical profiles, as collected together in Ref. [347].

faster, so that the average speed remains the same and mass is conserved. One thus gets a sequence of velocity profiles as shown in Fig. 2.5. Ultimately, the parabolic profile is approached and from there onwards the Poiseuille flow theory applies.

We can use this case to illustrate the dependence of the extent of the entry length on the Reynolds number. Defining X as the distance downstream from the entry at which u_{max} is within 5 per cent of its Poiseuille value [347],

$$\frac{X}{d} \approx \frac{Re}{30}.$$

(2.23)

This means, for example, that for flow at a Reynolds number of 10^4 (chosen as a high value at which this type of flow can occur) in a pipe of diameter 3 cm the entry length is 10 m. Evidently there will be many practical situations in which the Poiseuille flow pattern is never reached. Even if it is reached, it will often not occupy a sufficiently large fraction of the length for eqn (2.19) to be a good approximation to the relationship between pressure drop and mass flux.

Incidentally, the length at the other end over which the presence of the outlet has an effect is always relatively small.

2.6 Transition to turbulent flow

The above is an important, but perhaps rather uninteresting, limitation to the occurrence of Poiseuille flow. As the Reynolds number is increased further, there is a much more dramatic change in the flow. It undergoes transition to the type of motion known as turbulent, the flows considered so far being called laminar.

In laminar pipe flow the speed at a fixed position is always the same. Each element of fluid travels smoothly along a simple well-defined path. (In Poiseuille flow, this is a straight line at a constant distance from the axis; in the entry length it is a smooth curve.) Each element starting at the same place (at different times) follows the same path.

When the flow becomes turbulent, none of these features is retained. The flow develops a highly random character with rapid irregular fluctuations of velocity in both space and time. An example of the way in which (one component of) the velocity at a fixed position fluctuates is shown in Fig. 2.6. An element of fluid now follows a highly irregular distorted path. Different elements starting at the same place follow different paths, since the pattern of irregularities is changing all the time. These variations of the flow in time arise spontaneously, although the imposed conditions are all held steady.

Figure 2.7 shows the effect of the changeover from laminar to turbulent flow on dye introduced continuously near the entry of a pipe in a streak thin compared with the radius of the pipe. (This arrangement is, in its essentials, the one used by Reynolds [318] in the experiments that may be regarded as the genesis of systematic studies of transition to turbulence—although there had been some earlier work by Hagen.) When the flow is laminar, the dye just travels down the pipe in a straight or almost straight line as shown in the upper picture. As the flow rate is increased, thus increasing the Reynolds number, the pattern changes as

Fig. 2.6 Example of velocity variations in turbulent flow.

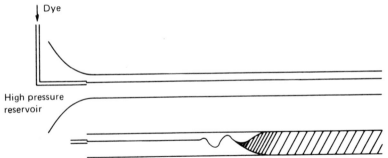

FIG. 2.7 Dye streaks in laminar and turbulent pipe flow. (Pipe length compressed relative to other dimensions.)

shown in the lower picture. The dye streak initially travels down the pipe in the same way as before, but, after some distance, it wavers and then suddenly the dye appears diffused over the whole cross-section of the pipe. The motion has become turbulent and the rapid fluctuations have mixed the dye up with the undyed fluid.

The distance fluids travels downstream before becoming turbulent varies with time, and, at any instant, there can be a laminar region downstream of a turbulent region. This comes about in the following way. The turbulence is generated initially over a small region. This region is actually localized radially (close to the wall) and azimuthally as well as axially. However, it quickly spreads over a cross-section of the pipe, and there is then a short length of turbulent flow with laminar regions both upstream and downstream of it (Fig. 2.8). This short length is known as a

FIG. 2.8 Photo sequence showing passage of turbulent slug past fixed observation point, similar to those in Refs. [254, 256]. Flow is downwards; time increases from left to right. Slug enters field of view at top of first frame (which also shows end of earlier slug at bottom) and leaves it at bottom of seventh frame. Flow visualization by addition of small amount of birefringent material—refractive index for polarized light is affected by shear (Refs. [36, 253]).

FIG. 2.9   Growth and transport of turbulent slug. Shaded regions are turbulent, unshaded laminar. The mean fluid speed is approximately midway between the speeds of the front and rear of the slug.

turbulent slug, a name that has superseded an earlier one of turbulent plug. (We postpone till Section 18.3 further consideration of the origin of turbulent slugs.) The turbulence then spreads—the laminar fluid next to each end of the turbulent slug is brought into turbulent motion—and the slug gets longer. The fluid is, of course, meanwhile travelling down the pipe. Hence, the development of a slug is as shown in Fig. 2.9; the shapes of the interfaces are indicated roughly [376, 414]. As the slug grows the interfaces soon occupy a short length of the pipe compared with the slug itself, and so laminar and turbulent regions are well demarcated.

After a while another slug is born in a similar way. By this time the previous slug has moved off downstream so there is again laminar fluid downstream of the new slug. The slugs sometimes originate randomly in time, sometimes periodically; the circumstances in which the two cases occur will be discussed further in Section 18.3. When the front interface

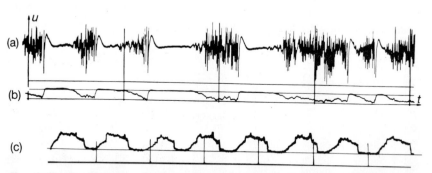

FIG. 2.10   Oscillograms of velocity fluctuations at the centre of a pipe. Traces (a) and (b) (from Ref. [327]) show random slug production for Re = 2550, $l/d = 322$; trace (a) was given by a.c. amplification of the signal and shows the velocity fluctuations in the slugs; trace (b) was given by d.c. amplification of the same signal and shows principally the local mean velocity change between laminar and turbulent flow. Trace (c) (from Ref. [296]) is the counterpart of trace (b) for periodic slug production for Re = 5000, $l/d = 290$. Note: velocity increases upwards in traces (a) and (b) but decreases upwards in trace (c).

of one slug meets the rear interface of another, as a result of their growth, the two simply merge to give a single longer slug.

As a result of these processes, a sensor at a fixed point in the pipe observes alternately laminar and turbulent motion. Figure 2.10 shows oscillograms of velocity fluctuations arising in this way, for random and for periodic slug production. The abrupt changes between laminar and turbulent motion illustrate again the sharpness of the interfaces.

The fraction of the time that the motion is turbulent—known as the intermittency factor—increases with distance downstream as a result of the growth of the slugs. Far enough downstream, in a long enough pipe, the laminar regions have all been absorbed and the flow is fully turbulent.

From the considerations earlier in this section, one would expect that there should be a critical value of the Reynolds number, below which the flow is wholly laminar, above which transition to turbulence occurs. In fact, the situation is more complicated than that. The transition process is extremely sensitive to the detailed geometry of the entry from the reservoir and to the level of small disturbances in the incoming fluid. As a result, transition has been observed to start at values of the Reynolds number ranging from 2×10^3 to 10^5. (This will be discussed further in Section 18.3.) The implication of this variation is not that the reasoning establishing the role of the Reynolds number was erroneous, but that dimensional parameters other than the four listed (d, u_{av}, ρ, and μ) are relevant to transition.

2.7 Relationship between flow rate and pressure gradient

The pressure difference needed to produce a given flow rate through a pipe is larger when the flow is turbulent than when it is laminar. This is shown in Fig. 2.11. Since the abscissa is the Reynolds number, it is appropriate that the ordinate should be a non-dimensional form of the average pressure gradient $(p_1 - p_2)/l$, chosen† here as $(p_1 - p_2)d^3\rho/\mu^2 l$. The dotted lines in Fig. 2.11 show this parameter plotted (logarithmically) against Reynolds number for Poiseuille flow and for wholly turbulent flow. The continuous line shows the behaviour for an actual case. At low Reynolds number it follows the Poiseuille flow line; as transition starts it rises above this, and, ultimately, when transition is

---

† A more conventional choice would be $(p_1 - p_2)d/\rho u_{av}^2 l$ (the above parameter divided by $(Re)^2$). However, because it does not contain u_{av}, our choice is more convenient for certain considerations in Section 18.3. It also shows the behaviour in a particular pipe more directly.

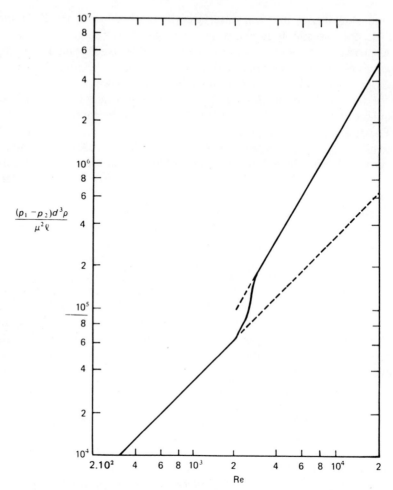

$$\frac{(p_1 - p_2)d^3\rho}{\mu^2 \ell}$$

Fig. 2.11 Variation of non-dimensional average pressure gradient with Reynolds number for pipe flow. Dotted lines: wholly laminar flow and wholly turbulent flow. Full line: example of actual case. Based on information in Refs. [44, 327].

complete in a short fraction of the total length, approaches the turbulent flow line. (The case chosen shows transition at a relatively low Reynolds number. With higher Reynolds number transition in a pipe of practicable length, there is likely to be some departure from the Poiseuille law before transition because of the importance of the entry length.)

3

FLOW PAST A CIRCULAR CYLINDER

3.1 Introduction

Relative motion between some object and a fluid is a common occurrence. Obvious examples are the motion of an aeroplane and of a submarine and the wind blowing past a structure such as a tall building or a bridge. Practical situations are, however, usually geometrically complicated. Here we wish to see the complexities of the flow that can arise even without geometrical complexity. We therefore choose a very simple geometrical arrangement, and one about which there is a lot of information available.

This is the two-dimensional flow past a circular cylinder. A cylinder of diameter d is placed with its axis normal to a flow of free stream speed u_0; that means that u_0 is the speed that would exist everywhere if the cylinder were absent and that still exists far away from the cylinder. The cylinder is so long compared with d that its ends have no effect; we can then think of it as an infinite cylinder with the same behaviour occurring in every plane normal to the axis. Also, the other boundaries to the flow (e.g. the walls of a wind-tunnel in which the cylinder is placed) are so far away that they have no effect.

An entirely equivalent situation exists when a cylinder is drawn perpendicularly to its axis through a fluid otherwise at rest. The only difference between the two situations is in the frame of reference from which the flow is being observed. (Aspects of relativity theory are already present in Newtonian mechanics.) The velocity at each point in one frame is given by the vectorial addition of u_0 onto the velocity at the geometrically similar point in the other frame (Fig. 3.1). This transformation does not change the accelerations involved or the velocity gradients giving rise to viscous forces; it thus has no effect on the dynamics of the situation. (These remarks apply only when u_0 is constant; if it is changing then one does have to distinguish between the cylinder accelerating and the fluid accelerating.)

It is, however, convenient to use a particular frame of reference for describing the flow. Except where otherwise stated, the following description will use the frame of reference in which the cylinder is at rest.

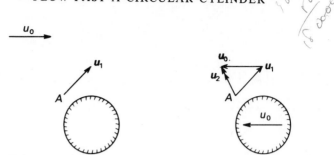

FIG. 3.1   Velocity vectors u_1 and u_2 at point A in two frames of reference.

3.2 The Reynolds number

One can have various values of d and u_0 and of the density, ρ, and viscosity, μ, of the fluid. For reasons like those applying to pipe flow (Section 2.4) the important parameter is the Reynolds number

$$\mathrm{Re} = \frac{\rho u_0 d}{\mu}. \tag{3.1}$$

This is the only dimensionless combination and one expects (and finds) that the flow pattern will be the same when Re is the same. We thus consider the sequence of changes that occurs to the flow pattern as Re is changed. We are concerned with a very wide range of Re. In practice this means that both u_0 and d have to be varied to make observations of the full range. For example, $\mathrm{Re} = 10^{-1}$ corresponds in air to a diameter of $10\ \mu\mathrm{m}$ with a speed of $0.15\ \mathrm{m\ s}^{-1}$ (or in glycerine to a diameter of $10\ \mathrm{mm}$ with a speed of $10\ \mathrm{mm\ s}^{-1}$); $\mathrm{Re} = 10^6$ corresponds in air to a diameter of $0.3\ \mathrm{m}$ with a speed of $50\ \mathrm{m\ s}^{-1}$. Thus experiments have been done with cylinders ranging from fine fibres to ones that can be used only in the largest wind-tunnels.

3.3 Flow patterns

The following description of the flow patterns is based almost entirely on experimental observations. Only for the lowest Reynolds numbers can the flow as a whole be determined analytically (Section 9.5), although there are theoretical treatments of aspects of the flow at other Reynolds numbers. Some of the flow patterns have also been studied computationally.

    Figure 3.2 shows the flow when $\mathrm{Re} \ll 1$. The lines indicate the paths of elements of fluid. The flow shows no unexpected properties, but two points are worth noting for comparison with higher values of the

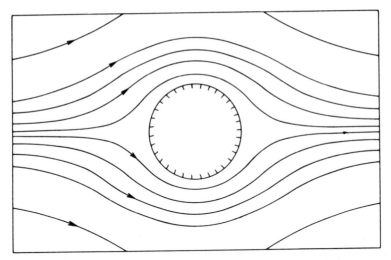

FIG. 3.2   Low Reynolds number flow past a circular cylinder.

Reynolds number. Firstly, the flow is symmetrical upstream and down-stream; the right-hand half of Fig. 3.2 is the mirror image of the left-hand half. Secondly, the presence of the cylinder has an effect over large distances; even many diameters to one side the velocity is appreciably different from u_0.

As Re is increased the upstream–downstream symmetry disappears. The particle paths are displaced by the cylinder for a larger distance behind it than in front of it. When Re exceeds about 4, this leads to the feature shown in the computed flow pattern of Fig. 3.3. Fluid that comes round the cylinder close to it moves away from it before reaching the rear point of symmetry. As a result, two 'attached eddies' exist behind the cylinder; the fluid in these circulates continuously, not moving off downstream. These eddies get bigger with increasing Re [131]; Fig. 3.4 shows a photograph of the flow at Re ≈ 40 just before the next flow development takes place. (For further computed flow patterns in this regime see Section 6.2.)

The tendency for the most striking flow features to occur downstream of the cylinder becomes even more marked as one goes to higher Reynolds numbers. This region is called the wake of the cylinder. For Re > 40 the flow in the wake becomes unsteady. As with transition to turbulence in a pipe (Section 2.6), this unsteadiness arises spontaneously even though all the imposed conditions are being held steady.

Figure 3.5 shows a sequence of patterns in a cylinder wake, produced by dye emitted through a small hole at the rear of the cylinder. The instability develops to give the flow pattern, known as a Kármán vortex

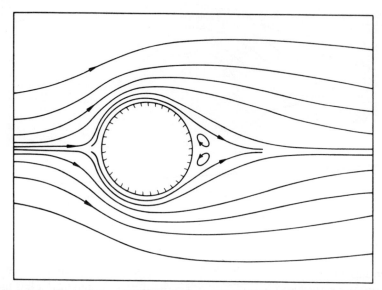

FIG. 3.3   Flow past a circular cylinder at Re = 10 (computed: Ref. [140]).

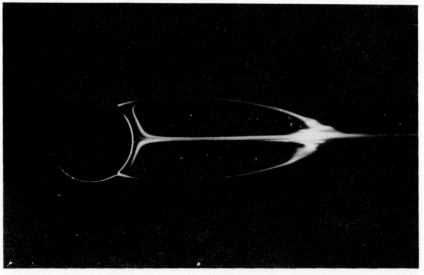

FIG. 3.4   Attached eddies on circular cylinder at Re = 41, exhibited by coating cylinder with dye (condensed milk). Ref. [364].

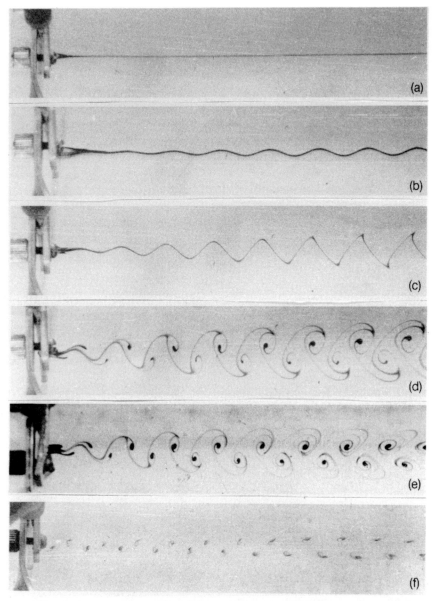

FIG. 3.5 Cylinder wakes exhibited by dye introduced through hole in cylinder. (Cylinder is obscured by bracket but can be seen obliquely, particularly in (a) and (f).) (a) Re ≈ 30; (b) Re ≈ 40; (c) Re = 47; (d) Re = 55; (e) Re = 67; (f) Re = 100.

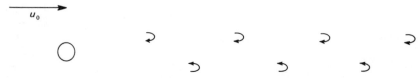

FIG. 3.6   Relative positions of vortices in Kármán vortex street.

street, shown schematically in Fig. 3.6. Concentrated regions of rapidly rotating fluid—more precisely regions of locally high vorticity (a term to be defined in Section 6.4)—form two rows on either side of the wake. All the vortices on one side rotate in the same sense, those on opposite sides in opposite senses. Longitudinally, the vortices on one side are midway between those on the other.

The whole pattern of vortices travels downstream, but with a speed rather smaller than u_0. This means that for the other frame of reference, a cylinder pulled through stationary fluid, the vortices slowly follow the cylinder. Figure 3.7 shows a photograph taken with this arrangement. A moderately long exposure shows the motion of individual particles in the fluid. Because the vortices are moving only slowly relative to the camera, the circular motion associated with them is shown well.

Vortex streets arise rather commonly in flow past obstacles (e.g. see also Sections 15.1, 17.8, 26.2, 26.8). Their basic cause is flow instability;

FIG. 3.7   Vortex street exhibited by motion of particles floating on water surface, through which cylinder has been drawn; Re = 200. From Ref. [380].

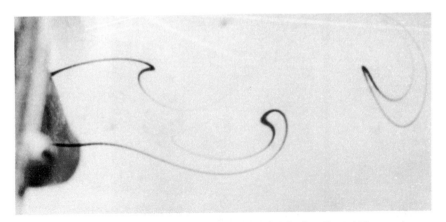

FIG. 3.8   Motion immediately behind cylinder; Re = 110.

the dynamics of this will be considered in Section 17.8. The process by which a vortex street is formed is often called 'eddy shedding', but this name may not always be appropriate. In the case of flow past a circular cylinder, it is applicable when the Reynolds number exceeds about 100 (Section 17.8). Then the attached eddies are periodically shed from the cylinder to form the vortices of the street. Whilst the eddy on one side is being shed, that on the other side is re-forming. Figure 3.8 shows a close-up of the region immediately behind the cylinder, with the same dye system as in Fig. 3.5, during this process.

Figure 3.9 shows a side-view of a vortex street, again shown by dye released at the cylinder. Sometimes the vortices are straight and closely parallel to the cylinder (indicating that the shedding occurs in phase all along the cylinder); sometimes they are straight but inclined to the cylinder, as in the top part of Fig. 3.9 (indicating a linear variation in the phase of shedding); and sometimes they are curved, as in the bottom part of Fig. 3.9 (indicating a more complicated phase variation). The exact behaviour is very sensitive to disturbances and so it is difficult to predict just what will occur at any given Reynolds number. One can, however, say that the lower the Reynolds number the greater is the likelihood of straight, parallel vortices.

It should be emphasized that in Figs. 3.5, 3.8, and 3.9 the introduction of dye was continuous. The gathering into discrete regions is entirely a function of the flow.

Figure 3.10 shows a sequence of oscillograms at two points fixed relative to a cylinder in a wind-tunnel as the main flow velocity is increased. In Fig. 3.10(a) and (b) it can be seen that the passage of the vortex street past the point of observation produces an almost sinusoidal variation. The frequency of this, n, is usually specified in terms of the

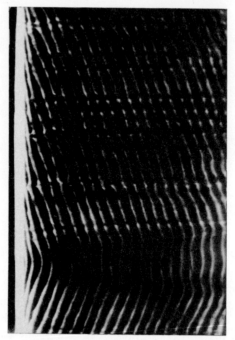

FIG. 3.9   Side-view of vortex street behind cylinder; Re = 150. From Ref. [89].

non-dimensional parameter

$$St = nd/u_0 \tag{3.2}$$

known as the Strouhal number. St is a function of Re (although a sufficiently slowly varying one that it may be said that St is typically 0.2).

The remainder of Fig. 3.10 illustrates the most important aspect of what happens as the Reynolds number is further increased. The complete regularity of the fluctuations is lost. Comparison of the oscillograms at the same Reynolds number shows that the irregularities become more marked as one goes downstream. Further instabilities are leading to the breakdown of the vortex street, producing ultimately a turbulent wake such as that shown in Fig. 3.11. The term turbulent implies the existence of highly irregular rapid velocity fluctuations, in the same way as in Section 2.6. The turbulence is confined to the long narrow wake region downstream of the cylinder. The character of a turbulent wake will be discussed in more detail in Sections 21.3 and 21.4.

The transition to turbulent motion is a consequence of further instabilities [95], which will be considered more fully in Section 17.8. In summary, there are two important types of instability: one occurring

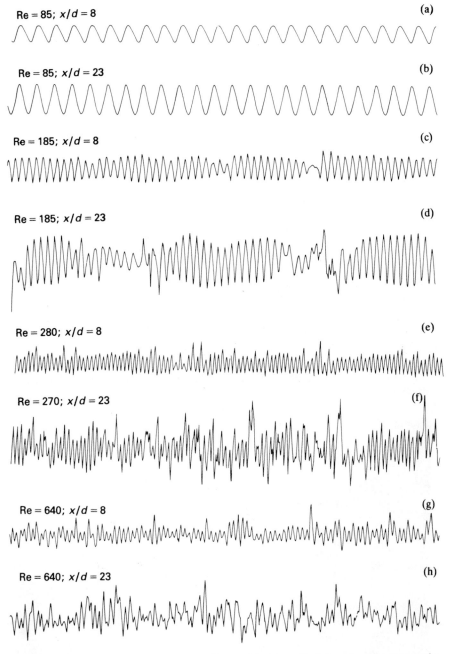

Re = 85; x/d = 8   (a)

Re = 85; x/d = 23   (b)

Re = 185; x/d = 8   (c)

Re = 185; x/d = 23   (d)

Re = 280; x/d = 8   (e)

Re = 270; x/d = 23   (f)

Re = 640; x/d = 8   (g)

Re = 640; x/d = 23   (h)

FIG. 3.10 Oscillograms of velocity fluctuations in cylinder wake. Traces are in pairs at same Re, but different distances, x, downstream from cylinder. For both positions, probe was slightly off-centre to be influenced mainly by vortices on one side of street. (Notes: relative velocity amplitudes are arbitrary; time scale is expanded by factor of about 3 in traces (g) and (h).)

FIG. 3.11 Turbulent wake exhibited by dye emitted from cylinder (out of picture ro right). From Ref. [186].

when the Reynolds number exceeds about 200 and acting three-dimensionally on the vortex street as a whole; the other occurring when Re exceeds about 400 and originating just downstream of the points of flow separation from the cylinder (Fig. 3.12). Because the latter occurs further upstream it is the primary cause of transition once Re > 400. In fact there is a wide range of Reynolds number, from this value up to about 3×10^5, in which, although there are changes in details of the flow [95, 175, 297], the broad picture, of a primary instability producing a vortex street and a secondary one disrupting this to give a turbulent wake, remains the same.

At Re $\approx 3 \times 10^5$, a dramatic development occurs. To understand this we must first consider developments at lower Re at the front and sides of the cylinder. There the phenomenon known as boundary layer formation

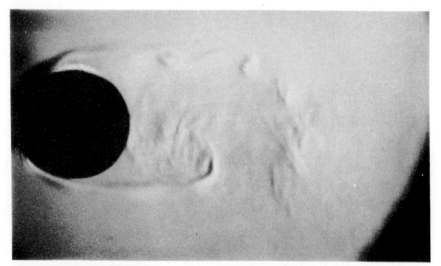

FIG. 3.12 Flow past cylinder at Re = 4800. High-speed schlieren (Section 25.4) photography with small density difference produced in air flow by introducing a small amount of carbon dioxide at cylinder surface. From Ref. [230].

occurs. This will be the subject of a full discussion later in the book (Section 8.3 and Chapter 11). For the moment, we may just say that there is a region, called the boundary layer, next to the wall of the cylinder in which all the changes to the detailed flow pattern occur. Outside this the flow pattern is independent of the Reynolds number. For these statements to mean anything, the boundary layer must be thin compared with the diameter of the cylinder; this is the case when Re is greater than about 100. (Boundary layer formation does not start by a sudden transition of the sort we have been considering previously; it is an asymptotic condition approached at high enough Re. The figure of 100 is just an order of magnitude.)

The change in the flow at $\mathrm{Re} \approx 3 \times 10^5$ results from developments in the boundary layer. Below this Reynolds number the motion there is laminar. Above, it undergoes transition to turbulence. At first, this transition takes a rather complicated form [157, 325]: laminar fluid close to the wall moves away from it as if it were entering the attached eddies; transition then occurs very quickly and the turbulent flow reattaches to the wall only a small distance downstream from the laminar separation. (See Section 12.6 for more precise consideration of these processes.) There are also complications due to the facts that the transition can occur asymmetrically between the two sides of the cylinder [157] and non-uniformly along its length [157, 203].

At higher values of the Reynolds number, above about 3×10^6, transition occurs in the boundary layer itself, thus eliminating the laminar separation and turbulent reattachment. The transition process is now similar to that described for pipe flow.

Whether or not it has previously undergone laminar separation and turbulent reattachment, the turbulent boundary layer itself separates; the fluid in it moves away from the wall of the cylinder and into the wake some distance before the rear line of symmetry. This occurs, however, much further round the cylinder than when the boundary layer remains laminar (Fig. 3.13). As a result the wake is narrower for $\mathrm{Re} > 3 \times 10^5$ than for $\mathrm{Re} < 3 \times 10^5$. When $\mathrm{Re} > 3 \times 10^5$, the fluid entering the wake is already turbulent and so the transition process immediately behind the cylinder is eliminated.

Markedly periodic vortex shedding remains a characteristic of the flow up to the highest Reynolds number ($\sim 10^7$) at which observations have been made. When this is not immediately apparent from oscillograms like Fig. 3.10, Fourier analysis reveals the presence of a dominant frequency amongst the other more random fluctuations. There are in fact two ranges, $200 < \mathrm{Re} < 400$ and $3 \times 10^5 < \mathrm{Re} < 3 \times 10^6$, in which the regularity of shedding decreases: in the former the Strouhal number shows a lot of scatter; in the latter the periodicity is lost except very close behind

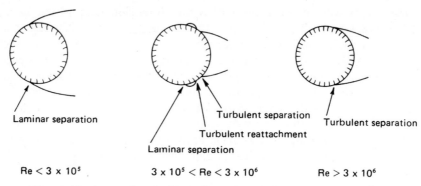

Laminar separation

Turbulent separation

Turbulent reattachment

Laminar separation

Turbulent separation

$\mathrm{Re} < 3 \times 10^5$ $3 \times 10^5 < \mathrm{Re} < 3 \times 10^6$ $\mathrm{Re} > 3 \times 10^6$

Fig. 3.13   Separation positions for various Reynolds number ranges.

the cylinder [90, 325]. The ends of these ranges can be identified with changes in the flow discussed above. At the top end of each range the change restores the regularity which had been lost at the bottom end; the Strouhal number becomes well defined again.

3.4   Drag

An important quantity associated with the relative motion between a body and a fluid is the force produced on the body. One has to apply a force in order to move a body at constant speed through a stationary fluid. Correspondingly an obstacle placed in a moving fluid would be carried away with the flow if no force were applied to hold it in place. The force in the flow direction exerted by the fluid on an obstacle is known as the drag. There is an equal and opposite force exerted by the obstacle on the fluid.

At high Reynolds numbers, this can be thought of as the physical mechanism of wake formation. Because of the force between it and the obstacle, momentum is removed from the fluid. The rate of momentum transport downstream must be smaller behind the obstacle than in front of it. There must thus be a reduction in the velocity in the wake region (Fig. 3.14). (At low Reynolds numbers, when the velocity and pressure are modified to large distances on either side of the obstacle, the situation is more complex.)

For the circular cylinder the important quantity is the drag per unit length; we denote this by D. It is conventional to present results for D in terms of the drag coefficient, defined as

$$C_D = \frac{D}{\frac{1}{2}\rho u_0^2 d}.$$
(3.3)

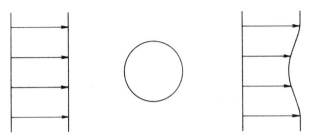

FIG. 3.14 Wake production: schematic velocity profiles upstream and down-stream of an obstacle.

This anticipates ideas to be introduced in Chapter 7. It is plausible ahead of that discussion that, since the Reynolds number is being used to specify the conditions, results for D should be presented non-dimensionally; C_D is non-dimensional. This procedure does in fact enable all conditions to be covered by a single curve, shown in Fig. 3.15. The curve is based primarily on experimental measurements, too numerous to show individual points. At the low Re end the experiments can be matched to theory.

The corresponding plot between Reynolds number and drag coefficient can also be given for a sphere, although with less precision because of the experimental problem of supporting the sphere. The curve, although different in detail, shows all the same principal features as the one for the cylinder.

A few features of the curves merit comment, some of these being points to which we shall return in later chapters.

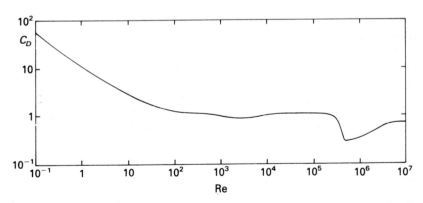

FIG. 3.15 Variation of drag coefficient with Reynolds number for circular cylinder. Curve is experimental, based on data in Refs. [139, 159, 325, 370, 386, 410]. See also Refs. [204, 211].

At low Reynolds numbers,

$$C_D \propto \frac{1}{\text{Re}}.$$ (3.4)

(This is actually an accurate representation for the sphere, an approximate one for the cylinder—see Sections 9.4 and 9.5.) For a given body in a given fluid (fixed d, ρ, and μ) this corresponds to

$$D \propto u_0.$$ (3.5)

Direct proportionality of the drag to the speed is a characteristic behaviour at low speeds.

Over a wide range of Reynolds number (10^2 to 3×10^5) the drag coefficient varies little. For a given body in a given fluid, constant C_D corresponds to

$$D \propto u_0^2.$$ (3.6)

Proportionality of the drag to the square of the speed is often a characteristic behaviour at high speeds.

However, there is a dramatic departure from this behaviour at $\text{Re} \approx 3 \times 10^5$. The drag coefficient drops by a factor of over 3. This drop occurs sufficiently rapidly that there is actually a range over which an increase in speed produces a decrease in drag. We have seen that this Reynolds number corresponds to the onset of turbulence in the boundary layer. The consequent delayed separation of the boundary layer results in a narrower wake. Since this in turn corresponds to less momentum extraction from the flow, one might expect the lower drag. The result is nonetheless a somewhat paradoxical one since transition to turbulence usually produces an increased drag. In fact, on the cylinder and sphere, the force exerted directly by the boundary layer does increase on transition. But this is more than counteracted by another effect: changes in the pressure distribution over the surface. This will be discussed more fully in Section 12.5.

4

FREE CONVECTION BETWEEN
PARALLEL WALLS

4.1 Introduction

Hot fluid rises. Thus the principle of free convection can be stated in what is almost an everyday expression—albeit somewhat loosely, since one should really say that hot fluid tends to rise and cold fluid tends to fall relative to each other. The principle is simple but its consequences are amongst the most complex in fluid dynamics; one could certainly not guess more than a small fraction of the phenomena of free convection from the basic notion.

It is therefore appropriate to make some aspect of free convection our third example of fluid motion in a simple configuration—in fact, a pair of configurations. We consider a layer of fluid between parallel walls maintained at different temperatures, and contrast the flows occurring in vertical and horizontal layers (Fig. 4.1). We shall see that a basic feature of this contrast is that, for the vertical layer, we know at the outset that motion will always occur (although not necessarily what pattern it will adopt), whereas, for the horizontal layer, the first question is when does the fluid convect. As a consequence of this difference, convection in horizontal layers has become one of the most basic and widely investigated topics in fluid dynamics; we shall return to it at greater length in Chapter 22. Section 4.3, on the other hand, is our only discussion of the vertical layer case.

However, some introductory remarks applicable to both orientations may be made. We suppose that the distance d between the two walls at temperatures T_1 and T_2 is small compared with the other dimensions of the layer; the other walls will then have comparatively little influence on what occurs in the layer. Heat is conducted into the fluid from the hot wall, transported through the layer and conducted out at the cold wall. If the space were occupied by a solid instead of a fluid, then the transport through the layer would also be by conduction. The temperature would fall linearly from the hot wall to the cold (cf. eqn (4.3) below). When the space is occupied by a fluid in motion, the temperature distribution is more complicated but it remains true that fluid close to the hot wall is at a higher temperature than fluid close to the cold wall. Associated with these temperature variations are density variations; the hotter fluid is

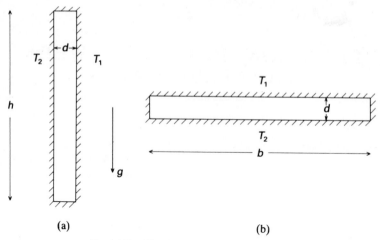

(a)   (b)

FIG. 4.1   Vertical and horizontal slots.

lighter. The cause of motion, in the cases we are going to consider, is the differing gravitational force per unit volume acting on these density differences. The fluid would remain at rest if $T_1 = T_2$, or if the density did not vary with the temperature, or if there were no gravitational field. As we shall discuss more systematically in Section 14.5, this is the definition of free convection.

4.2 The Rayleigh and Prandtl numbers

In this chapter we shall be concerned more with qualitative observations than with quantitative specification. However, the illustrations will be labelled with quantities that put them in the more complete context of Chapters 14 and 22. There is also one important quantitative result in Section 4.4. We therefore note briefly how the quantitative specification is formulated.

For the reasons discussed in Section 2.4 (and to be treated more fully in Chapter 7) we expect this to be done in terms of non-dimensional parameters. We cannot, however, use the Reynolds number for this purpose as there is no velocity involved in the specification of the problem. Its place is taken by a non-dimensional quantity known as the Rayleigh number:

$$\text{Ra} = g\alpha(T_2 - T_1)d^3/\nu\kappa, \tag{4.1}$$

where d, T_1, and T_2 are already defined, g is the acceleration due to gravity and α, ν, and κ are properties of the fluid, respectively its coefficient of expansion, kinematic viscosity and thermal diffusivity.† The derivation of this quantity is somewhat more complicated than that of the Reynolds number and we postpone a full explanation to Section 14.5.

In addition to the Rayleigh number, specification of a problem in free convection requires knowledge of the Prandtl number

$$Pr = v/\kappa. \tag{4.2}$$

This is the ratio of two fluid properties and thus itself determined by the choice of fluid. Again we postpone further explanation to Section 14.5, but values of Pr are indicated in figure captions in the present chapter.

4.3 Convection in vertical slots [243, 343]

We now turn specifically to the flow when the layer is vertical. As drawn in Fig. 4.2, the fluid on the left-hand side is hotter than that on the right-hand side. One anticipates that the fluid will rise on the left and fall on the right. The simplest possible type of flow is sketched in Fig. 4.2. There is no possibility of the fluid remaining at rest when $T_1 \neq T_2$. (A more formal justification of this statement, in terms of the governing equations, will be given in Section 14.5.)

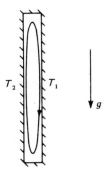

FIG. 4.2   Simplest form of convection in a vertical slot.

At low enough Rayleigh number the flow does indeed consist of a simple circulation like that in Fig. 4.2. However, just as increasing Reynolds number leads to a sequence of changes in the flows considered in previous chapters, so does increasing Rayleigh number here. Our knowledge of these developments comes not only from experimental observations, which we use to illustrate them here, but also from theoretical analyses and numerical solutions of the equations of motion.

† Thermal diffusivity is defined as $\kappa = k/\rho C_p$ (k = thermal conductivity; ρ = density; C_p = specific heat at constant pressure).

Nevertheless this knowledge is quite likely incomplete, because of the many possible conditions; not only do the developments have to be considered for different Prandtl number, but also a flow extending from top to bottom like that in Fig. 4.2 is influenced by h/d (Fig. 4.1(a)) even when this is large.

We can however, describe a typical sequence with increasing Ra, and this is illustrated in Figs. 4.3–4.5.†

Figure 4.3 shows changes in the distribution across the slot of the vertical velocity, whilst the overall flow pattern remains a single circulation as in Fig. 4.2. At first all the fluid participates in the circulation. As the Rayleigh number is raised, however, the motion becomes increasingly concentrated in 'boundary layers' (see Section 14.8) close to the hot and cold walls, leaving almost stationary fluid in the central region. The profiles in Fig. 4.3 were measured half-way up the apparatus; profiles at other heights were very similar. The temperature in the central region was almost constant over each horizontal plane but higher at the top of the slot than at the bottom.

The next development is more striking; the single circulation over the full height of the slot becomes unstable (cf. Chapter 17) and breaks up into a number of smaller circulating regions, stacked vertically. This flow is seen in Fig. 4.4(b), contrasted with a corresponding flow visualization of the previous flow. The motion may be described as cellular. The sense of circulation is the same in each cell (in contrast to the behaviour in horizontal layers we shall be considering shortly, Fig. 4.6), so that the fluid is always rising on the hot side and falling on the cold.

This pattern of motion involves large velocity gradients between neighbouring cells. The next development (Fig. 4.4(c), showing just a small part of the total height of the slot) may be seen as a consequence of this. Small cells of reverse circulation appear between the main cells, acting in a sense as 'idling wheels' between them. The small cells do not extend to the hot and cold walls and are probably nearly isothermal; temperature differences are driving them only indirectly via the main cells.

Although it has become quite complicated, the flow has so far remained steady. At the highest values of the Rayleigh number it becomes spontaneously unsteady. The cellular pattern is disrupted. Waves are generated in the lower part of this hot upgoing flow and the upper part of the cold downgoing flow. These lose their regularity as they travel up or down, producing a turbulent core in the centre of the slot.

† Figure 4.3 is more characteristic of high Prandtl number than low, because in the latter case the development shown in Fig. 4.4(b) occurs before the profile has changed greatly [401]. The cellular structure of Fig. 4.4(b) is to be found, with some differences in detail, over some Rayleigh number range, for a remarkably wide range of Prandtl number [243].

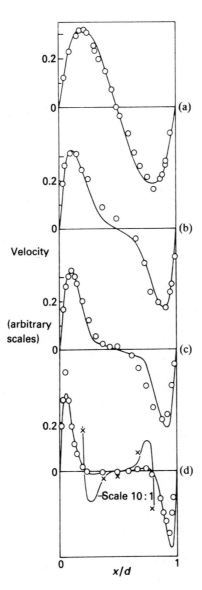

FIG. 4.3 Distribution of vertical velocity at midheight across a vertical slot containing paraffin for various values of the Rayleigh number: (a) 3.1×10^4; (b) 2.95×10^5; (c) 6.6×10^5; (d) 3.6×10^6. From Ref. [153].

FIG. 4.4 Convection patterns in a vertical slot, shown by suspended beads illuminated in a narrow sheet. $h/d = 15$; $Pr = 480$; Ra based on d: (a) 1.3×10^5; (b) 2.9×10^5; (c) 1.5×10^6. From Ref. [343].

Figure 4.5 shows schematically the principal features of the convective motion at high Rayleigh number.

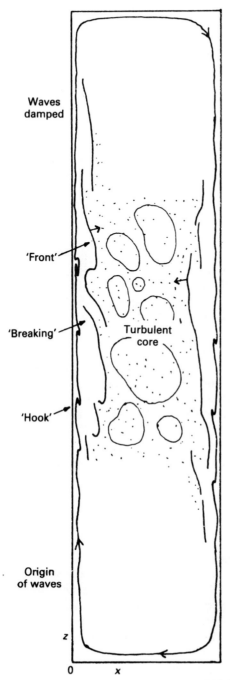

FIG. 4.5 Schematic representation of principal features of high Rayleigh number ($> \sim 10^6$) convection in a vertical slot. From Ref. [154].

4.4 Convection in horizontal layers

If the fluid remains at rest the temperature distribution is just the linear one associated with conduction,

$$T = T_2 - (T_2 - T_1)z/d, \tag{4.3}$$

where z is the distance from the wall at T_2. If the layer is horizontal this involves no horizontal variations; i.e. no lighter and heavier regions which must rise and fall relative to one another. (Again see Section 14.5 for a more formal treatment.) The rest configuration is an equilibrium one. If $T_2 < T_1$, with T_2 now specifically the temperature of the bottom surface, the fluid always remains at rest. On the other hand, when†

$$T_2 > T_1 \tag{4.4}$$

heavy cold fluid is situated above light hot fluid; if the former moves downward and the latter upwards, there is a release of potential energy which can provide kinetic energy for the motion. There is thus a possibility that the equilibrium will be unstable. Thus when flow arises, it does so not because of the absence of equilibrium (i.e. the absence of any solutions of the governing equations with the fluid at rest) but because the equilibrium is unstable.

Because stability considerations are very important in fluid dynamics and because this is one of the simplest configurations for which they are needed, the case of a horizontal fluid layer heated from below and cooled from above has received a great deal of attention, both theoretically and experimentally. We shall return to it in Chapter 22. It is known as the Bénard (or Rayleigh–Bénard) configuration. (Historically, the name is inaccurate; Bénard's pioneering observations, although for long believed to relate to this configuration, were actually mostly of another phenomenon—see Section 4.5—that gives rise to similar effects. However, the name is so well established that its use in this way causes no confusion.)

The inequality (4.4) is a necessary criterion for motion to occur, but it is not the only one. The instability is opposed by the frictional action of viscosity. It is also opposed by the action of thermal conductivity, which tends to remove the temperature difference between the hot rising regions and the cold falling ones. Motion occurs only when the destabilizing action of the temperature difference is strong enough to overcome these. This in fact means when the Rayleigh number, eqn (4.1), is large enough. Although we have not yet considered the Rayleigh number systematically we may notice here that the factors that

---

† This assumes, as we have been doing throughout, that the coefficient of expansion is positive. In the few cases when it is negative, such as water below 4°C, the criterion (4.4) is of course reversed.

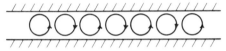

FIG. 4.6 Schematic representation of Bénard convection at a Rayleigh number a little above critical.

drive the motion—the combination of the temperature difference, the consequent expansion, and gravity—appear in its numerator and the factors mentioned above as opposing it in its denominator.

Instability occurs when the Rayleigh number exceeds a critical value of about 1700, independent of the Prandtl number (see Section 22.2). Below this the fluid remains at rest. Above, it comes into motion. The fluid establishes hot rising regions and cold falling regions with horizontal motion at top and bottom to maintain continuity, as shown schematically in Fig. 4.6. The rising fluid loses its heat by thermal conduction when it gets near to the cold top wall and can thus move downwards again. Similarly, the cold downgoing fluid is warmed in the vicinity of the hot bottom wall and can rise again. When the flow is established as a steady pattern, the continuous release of potential energy is balanced by viscous dissipation of mechanical energy. The potential energy is provided by the heating at the bottom and the cooling at the top. From a thermodynamic point of view the system is thus a heat engine (cf. Section 14.7).

Figure 4.6 is, of course, just a vertical section through a three-dimensional pattern. Figures 4.7–4.9 illustrate a few of the planforms that have been observed, and make the point that these are quite varied.

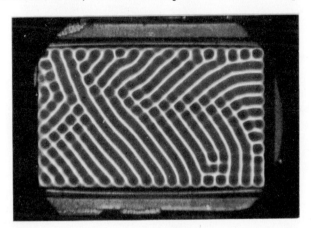

FIG. 4.7 Shadowgraph (Section 25.4) of Bénard convection at $Ra = 2.2 \times 10^3$, $Pr = 70$. Photo by V. Croquette, A. Pocheau, P. Le Gal, and C. Poitou of C.E.N., Saclay.

FIG. 4.8   Bénard convection at Ra = 2.6 × 10⁴, Pr = 100; aluminium powder flow visualization (Section 25.4). From Ref. [326].

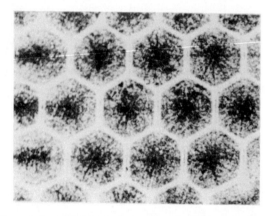

FIG. 4.9   Shadowgraph of Bénard convection in fluid of which viscosity varies strongly with the temperature. Rayleigh number is factor of 1.45 above its critical value, and the ratio of maximum to minimum viscosity is 7. From Ref. [319].

FIG. 4.10   Temperature variations in Bénard convection observed by moving and stationary probes. (a) Ra = 5.0 × 10³, (b) 1.21 × 10⁴, (c) 3.03 × 10⁴, (d) 5.61 × 10⁴, (e) 1.55 × 10⁶; Pr = 0.72 (air) throughout. Abscissa scales represent x/d and time (minutes) for the two cases; ordinate scales represent the difference between the instantaneous temperature and the mean temperature divided by $(T_2 - T_1)$. From Ref. [411].

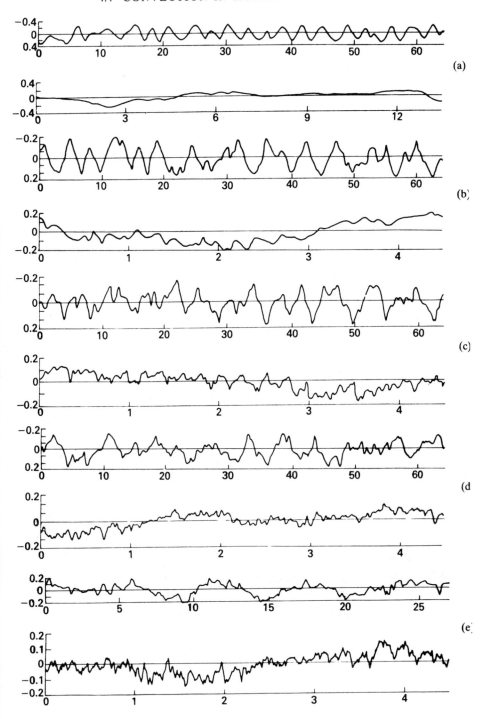

There are two aspects to this variety. Principally, as with vertical slot convection, there are the changes that occur as the Rayleigh and Prandtl numbers are varied (e.g. compare Figs. 4.7 and 4.8). Additionally, the flow pattern is sensitive to deviations from the ideal configuration such as variations of the fluid properties, notably viscosity, with temperature. Figure 4.9 shows a very regular hexagonal pattern that has arisen as a result of such viscosity variations. It was once thought that such hexagonal arrays were the most basic type of Bénard convection; but it is now known that roll-like cells, as seen in Fig. 4.7, are the ones least influenced by deviations from the ideal—when the Rayleigh number is not too much above its critical value.

The sequence of events for a given fluid as the Rayleigh number is increased from the critical value at which motion first occurs to much higher values is of particular interest. We postpone detailed consideration to Chapter 22, but outline the main features here. For values of Ra not too much in excess of the critical value, the convection always occurs in a regular or fairly regular pattern as implied by Fig. 4.6. The individual elements in such a pattern are known as convection cells, or Bénard cells. The flow within this pattern is steady. Increase of the Rayleigh number leads to time-varying flows and loss of the spatial regularity. As in the sequences considered in Sections 2.5, 3.3, and 4.3, the ultimate result is turbulent motion.

The overall picture is summarized by Fig. 4.10, showing observations of temperature variations in Bénard convection in air. At each Rayleigh number, one trace shows the variation with time at a fixed position, the other the variation with position observed by traversing a probe through the fluid. At low Rayleigh numbers, the temperature at a fixed position is effectively constant, whilst the traverse produces a roughly periodic variation associated with the passage of the probe through the cells. At higher Rayleigh numbers unsteadiness is apparent. At first, the time variations are a perturbation of the previous pattern, but with increasing Ra, they become increasingly dominant. At the highest Ra, the fluctuations have a general resemblance to Fig. 2.6.

4.5 Convection due to surface tension variation [47]

In the last section we remarked that Bénard's original experimental observations were not, for the most part, of the phenomenon that bears his name. What then did he observe? Although, as a free surface effect, it is strictly outside the scope of this book, we should perhaps not leave that question unanswered.

The flow is again due to an instability of a rest configuration, with

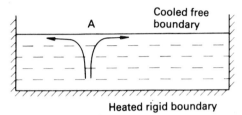

FIG. 4.11 Mechanism of surface tension instability; effect of local increase in temperature at *A*.

temperature variation of surface tension being the crucial feature. Figure 4.11 illustrates the process. A layer of liquid is bounded by a solid wall from which it receives heat and a free surface at which it loses heat. The surface tension of a liquid decreases with increasing temperature. Consequently, if part of the free surface should become locally hotter than the rest, as a result of some small disturbance, fluid is drawn away from the region by the action of surface tension. Other fluid comes from below the surface to replace it. This fluid has been closer to the hot solid boundary and will thus be hotter than the fluid at the surface. The original disturbance in the temperature distribution is thus amplified.

A steady state can be established with hot fluid moving towards the free surface, losing heat as it travels parallel to and close to the surface and then moving away again as cold fluid. In this way a distribution of surface tension can be maintained that drives the motion against the retarding action of viscous forces. This is another situation which can give rise to a very regular array of hexagonal cells, similar to that occurring in Bénard convection with large viscosity variations (Fig. 4.9). To-day's name for this type of flow is Marangoni convection.

EQUATIONS OF MOTION

5.1 Introduction

We turn now to the formulation of the basic equations—the starting point of the theories that will lead, one hopes, to an understanding of phenomena such as those described in Chapters 2 to 4. These equations are formulations appropriate to a fluid in motion of the usual laws of mechanics—conservation of mass and Newton's laws of motion. In some situations other physical processes may be present, thermodynamic processes for instance, and equations for these are similarly formulated, as we shall examine more fully in Chapter 14.

5.2 Fluid particles and continuum mechanics

Before we can proceed with this formulation we need certain preliminary ideas, the most important being the concept of a fluid particle.

The equations concern physical and mechanical quantities, such as velocity, density, pressure, temperature, which will be supposed to vary continuously from point to point throughout the fluid. How do we define these quantities at a point? To do so we have to make what is known as the assumption of the applicability of continuum mechanics or the continuum hypothesis. We suppose that we can associate with any volume of fluid, no matter how small, those macroscopic properties that we associate with the fluid in bulk. We can then say that at each point there is a particle of fluid and that a large volume of fluid consists of a continuous aggregate of such particles, each having a certain velocity, temperature, etc.

Now we know that this assumption is not correct if we go right down to molecular scales. We have to consider why it is nonetheless plausible to formulate the equations on the basis of the continuum hypothesis. It is simplest to think of a gas, although the considerations for a liquid are very similar.

The various macroscopic properties are defined by averaging over a large number of moelcules. Consider velocity for example. The molecules of a gas have high speeds associated with their Brownian motion, but these do not result in a bulk transfer of gas from one place to another. The flow velocity is thus defined as the average velocity of many molecules. Similarly, the temperature is defined by the average energy of the Brownian motion. The density is defined by the mass of the average number of molecules to be found in a given volume. Other macroscopic properties, such as pressure and viscosity, likewise result from the average action of many molecules.

None of these averaging processes is meaningful unless the averaging is carried out over a large number of molecules. A fluid particle must thus be large enough to contain many molecules. It must still be effectively at a point with respect to the flow as a whole. Thus the continuum hypothesis can be valid only if there is a length scale, L_2, which we can think of as the size of a fluid particle, such that

$$L_1 \ll L_2 \ll L_3 \qquad (5.1)$$

where the meanings of L_1 and L_3 are illustrated by Fig. 5.1. This figure uses the example of temperature, rather than the natural first choice of velocity, because it is easier to discuss a scalar. It shows schematically the average Brownian energy of the molecules in a volume L^3 plotted against the length scale L (on a logarithmic scale). The centre of the volume may be supposed fixed as its size is varied. When the volume is so small that it contains only a few molecules, there are large random fluctuations; the change produced by increasing the volume depends on the particular speeds of the new molecules then included. As the volume becomes large enough to contain many molecules, the fluctuations become negligibly small. A temperature can then meaningfully be defined. L_1 is proportional to, but an order of magnitude or so larger than, the average

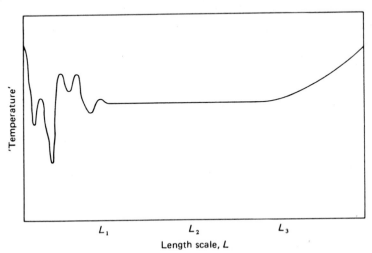

FIG. 5.1 Schematic variation of average energy of molecules with length scale. See text.

distance from a molecule to its nearest neighbour. At the other extreme, the volume may become so large that it extends into regions where the temperature is significantly different. This will result in an increase or decrease in the average. L_3 is a typical length scale associated with the flow; that is, a typical distance over which the macroscopic properties vary appreciably.

The applicability of the continuum hypothesis depends on there being a significant plateau between L_1 and L_3 as shown. One may regard L_2 as being an infinitesimal distance so far as macroscopic effects are concerned, and formulate the equations (as differential equations implicitly involving the limit of small separations) ignoring the behaviour on still smaller length scales.

The same fluid particle does not consist of just the same molecules at all times. The interchange of molecules between fluid particles is taken into account in the macroscopic equations by assigning to the fluid diffusive properties such as viscosity and thermal conductivity. For example (again considering a gas for simplicity) the physical process by which the velocity distribution of Fig. 1.1 generates the stress shown is the Brownian movement of molecules across AB; those crossing in, say, the $+y$-direction have on average less x-momentum and so tend to reduce the momentum of the fluid above AB. The same fluid particle may be identified at different times, once the continuum hypothesis is accepted, through the macroscopic formulation. This specifies (in principle) a trajectory for every particle and thus provides meaning to the statement that the fluid at one point at one time is the same as that at another point at another time. For example, for a fluid macroscopically at rest, it is obviously sensible to say that the same fluid particle is always in the same place—even though, because of the Brownian motion, the same molecules will not always be at that place.

However, for the continuum hypothesis to be plausible, it is evidently necessary for the molecules within a fluid particle to be strongly interacting with one another. If each molecule acted just as if the others were not there, there would be little point in identifying the aggregate as a particle. Thus, if λ is the molecular-mean-free path, continuum mechanics can be applied only if

$$\lambda \ll L_2 \qquad\qquad (5.2)$$

so that each molecule undergoes many collisions whilst traversing a distance that can still be regarded as infinitesimal. Since λ can be large compared with L_1 as defined above, this is an additional requirement to (5.1).

Once the continuum hypothesis has been introduced, we can formulate the equations of motion on a continuum basis, and the molecular structure of the fluid need not be mentioned any more. Hence, although the concepts developed above underlie the whole formulation, we shall not have much occasion to refer back to them. Velocity, henceforth, will be either a mathematical quantity or something (hopefully equivalent) that one measures experimentally. So will all the other parameters. Their definitions as averages over molecules provide answers to the implicit, but rarely explicit, questions: 'What is the real physical meaning of this mathematical quantity?' 'What quantity does one ideally wish to measure?'

The continuum hypothesis is only a hypothesis. The above discussion suggests that it is plausible, but nothing more. The real justification for it comes subsequently, through the experimental verification of predictions of the equations developed on the basis of the hypothesis.

5.3 Eulerian and Lagrangian coordinates

In setting up the equations governing the dynamics of a fluid particle, we evidently need to decide whether we should use coordinates fixed in space or coordinates that move with the particle. These two procedures are known respectively as the Eulerian and Lagrangian specifications. The equations are much more readily formulated using the former because the Lagrangian specification does not immediately indicate the instantaneous velocity field on which depend the stresses acting between fluid particles. Throughout this book we use only the Eulerian specification; i.e. we write the velocity

$$\boldsymbol{u} = \boldsymbol{u}(\boldsymbol{r}, t) \tag{5.3}$$

where $\boldsymbol{r}$ is the position in an inertial frame of reference and t is time. Values of $\boldsymbol{u}$ at the same $\boldsymbol{r}$ but different t do not, of course, correspond to the same fluid particle.

It is not always easy to relate Lagrangian aspects, such as the trajectories of fluid particles, to an Eulerian specification. In the context of this book, this is particularly relevant to the interpretation of flow visualization experiments in which dye marks certain fluid particles. The relationship of the observed patterns to the corresponding Eulerian velocity field may not be simple.

5.4 Continuity equation

We are now ready to start on the actual formulation of the equations. We consider first the representation of mass conservation, often called continuity.

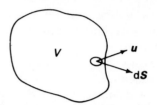

FIG. 5.2 Definition sketch for derivation of continuity equation.

Consider an arbitrary volume V fixed relative to Eulerian coordinates and entirely within the fluid (Fig. 5.2). Fluid moves into or out of this volume at points over its surface. If dS is an element of the surface (the magnitude of dS being the area of the element and its direction the outward normal) and u is the velocity at the position of this element, it is the component of u parallel to dS that transfers fluid out of V. Thus, the outward mass flux (mass flow per unit time) through the element is $\rho u \cdot dS$, where ρ is the fluid density. Hence,

$$\text{rate of loss of mass from } V = \int_S \rho u \cdot dS. \qquad (5.4)$$

(This is, of course, negative if the mass in V is increasing.) We have also

$$\text{total mass in volume } V = \int_V \rho \, dV. \qquad (5.5)$$

Hence

$$\frac{d}{dt} \int_V \rho \, dV = \int_V \frac{\partial \rho}{\partial t} dV = -\int_S \rho u \cdot dS. \qquad (5.6)$$

We are interested in the mass balance at a point, rather than that over an arbitrary finite volume. Hence, we allow V to shrink to an infinitesimal volume; the integration in $\int (\partial \rho / \partial t) \, dV$ is redundant and we have

$$\frac{\partial \rho}{\partial t} = -\lim_{V \to 0} \left[\int \rho u \cdot dS / V \right]. \qquad (5.7)$$

That is,

$$\frac{\partial \rho}{\partial t} = -\operatorname{div} \rho u \qquad (5.8)$$

by definition of the operator div. This gives the general expression representing mass conservation for a fluid in which both u and ρ are functions of position,

$$\frac{\partial \rho}{\partial t} + \nabla \cdot (\rho u) = 0. \qquad (5.9)$$

This is known as the continuity equation.

For the important special case when ρ is constant (see Section 1.2 (3) and Section 5.8) the continuity equation reduces to the very simple form

$$\nabla \cdot \boldsymbol{u} = 0. \tag{5.10}$$

The velocity field is solenoidal vector. We notice that this does not assume steady flow. The time-variation does not appear explicitly in the continuity equation of a constant density fluid even when the flow is unsteady.

5.5 The substantive derivative

The next equation to be derived is the representation of Newton's second law of motion; i.e. the rate of change of momentum of a fluid particle is equal to the net force acting on it. We need first of all an expression for the rate of change of momentum of a fluid particle. It would not be correct to equate the rate of change of momentum at a fixed point to the force, because different particles are there at different times. Even in steady flow, for example, a fluid particle can change its momentum by travelling to a place where the velocity is different; this acceleration requires a force to produce it.

This is one example of a general problem. On occasion, one needs to know the rate of change of other quantities whilst following a fluid particle. In problems where thermal effects are important, for example, various physical processes may heat or cool the fluid. These determine the rate of change of the temperature of a fluid particle, not the rate of change at a fixed point.

In this section, therefore, we examine the general question of rates of change following the fluid. It is easier in the first place to consider a scalar quantity, and we denote this by T, thinking of the example of temperature.

Quite generally, the small change δT produced by a small change δt in time and small changes δx, δy, δz in Cartesian position coordinates is

$$\delta T = \frac{\partial T}{\partial t} \delta t + \frac{\partial T}{\partial x} \delta x + \frac{\partial T}{\partial y} \delta y + \frac{\partial T}{\partial z} \delta z \tag{5.11}$$

and a rate of change can be formulated by dividing by δt:

$$\frac{\delta T}{\delta t} = \frac{\partial T}{\partial t} + \frac{\partial T}{\partial x} \frac{\delta x}{\delta t} + \frac{\partial T}{\partial y} \frac{\delta y}{\delta t} + \frac{\partial T}{\partial z} \frac{\delta z}{\delta t}. \tag{5.12}$$

If now we choose δx, δy, and δz to be the components of the small distance moved by a fluid particle in time δt, then (in the limit $\delta t \to 0$)

this is the rate of change of T of that particle. Also $\delta x/\delta t$, $\delta y/\delta t$ and $\delta z/\delta t$ are then (in the same limit) the three components of the velocity of the particle (u, v, and w). We thus have

$$\frac{DT}{Dt} = \frac{\partial T}{\partial t} + u\frac{\partial T}{\partial x} + v\frac{\partial T}{\partial y} + w\frac{\partial T}{\partial z}. \tag{5.13}$$

In general D/Dt denotes the rate of change (of whatever quantity it operates on) following the fluid. This operator is known as the substantive derivative.

We can rewrite eqn (5.13) as

$$\frac{DT}{Dt} = \frac{\partial T}{\partial t} + \boldsymbol{u}\cdot\nabla T \tag{5.14}$$

and the operator in general as

$$\frac{D}{Dt} = \frac{\partial}{\partial t} + \boldsymbol{u}\cdot\nabla \tag{5.15}$$

(exhibiting the physically obvious fact that it does not depend on the particular coordinates used).

We see that the relationship combines the two ways in which the temperature of a fluid particle can change. It can change because the whole temperature field is changing—a process present even if the particle is at rest. And it can change by moving to a position where the temperature is different—a process present even if the temperature field as a whole is steady. As one would expect, this latter process depends on the magnitude of the spatial variations of the temperature and on the velocity, determining how quickly the fluid moves through the spatial variations.

Nothing in the above analysis restricts it to scalar quantities, and we can similarly write that the rate of change of a vector quantity $\boldsymbol{B}$ following a fluid particle is

$$\frac{D\boldsymbol{B}}{Dt} = \frac{\partial \boldsymbol{B}}{\partial t} + u\frac{\partial \boldsymbol{B}}{\partial x} + v\frac{\partial \boldsymbol{B}}{\partial y} + w\frac{\partial \boldsymbol{B}}{\partial z} = \frac{\partial \boldsymbol{B}}{\partial t} + \boldsymbol{u}\cdot\nabla\boldsymbol{B}. \tag{5.16}$$

Whereas $\boldsymbol{u}\cdot\nabla T$ is the scalar product of vectors $\boldsymbol{u}$ and ∇T, $\boldsymbol{u}\cdot\nabla\boldsymbol{B}$ cannot be similarly split up. It is meaningful only as a whole. $(\boldsymbol{u}\cdot\nabla)$ operating on a vector must be thought of as a new operator (defined through its Cartesian expansion).

The particular case, $\boldsymbol{B}=\boldsymbol{u}$, gives the rate of change of velocity following a fluid particle,

$$\frac{D\boldsymbol{u}}{Dt} = \frac{\partial \boldsymbol{u}}{\partial t} + \boldsymbol{u}\cdot\nabla\boldsymbol{u}. \tag{5.17}$$

The velocity u now enters in two ways, both as the quantity that changes as the fluid moves and as the quantity that governs how fast the change occurs. Mathematically, however, it is just the same quantity in both its roles.

Returning finally to the information that we require for the dynamical equation, we have that the rate of change of momentum per unit volume following the fluid is

$$\rho\frac{\mathrm{D}u}{\mathrm{D}t} = \rho\frac{\partial u}{\partial t} + \rho u \cdot \nabla u. \tag{5.18}$$

Why, the reader may ask, is it $\rho\,\mathrm{D}u/\mathrm{D}t$ and not $\mathrm{D}(\rho u)/\mathrm{D}t$? (The distinction is important in the general case when both u and ρ are variables.) The only reason why a particular bit of fluid is changing its momentum is that it is changing its velocity. If it is simultaneously changing its density, this is not because it is gaining or losing mass, but because it is changing the volume it occupies. This change is therefore irrelevant to the momentum change. Expressing the distinction verbally instead of algebraically, we may say that '(the rate of change of momentum) per unit volume' is different from 'the rate of change of (momentum per unit volume)', and the former is the relevant quantity.

5.6 The Navier–Stokes equation

From above, the left-hand side of the dynamical equation, representing Newton's second law of motion, is $\rho\,\mathrm{D}u/\mathrm{D}t$. The right-hand side is the sum of the forces (per unit volume) acting on the fluid particle. We now consider the nature of these forces in order to complete the equation.

Some forces are imposed on the fluid externally, and are part of the specification of the particular problem. One may need, for example, to specify the gravity field in which the flow is occurring. On the other hand, the forces due to the pressure and to viscous action, of which simple examples have been given in Section 2.2, are related to the velocity field. They are thus intrinsic parts of the equations of motion and have to be considered here. Both the pressure and viscous action generate stresses acting across any arbitrary surface within the fluid; the force on a fluid particle is the net effect of the stresses over its surface.

The generalization of the pressure force from the simple case considered in Section 2.2 is straightforward. We saw there that the net force per unit volume in the x-direction as a result of a pressure change in that direction is $-\partial p/\partial x$. For a general pressure field, similar effects act in all directions and the total force per unit volume is $-\mathrm{grad}\, p$. This term always appears in the dynamical equation; when a fluid is brought into motion, the pressure field is changed from that existing when it is at rest (the hydrostatic pressure). We can regard this for the moment as an experimental result. We shall be seeing that it is necessary to have the pressure as a variable in order that the number of variables matches the number of equations.

The general form of the viscous force is not so readily inferred from any simple example. The mathematical formulation is outlined as an appendix to this chapter. Here we shall look at some of the physical concepts underlying viscous action, and then quote the expression for the viscous term in the dynamical equation that is given by rigorous formulation of these concepts. (Some further discussion of the physical action of viscosity will be given in the context of the particular example of jet flow in Section 11.7.)

Viscous stresses oppose relative movements between neighbouring fluid particles. Equivalently, they oppose the deformation of fluid particles. The difference between these statements lies only in the way of verbalizing the rigorous mathematical concepts, as is illustrated by Fig. 5.3. The change in shape of the initially rectangular region is produced by the ends of one diagonal moving apart and the ends of the other moving together. As the whole configuration is shrunk to an infinitesimal one, it may be said either that the particle shown is deforming or that particles on either side of AB are in relative motion. The rate of deformation depends on the velocity gradients in the fluid. The consequence of this behaviour is the generation of a stress (equal and opposite forces on the two sides) across a surface such as AB; this stress depends on the properties of the fluid as well as on the rate of deformation.

The stress can have any orientation relative to the surface across which it acts. The special case, considered in Sections 1.2(4) and 2.2, in which the stress is in the plane of the surface is often thought of as the 'standard' case. We therefore look for a moment at a simple situation in which viscous stresses normal to the surface govern the behaviour. This is the falling column produced for example when a viscous liquid is poured from a container (Fig. 5.4). In the absence of side-walls transverse viscous stresses cannot be generated as they are in channel flow. The

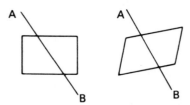

FIG. 5.3 Deformation of rectangular element and relative motion of fluid on either side of arbitrary line through the element.

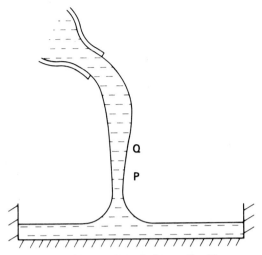

FIG. 5.4 Pouring of viscous liquid.

reason the fluid at, say, P does not fall with an acceleration of g is the viscous interaction with the more slowly falling fluid at Q.

In the general case, the stress is a quantity with a magnitude and two directions, the direction in which it acts and the normal to the surface, associated with it. It is thus a second-order tensor. (The stresses acting across surfaces of different orientations through the same point are not, or course, independent of one another.) The rate of deformation is also expressed by a second-order tensor—the rate-of-strain tensor. From the considerations above we expect this to involve the velocity gradients. However, not all distributions of velocity variation lead to deformation; a counter-example is the rotation of a body of fluid as if it were rigid (see Section 6.4). The rate-of-strain tensor selects the appropriate features of the velocity field.

One expects the stress tensor to depend on the rate-of-strain tensor and on the properties of the fluid. A Newtonian fluid (see Section 1.2(4)) can now be defined rigorously as one in which the stress tensor and the rate-of-strain tensor are linearly related.

The remaining ideas contained in the derivation of the viscous term of the dynamical equation are simply symmetry considerations. For example, a mirror-image flow pattern must generate a mirror-image stress distribution. And the analysis of a flow configuration using different coordinates must give the same result.

In Cartesian coordinates, the x-component of the viscous force per unit volume (see appendix to this chapter) is

$$\frac{\partial}{\partial x}\left[2\mu\frac{\partial u}{\partial x} + \lambda\nabla\cdot\boldsymbol{u}\right] + \frac{\partial}{\partial y}\left[\mu\left(\frac{\partial u}{\partial y} + \frac{\partial v}{\partial x}\right)\right] + \frac{\partial}{\partial z}\left[\mu\left(\frac{\partial w}{\partial x} + \frac{\partial u}{\partial z}\right)\right]. \quad (5.19)$$

Similar expressions for the y- and z-components are given by appropriate permutations.

Here μ is the coefficient of viscosity, defined through the special case considered in Section 1.2(4). λ is a second viscosity coefficient. One would expect there to be a second such coefficient, independent of the first, by analogy with the fact that there are two independent elastic moduli. This is a valid analogy. However, it has often been the practice to introduce a relationship between μ and λ ($\lambda = -2\mu/3$). This is done at the cost of redefining the pressure so that it is not the thermodynamic pressure, and the second independent parameter then appears in the relationship between the two pressures [26,324]. λ is difficult to measure experimentally and is not known for the variety of fluids for which there are values of μ. Hence, the statement that a fluid is Newtonian usually means that μ is observed to be independent of the rate of strain and that λ is assumed to be so too.

However, for a fluid of constant density, the continuity equation (5.10) causes the term involving λ to drop out. If, additionally, μ is taken to be a constant, expression (5.19) reduces (with a further use of (5.10)) to simply

$$\mu\nabla^2 u. \qquad (5.20)$$

The y- and z-components are correspondingly $\mu\nabla^2 v$ and $\mu\nabla^2 w$, and the vectorial viscous force per unit volume is $\mu\nabla^2\boldsymbol{u}$.†

Because (5.20) is so much simpler than (5.19), one prefers to use the former whenever possible; one tends to make this approximation even when the density and/or the viscosity do vary appreciably (see also appendix to Chapter 14). That will be the procedure throughout this book. But it should be remembered that situations may arise in the laboratory or in a practical application which are properly described only by the full expression.

Collecting together the various contributions mentioned, we have

$$\rho\frac{D\boldsymbol{u}}{Dt} = -\nabla p + \mu\nabla^2\boldsymbol{u} + \boldsymbol{F}. \qquad (5.21)$$

This is known as the Navier–Stokes equation. It is the basic dynamical equation expressing Newton's second law of motion for a fluid of constant density.

The term $\boldsymbol{F}$ represents the contribution of those forces (such as gravity) mentioned at the beginning of this section that have to be included in the specification of the problem. This is often known as the body force term, because such forces act on the volume of a fluid

† The meaning of ∇^2 operating on a vector is defined through its Cartesian expansion.

particle, not over its surface in the way the stresses between fluid particles act. The reaction to a body force is remote from the fluid particle concerned: usually outside the fluid region, although occasionally on distant fluid particles.

We shall often be considering problems in which $F = 0$, the cause of motion being either imposed pressure differences or relative movement of boundaries. No body forces are applied. It might be objected that, although one can well imagine all other sources of body force being eliminated, almost every flow will take place in a gravity field. It can be shown that gravitational body forces act significantly only on density differences. If the density is uniform, the gravitational force is balanced by a vertical pressure gradient which is present whether or not the fluid is moving and which does not interact with any flow. This hydrostatic balance can be subtracted out of the dynamical equation and the problem reduced to one without body forces. This assumes, of course, that the fluid region is supported at the bottom; flow under gravity down a vertical pipe provides an obvious example where this is not so and where the above remarks do not apply. We will consider the justification for subtracting out the hydrostatic balance more formally in Section 14.2.

The Navier–Stokes equation may be rewritten

$$\frac{\partial \boldsymbol{u}}{\partial t} + \boldsymbol{u} \cdot \nabla \boldsymbol{u} = -\frac{1}{\rho} \nabla p + \nu \nabla^2 \boldsymbol{u} + \frac{1}{\rho} \boldsymbol{F}, \qquad (5.22)$$

where $\nu = \mu/\rho$ and is a property of the fluid called the kinematic viscosity.

We notice that the equation is a non-linear partial differential equation in $\boldsymbol{u}$. The non-linearity arises from the dual role of the velocity in determining the acceleration of a fluid particle, as mentioned in Section 5.5. This non-linearity is responsible for much of the mathematical difficulty of fluid dynamics, and is the principal reason why our knowledge of the behaviour of fluids in motion is obtained in many cases from observation (both of laboratory experiments and of natural phenomena) rather than from theoretical prediction. The physical counterpart of the mathematical difficulty is the variety and complexity of fluid dynamical phenomena; without the non-linearity the range of these would be much more limited.

The continuity eqn (5.10) and the Navier–Stokes eqn (5.21) constitute a pair of simultaneous partial differential equations. Both represent physical laws which will always apply to every fluid particle. Together they provide one scalar equation and one vector equation—effectively four simultaneous equations—for one scalar variable (the pressure) and one vector variable (the velocity)—effectively four unknown quantities. The number of unknowns is thus correctly matched to the number of equations. We see that the pressure must necessarily be an intrinsic variable in fluid dynamical problems for there to be enough variables to satisfy the basic laws of mechanics.

For many particular problems it is convenient to use the equations referred to a coordinate system rather than the vectorial forms. Most often one uses Cartesian coordinates, but sometimes a curvilinear system is suggested by the geometry. Listed below are the forms taken by the continuity eqn (5.10) and the three components of the Navier–Stokes eqn (5.21) in Cartesian, cylindrical polar, and spherical polar co-ordinates:

1. Cartesian coordinates:

$$\frac{\partial u}{\partial x} + \frac{\partial v}{\partial y} + \frac{\partial w}{\partial z} = 0 \tag{5.23}$$

$$\rho\left[\frac{\partial u}{\partial t} + u\frac{\partial u}{\partial x} + v\frac{\partial u}{\partial y} + w\frac{\partial u}{\partial z}\right] = -\frac{\partial p}{\partial x} + \mu\left[\frac{\partial^2 u}{\partial x^2} + \frac{\partial^2 u}{\partial y^2} + \frac{\partial^2 u}{\partial z^2}\right] + F_x \tag{5.24}$$

together with similar equations for v and w.

2. Cylindrical polar coordinates (r = distance from axis, ϕ = azimuthal angle about axis, z = distance along axis):

$$\frac{\partial u_r}{\partial r} + \frac{u_r}{r} + \frac{1}{r}\frac{\partial u_\phi}{\partial \phi} + \frac{\partial u_z}{\partial z} = 0 \tag{5.25}$$

$$\rho\left[\frac{\partial u_r}{\partial t} + u_r\frac{\partial u_r}{\partial r} + \frac{u_\phi}{r}\frac{\partial u_r}{\partial \phi} + u_z\frac{\partial u_r}{\partial z} - \frac{u_\phi^2}{r}\right] = -\frac{\partial p}{\partial r}$$

$$+ \mu\left[\frac{\partial^2 u_r}{\partial r^2} + \frac{1}{r}\frac{\partial u_r}{\partial r} - \frac{u_r}{r^2} + \frac{1}{r^2}\frac{\partial^2 u_r}{\partial \phi^2} + \frac{\partial^2 u_r}{\partial z^2} - \frac{2}{r^2}\frac{\partial u_\phi}{\partial \phi}\right] + F_r \tag{5.26}$$

$$\rho\left[\frac{\partial u_\phi}{\partial t} + u_r\frac{\partial u_\phi}{\partial r} + \frac{u_r u_\phi}{r} + \frac{u_\phi}{r}\frac{\partial u_\phi}{\partial \phi} + u_z\frac{\partial u_\phi}{\partial z}\right] = -\frac{1}{r}\frac{\partial p}{\partial \phi}$$

$$+ \mu\left[\frac{\partial^2 u_\phi}{\partial r^2} + \frac{1}{r}\frac{\partial u_\phi}{\partial r} - \frac{u_\phi}{r^2} + \frac{1}{r^2}\frac{\partial^2 u_\phi}{\partial \phi^2} + \frac{\partial^2 u_\phi}{\partial z^2} + \frac{2}{r^2}\frac{\partial u_r}{\partial \phi}\right] + F_\phi \tag{5.27}$$

$$\rho\left[\frac{\partial u_z}{\partial t} + u_r\frac{\partial u_z}{\partial r} + \frac{u_\phi}{r}\frac{\partial u_z}{\partial \phi} + u_z\frac{\partial u_z}{\partial z}\right] = -\frac{\partial p}{\partial z}$$

$$+ \mu\left[\frac{\partial^2 u_z}{\partial r^2} + \frac{1}{r}\frac{\partial u_z}{\partial r} + \frac{1}{r^2}\frac{\partial^2 u_z}{\partial \phi^2} + \frac{\partial^2 u_z}{\partial z^2}\right] + F_z. \tag{5.28}$$

3. Spherical polar coordinates (r = distance from origin, θ = angular displacement from reference direction, ϕ = azimuthal angle about line

$\theta = 0$):

$$\frac{\partial u_r}{\partial r} + \frac{2u_r}{r} + \frac{1}{r}\frac{\partial u_\theta}{\partial \theta} + \frac{u_\theta \cot \theta}{r} + \frac{1}{r \sin \theta}\frac{\partial u_\phi}{\partial \phi} = 0 \qquad (5.29)$$

$$\rho\left[\frac{\partial u_r}{\partial t} + u_r\frac{\partial u_r}{\partial r} + \frac{u_\theta}{r}\frac{\partial u_r}{\partial \theta} + \frac{u_\phi}{r \sin \theta}\frac{\partial u_r}{\partial \phi} - \frac{u_\theta^2}{r} - \frac{u_\phi^2}{r}\right]$$

$$= -\frac{\partial p}{\partial r} + \mu\left[\frac{\partial^2 u_r}{\partial r^2} + \frac{2}{r}\frac{\partial u_r}{\partial r} - \frac{2u_r}{r^2} + \frac{1}{r^2}\frac{\partial^2 u_r}{\partial \theta^2} + \frac{\cot \theta}{r^2}\frac{\partial u_r}{\partial \theta}\right.$$

$$\left. + \frac{1}{r^2 \sin^2 \theta}\frac{\partial^2 u_r}{\partial \phi^2} - \frac{2}{r^2}\frac{\partial u_\theta}{\partial \theta} - \frac{2u_\theta \cot \theta}{r^2} - \frac{2}{r^2 \sin \theta}\frac{\partial u_\phi}{\partial \phi}\right] + F_r \quad (5.30)$$

$$\rho\left[\frac{\partial u_\theta}{\partial t} + u_r\frac{\partial u_\theta}{\partial r} + \frac{u_r u_\theta}{r} + \frac{u_\theta}{r}\frac{\partial u_\theta}{\partial \theta} + \frac{u_\phi}{r \sin \theta}\frac{\partial u_\theta}{\partial \phi} - \frac{u_\phi^2 \cot \theta}{r}\right]$$

$$= -\frac{1}{r}\frac{\partial p}{\partial \theta} + \mu\left[\frac{\partial^2 u_\theta}{\partial r^2} + \frac{2}{r}\frac{\partial u_\theta}{\partial r} - \frac{u_\theta}{r^2 \sin^2 \theta} + \frac{1}{r^2}\frac{\partial^2 u_\theta}{\partial \theta^2}\right.$$

$$\left. + \frac{\cot \theta}{r^2}\frac{\partial u_\theta}{\partial \theta} + \frac{1}{r^2 \sin^2 \theta}\frac{\partial^2 u_\theta}{\partial \phi^2} + \frac{2}{r^2}\frac{\partial u_r}{\partial \theta} - \frac{2\cot \theta}{r^2 \sin \theta}\frac{\partial u_\phi}{\partial \phi}\right] + F_\theta \quad (5.31)$$

$$\rho\left[\frac{\partial u_\phi}{\partial t} + u_r\frac{\partial u_\phi}{\partial r} + \frac{u_r u_\phi}{r} + \frac{u_\theta}{r}\frac{\partial u_\phi}{\partial \theta} + \frac{u_\theta u_\phi \cot \theta}{r} + \frac{u_\phi}{r \sin \theta}\frac{\partial u_\phi}{\partial \phi}\right]$$

$$= -\frac{1}{r \sin \theta}\frac{\partial p}{\partial \phi} + \mu\left[\frac{\partial^2 u_\phi}{\partial r^2} + \frac{2}{r}\frac{\partial u_\phi}{\partial r} - \frac{u_\phi}{r^2 \sin^2 \theta} + \frac{1}{r^2}\frac{\partial^2 u_\phi}{\partial \theta^2} + \frac{\cot \theta}{r^2}\frac{\partial u_\phi}{\partial \theta}\right.$$

$$\left. + \frac{1}{r^2 \sin^2 \theta}\frac{\partial^2 u_\phi}{\partial \phi^2} + \frac{2}{r^2 \sin \theta}\frac{\partial u_r}{\partial \phi} + \frac{2\cot \theta}{r^2 \sin \theta}\frac{\partial u_\theta}{\partial \phi}\right] + F_\phi. \quad (5.32)$$

(One sometimes requires also the individual stress components in polar coordinates. These will be introduced as required. For a more systematic treatment the reader is referred to Refs. [11] and [22].)

5.7 Boundary conditions

Since the governing equations of fluid motion are differential equations, the specification of any problem must include the boundary conditions. We would expect this on physical grounds; the motion throughout a fluid region is evidently influenced by the presence and motion of walls or other boundaries. We examine now the form taken by the conditions on the velocity field applying at boundaries.

There are obviously various types of boundary, giving rise to different possible conditions. However, the only case that we need consider in any detail for the purposes of this book is the most common type of boundary to a fluid region—the rigid impermeable wall.

One condition applying at such a wall is obviously provided by the requirement that no fluid should pass through the wall. If the wall is moving with velocity U and the velocity of a fluid particle right next to the wall is u, then this means that the normal components of these two velocities must be the same:

$$u \cdot \hat{n} = U \cdot \hat{n}, \qquad (5.33)$$

where $\hat{n}$ is the unit normal to the surface (Fig. 5.5). One often chooses a frame of reference in which the boundaries are at rest, giving $U = 0$; this boundary condition then becomes

$$u \cdot \hat{n} = 0 \qquad (5.34)$$

or in Cartesian coordinates with x and z in the local tangential plane to the wall and y normal to it,

$$v = 0. \qquad (5.35)$$

Another condition is provided by the no-slip condition, already mentioned in Section 2.2, that there should be no relative tangential velocity between a rigid wall and the fluid immediately next to it. Formally,

$$u \times \hat{n} = U \times \hat{n} \qquad (5.36)$$

and, when $U = 0$,

$$u \times \hat{n} = 0 \qquad (5.37)$$

or

$$u = w = 0. \qquad (5.38)$$

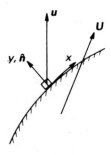

FIG. 5.5   Velocity vectors of solid and of fluid particle immediately next to its surface. (Note: U and u are shown different for definition purposes, although the text subsequently shows them to be the same.)

It is apparent that some such condition must pertain. For example, without it, there would be no boundary condition on u (eqns (2.7) and (2.16)) for pipe or channel flow and the solutions (eqns (2.8) and (2.17)) could not be obtained; viscous action would place no limit on the amount of fluid per unit time that could pass through a pipe under a given pressure gradient. However, it is not so apparent that the condition should take this exact form. The notion underlying the no-slip condition is that the interaction between a fluid particle and a wall is similar to that between neighbouring fluid particles. Within a fluid there cannot be any finite discontinuity of velocity. This would involve an infinite velocity gradient and so produce an infinite viscous stress that would destroy the discontinuity in an infinitesimal time. If, therefore, the wall acts like further fluid, the action of viscosity prevents a discontinuity in velocity between the wall and fluid; the no-slip condition must apply. However, this concept that the wall acts like further fluid is itself an assumption and the justification for the no-slip condition lies ultimately in experimental observation. This experimental justification takes two forms. The first is direct observation; dye or smoke introduced very close to a wall does stay at rest relative to the wall. The second is an *a posteriori* justification; the no-slip condition is assumed and the solutions of the equations found in simple cases; agreement between theory and experiment then justifies the original assumption. The no-slip condition is found to be violated only when the molecular-mean-free path becomes comparable with the distances involved; then the continuum equations are ceasing to be applicable anyway.

Conditions (5.33) and (5.36) in combination do, of course, give

$$u = U \tag{5.39}$$

or, in Cartesian coordinates on a wall at rest,

$$u = v = w = 0. \tag{5.40}$$

The total boundary condition is simply that there is no relative motion between a wall and the fluid next to it. It is, however, important to note (and it will be of significance subsequently, in Section 8.3) that the physical origins of the two parts of the condition are quite different. The no-slip condition depends essentially on the action of viscosity, whilst the impermeability condition does not.

This is a convenient point to mention the forces exerted by a moving fluid on a rigid boundary, a matter of obvious practical importance. This again uses the notion that the fluid acts on a wall in the same way as it acts on further fluid. However, no assumption is involved here; the stresses must be continuous or a fluid particle at a wall would experience infinite acceleration. Thus we may use the expressions in Section 5.6 for stresses in the interior of the fluid. We use Cartesian coordinates as above with the wall at $y = 0$. Then the viscous stresses in the x-, y- and z-directions can be extracted from expression (5.19) and are given in the first place as

$$\mu\left(\frac{\partial u}{\partial y} + \frac{\partial v}{\partial x}\right)_{y=0}; \quad 2\mu\left(\frac{\partial v}{\partial y}\right)_{y=0}; \quad \mu\left(\frac{\partial w}{\partial y} + \frac{\partial v}{\partial z}\right)_{y=0}. \qquad (5.41)$$

Since, from the continuity equation,

$$\frac{\partial v}{\partial y} = -\frac{\partial u}{\partial x} - \frac{\partial w}{\partial z} \qquad (5.42)$$

and since, also,

$$u = v = w = 0 \quad \text{at} \quad y = 0 \quad \text{for all } x \text{ and } z \qquad (5.43)$$

these reduce to

$$\mu\left(\frac{\partial u}{\partial y}\right)_{y=0}; \quad 0; \quad \mu\left(\frac{\partial w}{\partial y}\right)_{y=0}. \qquad (5.44)$$

The first and third quantities are the tangential forces per unit area on the wall. There is no normal viscous force, but there is a pressure force of p per unit area in the $-y$-direction.

There are other important types of boundary besides the rigid impermeable wall. The free surface of a liquid is an obvious example. And a rigid wall with suction or injection of fluid through it has important practical applications, such as some aircraft wings where the flow is controlled by suction of the air. However, we do not need to formulate the corresponding boundary conditions for the purposes of this book.

One other case does need mention—when the boundary condition is applied at infinity. Often any boundaries are far from the region of interest, as in the examples of an aeroplane well above the ground or a small model placed in a wind-tunnel. The motion far from such an obstacle is the same as in the absence of the obstacle, and one has a boundary condition of the form

$$u \to u_0 \quad \text{as} \quad r \to \infty. \qquad (5.45)$$

5.8 Condition for incompressibility

We have seen that both the continuity equation and the dynamical equation simplify greatly if one can treat the density, ρ, as a constant. Fortunately, there are many situations in which it is a good approximation to do so. Brief consideration has been given to this matter in Section

1.2(3) and we now examine more fully the conditions in which one may make this approximation.

The status of the equation

$$\rho = \text{const.} \qquad (5.46)$$

is that of an equation of state. That is to say, in circumstances where it is not a good approximation, one needs instead an equation of state giving the density as a function of pressure and temperature.

Correspondingly, when one does take the density as being constant, one is saying that the density variations produced by the pressure and temperature variations are sufficiently small to be unimportant. In this section we shall be principally concerned with the effect of pressure variations, as we have already seen that such variations are intrinsic to any flow. We shall derive a criterion for these to have a negligible effect. Non-fulfilment of this criterion is the most familiar reason for departures from eqn (5.46); consequently this equation is called the incompressibility condition, although the name is not a complete summary of the requirements for it to be applicable.

Temperature variations are also in principle intrinsic to any flow. (They may also be introduced specifically but we leave to Section 14.2 and the appendix to Chapter 14 the corresponding considerations for that case.) Firstly, the expansions and contractions as the fluid moves through the pressure field involve temperature changes. The effect of these will, however, be covered by the following discussion of the pressure effect; nowhere will it be assumed that the changes are isothermal. Secondly, viscous action involves the dissipation of mechanical energy (see Section 11.7), which reappears as heat; we will consider this briefly at the end of this section.

Liquids are known to change their density very little even for large pressure changes. One would expect to be able to treat these as incompressible. It is less apparent that there are important circumstances in which gases can be so treated, although the result is perhaps not wholly unexpected if one recalls that the fractional change in the atmospheric pressure (and so the fractional change in the air density) is small even when strong winds are blowing. The following derivation of the criterion is thus of importance primarily for gases, although it is in fact quite general.

We can write the density

$$\rho = \rho_0 + \Delta\rho \qquad (5.47)$$

where ρ_0 is a reference density—for example, the density at some arbitrarily chosen point—and $\Delta\rho$ is the local departure from this. If

$$\Delta\rho/\rho_0 \ll 1 \qquad (5.48)$$

then, for example, the term $\rho \, Du/Dt$ in the dynamical equation can be approximated by $\rho_0 \, Du/Dt$. A similar comment applies to the other places where ρ appears in the continuity and dynamical equations. Thus the equations with ρ const. may be used when relationship (5.48) is satisfied.

Since the density changes under consideration result from pressure variations, in order to estimate the typical size of $\Delta\rho$ we need to know the typical size ΔP of these pressure variations. We get this information from the requirement that the pressure force must be balanced by other terms in the Navier–Stokes equation and thus will be of the same order of magnitude as at least one other term. (If the flow is produced by imposed pressure differences, then at the start of the motion the fluid will accelerate until terms involving the velocity become comparable with the pressure force. If the flow is produced by imposed velocity differences, then the flow will generate pressure differences of an appropriate size.) We confine attention to steady flow without body forces, and so, either

$$\nabla p \sim \rho \boldsymbol{u} \cdot \nabla \boldsymbol{u} \qquad (5.49)$$

or

$$\nabla p \sim \mu \nabla^2 \boldsymbol{u} \qquad (5.50)$$

or both (with the symbol $\sim$ meaning 'is of the same order of magnitude as'). We shall pursue the consequences of (5.49). We shall see in Chapter 8 that the only circumstances when (5.50) applies whilst (5.49) does not are when the Reynolds number is low, and the following analysis does not then apply.

Provided that the x-axis is chosen in a direction in which significant variations occur, (5.49) can be written

$$\frac{\partial p}{\partial x} \sim \rho u \frac{\partial u}{\partial x} = \tfrac{1}{2}\rho \frac{\partial u^2}{\partial x}. \qquad (5.51)$$

This indicates that

$$\Delta P/L \sim \rho \Delta(U^2)/L \qquad (5.52)$$

where ΔP and $\Delta(U^2)$ are typical differences in p and u^2 between points a distance L apart. This means that, if one arbitrarily chose many such pairs of points, the average difference in p would be of the general size ΔP. Since ΔP and $\Delta(U^2)$ are defined only as order-of-magnitude quantities they do not require more precise definition than that.

If L is the general length scale of the flow, ΔP and $\Delta(U^2)$ are the orders of magnitude of the variations of p and u^2 within the fluid; and are related by

$$\Delta P \sim \rho \Delta(U^2). \qquad (5.53)$$

We do not need to maintain the distinction between a typical difference in the (square of the) velocity and a typical value of the (square of the) velocity itself; i.e. we can write

$$\Delta U \sim U; \quad \Delta(U^2) \sim U^2 \tag{5.54}$$

where U is a velocity scale. The reason for this is that one always can (and normally will) choose a frame of reference in which some points of the flow, for example those at a boundary, are at rest. (It would be perverse to analyse the dynamics of a low-speed aeroplane from the frame of reference of a high-speed aeroplane.) On the other hand, it is necessary to maintain the distinction between the pressure difference scale and the pressure itself. Since the pressure appears only in the form ∇p in the governing equations, the absolute pressure can be increased indefinitely without directly† altering the dynamics; only pressure differences are relevant.

Thus we have

$$\Delta P \sim \rho U^2 \tag{5.55}$$

which indicates the typical pressure variation in a flow of typical speed U. We now use this to determine the typical density variation. This depends on the fluid and in particular on its compressibility, β;

$$\Delta \rho / \rho \sim \beta \Delta P. \tag{5.56}$$

(For order of magnitude considerations it does not matter whether β is the isothermal compressibility, the adiabatic compressibility or what.) The final result is given in a convenient form if we now introduce the speed of sound, a, in the fluid:

$$a^2 = \left(\frac{\partial p}{\partial \rho}\right)_s \sim \frac{1}{\rho \beta} \tag{5.57}$$

(S = entropy). Here a is introduced simply as a property of the fluid under consideration, a measure of its compressibility.

Combining the various relationships, we have

$$\frac{\Delta \rho}{\rho} \sim \frac{\Delta P}{\rho a^2} \sim \frac{U^2}{a^2}. \tag{5.58}$$

Thus criterion (5.48) is fulfilled if

$$(\text{Ma})^2 = U^2/a^2 \ll 1. \tag{5.59}$$

Flows at speeds low compared with the speed of sound in the fluid thus behave as if the fluid were incompressible.

† It can do so indirectly by changing fluid properties.

The ratio U/a is known as the Mach number of the flow, and incompressible flows thus occur at low Mach number. The fact that $(\text{Ma})^2$ is involved in relationship (5.59) means that Ma does not have to be very small; when Ma is less than about 0.2, density variations are only a few per cent, bringing the accuracy of the incompressibility assumption to within the sort of accuracy attainable in many fluid dynamical investigations.

Many important gas flows do occur at low Mach number. For example, the speed of sound in air under atmospheric conditions is around $300\,\text{m s}^{-1}$. Evidently, one will often be concerned with speeds low compared with this.

The fact that liquids are much less compressible than gases is contained in this analysis by the fact that they have much higher sound speeds, thus giving lower Mach numbers at the same U.

One general comment about the nature of the above argument may be made. The way in which ρ enters the governing equations is important in deciding that relationship (5.48) is a justification for treating ρ as a constant. Pressure provides an immediate counter-example in the present context. The fractional pressure change may also be small; in fact, for a gas,

$$\Delta P/p \sim \Delta\rho/\rho. \qquad (5.60)$$

This does not mean that the pressure can be treated as a constant. As we have already noted, the pressure appears only in ∇p and the absolute pressure can be altered at will without changing the equations; comparison with it is thus irrelevant.

It was remarked that the above analysis does not apply at low Reynolds number. Then one has to use (5.50) instead of (5.49). The corresponding analysis then gives the criterion for incompressibility as

$$(\text{Ma})^2 \ll \text{Re} \qquad (5.61)$$

where Re is the Reynolds number—a somewhat academic result, rarely relevant to real situations. (But see Section 26.5.)

A similar treatment of viscous dissipation shows that it also frequently has negligible effect. (It is, for example, not noticeably warmer swimming at the bottom of a waterfall than at the top.) The details will not be given here; the corresponding matter in the topic of free convection is discussed in the appendix to Chapter 14. As in that case, the criterion for the resulting density changes to be negligible can be expressed in a form not involving the viscosity coefficient and related to the criterion for incompressibility. This effect can be ignored wherever (5.59) is fulfilled. Hence, this provides an adequate criterion for the use of the incompressible flow equations.

In general, the character of low Mach number flow may be sum-marized by saying that, from a thermodynamic point of view, the whole flow is only a perturbation. This remark is most directly applicable to gases—the corresponding considerations for liquids are more complicated—and in fact we use the properties of a perfect gas to illustrate it. Then (by putting the appropriate equation of state into (5.57))

$$a^2 = \gamma R T \qquad (5.62)$$

where γ is the ratio of specific heat capacities, R the gas constant (the universal gas constant divided by the molecular weight, see comment in the Notation section, p. 468) and T the temperature. Also the internal energy per unit mass (often denoted by U in texts in thermodynamics) is

$$E = C_V T. \qquad (5.63)$$

γR and C_V are of the same order of magnitude. Hence, when $U^2 \ll a^2$, the flow kinetic energy per unit volume is small compared with the internal energy per unit volume

$$\tfrac{1}{2} \rho U^2 \ll \rho E. \qquad (5.64)$$

Thus, for example, if all the energy of a flow were dissipated viscously, the net change in internal energy would be fractionally small. More generally none of the processes involved in low Mach number flow involve major changes in thermodynamic state. This is just what is required for a gas to have only small fractional changes in density.

Incidentally, it follows from the above that the flow velocity in low Mach number flow of a gas is small compared with the typical Brownian velocity of a molecule. Any individual molecule is moving much faster as a result of its thermal motions than as a result of the flow. It is only because the Brownian velocity averages very closely indeed to zero when one considers a large number of molecules that the flow emerges as a significant net effect.

Thus there is a sense in which all incompressible flows are 'only a perturbation'. But that remark will scarcely impress someone standing in a wind of 100 km/hour (Ma ≈ 0.1)!

Appendix: Derivation of viscous term of dynamical equation

This appendix presents the main points of the mathematical formulation of the ideas described physically in Section 5.6. For a more complete treatment, see, for example, Refs. [11, 14, 19, 20, 26]. We consider briefly each of the stress tensor and rate of strain tensor, and then

EQUATIONS OF MOTION

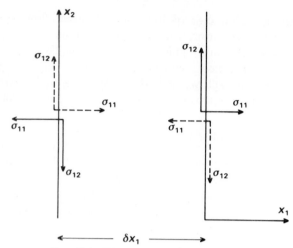

FIG. 5.6 The stress tensor: see text.

determine the consequences of a linear relationship between them. The summation convention for repeated algebraic suffixes applies throughout.

Two of the nine components of the stress tensor σ_{ij} in Cartesian coordinates (x_1, x_2, x_3) are shown in Fig. 5.6. We consider for the moment just the left-hand half of this figure, where the stresses acting across a surface normal to the x_1 direction are shown (σ_{13} is normal to the page). The same convention as in Fig. 1.1 is adopted, that the side of the surface of the fluid on which the stress acts is indicated by a slight displacement of the roots of the arrows; stresses on opposite sides are, of course, exactly equal and opposite. The first suffix on each component indicates the orientation of the surface across which it acts and the second indicates the direction in which it acts (with the sign convention that σ_{ij} is positive when the stress on the lower x_i side is in the positive x_j direction). Some authors use the suffixes the other way round; since the stress tensor can be shown to be symmetric (otherwise infinite angular accelerations would arise) this does not much matter. Definition of nine components in this way provides a complete specification of the stress; components acting across surfaces of other orientations are given by the usual rules for the effect of coordinate rotation.

The total stress acting in a fluid has contributions from both pressure and viscous effects,

$$\sigma_{ij} = -p\delta_{ij} + \tau_{ij}. \tag{5.65}$$

Inclusion of the right-hand half of Fig. 5.6 extends it analogously to the extension from Fig. 1.1 to Fig. 2.3. The net force on a fluid particle is given by the differences in the stresses acting across opposite faces. The solid arrows in Fig. 5.6 are those acting on the fluid between the two surfaces. For example, a net force in the x_1-direction arises from the difference in the two values of σ_{11}. However, forces in this direction arise also from variations of σ_{21} in the x_2-direction and of σ_{31} in the x_3-direction. Consequently, the total force per unit volume in the x_j-direction is

$$\frac{\partial \sigma_{ij}}{\partial x_i} = -\frac{\partial p}{\partial x_j} + \frac{\partial \tau_{ij}}{\partial x_i}. \tag{5.66}$$

What determines the viscous stress τ_{ij}? Figure 5.7 illustrates the effect of velocity gradients in the fluid. Two material points A and B are instantaneously at vectorial positions x_i and $x_i + \delta x_i$ and thus separated by δx_i. They are moving with velocities u_i and $u_i + \delta u_i$ so that after a time δt they are at A' and B' separated by $\delta x_i + \delta u_i\, \delta t$. The strain is related to the change $\delta u_i\, \delta t$ normalized by the original separation δx_i. (Doubling δx_i will double δu_i provided that both are infinitesimal.) Hence, rates of strain are related to the velocity gradient tensor

$$\zeta_{ij} = \partial u_i / \partial x_j. \tag{5.67}$$

However, there are some velocity gradient fields that involve no changes in the length of any material line, and thus no distortions and no viscous effects. Suppose the length of AB is δl,

$$(\delta l)^2 = \delta x_i\, \delta x_i \tag{5.68}$$

and so

$$\frac{D(\delta l)^2}{Dt} = 2\, \delta x_i \frac{D(\delta x_i)}{Dt} = 2\delta x_i\, \delta u_i$$

$$= 2\, \delta x_i\, \delta x_j \frac{\partial u_i}{\partial x_j} = \delta x_i\, \delta x_j \left(\frac{\partial u_i}{\partial x_j} + \frac{\partial u_j}{\partial x_i} \right) = \delta x_i\, \delta x_j (\zeta_{ij} + \zeta_{ji}). \tag{5.69}$$

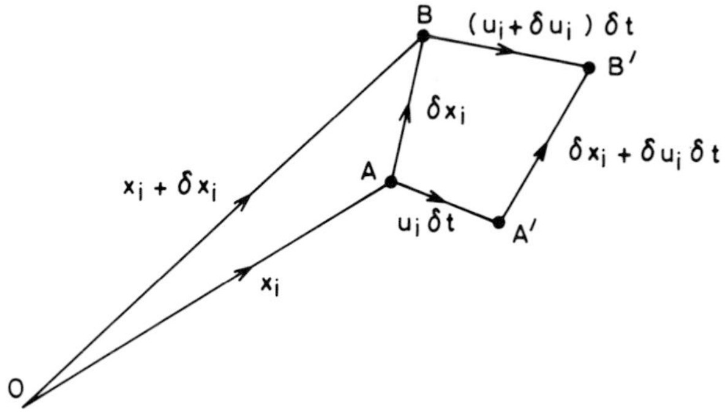

FIG. 5.7 Relative movements in flow: see text.

It is thus the symmetrical combinations of the velocity gradients that give rise to rates of strain. (A motion in which

$$\frac{\partial u_i}{\partial x_j} = -\frac{\partial u_j}{\partial x_i} \tag{5.70}$$

for all i and j does not involve any distortions. It is in fact some combination of uniform translation and rigid body rotation. The step from the fourth to the fifth expression in (5.69) makes explicit the otherwise not immediately apparent fact that the summation over i and j in the fourth expression involves terms that cancel when (5.70) is true.) Hence, the rate of strain tensor e_{ij} is the symmetric part of ζ_{ij},

$$e_{ij} = \tfrac{1}{2}(\zeta_{ij} + \zeta_{ji}). \tag{5.71}$$

(Parenthetically, the antisymmetric part

$$\eta_{ij} = \tfrac{1}{2}(\zeta_{ij} - \zeta_{ji}) \tag{5.72}$$

corresponds to the vorticity—see Section 6.4;

$$\eta_{ij} = -\tfrac{1}{2}\,\epsilon_{ijk}\omega_k.) \tag{5.73}$$

For a Newtonian fluid τ_{ij} is linearly related to e_{ij}:

$$\tau_{ij} = \Lambda_{ijkl}e_{kl}. \tag{5.74}$$

The physical processes must be independent of the orientation and handedness of the axes. Λ_{ijkl} must thus be an isotropic tensor and the most general form it can take is [212]

$$\Lambda_{ijkl} = \lambda\,\delta_{ij}\,\delta_{kl} + \xi\,\delta_{ik}\,\delta_{jl} + \chi\,\delta_{il}\,\delta_{jk}. \tag{5.75}$$

This gives

$$\tau_{ij} = \lambda\,\delta_{ij}e_{kk} + (\xi + \chi)e_{ij} \tag{5.76}$$

(since $e_{ij} = e_{ji}$). There are thus two arbitrary constants involved; these are physical properties of the particular fluid. From the particular case $e_{12} = \tfrac{1}{2}\,\partial u/\partial y$ and all other e_{ij} equal to zero, we can identify that

$$\xi + \chi = 2\mu \tag{5.77}$$

where μ is the coefficient of viscosity introduced in Section 1.2. Also,

$$e_{kk} = \frac{\partial u_k}{\partial x_k} = \text{div }\boldsymbol{u} \tag{5.78}$$

and so

$$\tau_{ik} = \mu\left(\frac{\partial u_i}{\partial x_j} + \frac{\partial u_j}{\partial x_i}\right) + \lambda\delta_{ij}\,\text{div }\boldsymbol{u}. \tag{5.79}$$

From (5.66) the viscous force per unit volume is $\partial\tau_{ij}/\partial x_i$. Putting (5.79) into this and writing it in expanded form gives expression (5.19).

6

FURTHER BASIC IDEAS

6.1 Streamlines, streamtubes, particle paths and streaklines [388]

A streamline is defined as a continuous line within the fluid of which the tangent at any point is in the direction of the velocity at that point. Its relationship to the velocity field is thus analogous to the relationship of a line of force to an electric field. Patterns of streamlines are useful (particularly in two-dimensional flow) in providing a pictorial representation of a flow.

The streamlines for a known velocity field (u, v, w) are given as solutions of the pair of differential equations

$$\frac{dx}{u} = \frac{dy}{v} = \frac{dz}{w}. \tag{6.1}$$

Two streamlines cannot intersect except at a position of zero velocity; otherwise one would have the meaningless situation of a velocity with two directions.

A streamtube is a tubular region within the fluid bounded by streamlines. Because streamlines cannot intersect, the same streamlines pass through a streamtube at all stations along its length.

Consider two stations along a streamtube of cross-sectional areas S_1 and S_2 as in Fig. 6.1. We suppose that the cross-sections are small enough that there is negligible variation of physical quantities over them and we can say that the densities and speeds at the two stations are ρ_1 and q_1 and ρ_2 and q_2 ($q = |\boldsymbol{u}|$; we can use the scalar quantity, as the direction is by definition along the streamtube). The rate at which mass is entering the volume between the two stations is $\rho_1 q_1 S_1$; the rate at which it is leaving is $\rho_2 q_2 S_2$. If the flow is either steady or incompressible (or, but not necessarily, both) the mass in this region is not changing, and so

$$\rho_1 q_1 S_1 = \rho_2 q_2 S_2. \tag{6.2}$$

If the flow is incompressible, $\rho_1 = \rho_2$ and

$$q_2/q_1 = S_1/S_2 \tag{6.3}$$

or, for a general station along the streamtube,

$$q \propto 1/S. \tag{6.4}$$

The speed is inversely proportional to the cross-sectional area. Hence

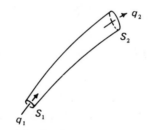

FIG. 6.1   Section of a streamtube.

where the streamlines are close together the speed is high; where they are far apart it is low. Convergence of streamlines indicates acceleration, divergence deceleration. These facts are useful in the interpretation of streamline patterns. It is important to note that we have derived them only for incompressible flow. Result (6.4) is, of course, directly related to the continuity equation, div $u = 0$, just as the parallel interpretation that the electric field is large where there is a high density of lines of force applies only when there are no space charges and so div $E = 0$.

One cannot directly observe a streamline by flow visualization. One can synthesize a streamline pattern from pictures, such as Fig. 3.7, showing the change in position of many particles during a short time interval.

In the most common types of flow visualization, dye (or smoke) is introduced at a point. One then sees either a particle path or a streakline: the former if the dye is introduced instantaneously and observed continuously or photographed with a long exposure; the latter if the dye is introduced continuously and observed or photographed instantaneously.

A particle path is effectively defined by its name. It is the trajectory of an individual element of fluid.

A streakline is the locus of all fluid elements that have previously passed through a particular fixed point.

In steady flow, streamlines, particle paths, and streaklines are all identical. A particle is instantaneously moving along a streamline; if that streamline is unchanged at a slightly later time, the particle will continue to move along it, and so it traverses the streamline. Similarly, particles starting from the same point at different times will all follow the same path, and so a streakline will consist of that path.

In unsteady flow, streamlines, particle paths, and streaklines are all different; the remarks in the last paragraph no longer apply. It is then that the interpretation of patterns obtained by flow visualization calls for care; it may not be easy to infer instantaneous flow patterns from the observations [236].

Thus, in Chapter 3, the steady flow patterns at low Reynolds number are quite fully specified by the illustrations. Although the lines in Figs. 3.2 and 3.3 were there defined as particle paths, we can now say that these figures show streamline patterns. Further, Fig. 3.4, which is in principle a sort of streakline pattern, can be used to illustrate one of these flows. When the flow becomes unsteady, any individual illustration tells a more partial story. For example, the streakline patterns in Fig. 3.5 and the streamline pattern deducible from Fig. 3.7 complement one another.

The relationships between the streamlines, particle paths, and streaklines in an unsteady flow—and the fact that they can look quite different from one another—are best seen through an example. Section 6.2 will consider one case in some detail, and we will take up the introduction of the basic ideas of fluid dynamics again in Section 6.3.

6.2 Computations for flow past a circular cylinder

In this section we look at some computed flow patterns that illustrate the relationship between streamlines, particle paths and streaklines. They concern a flow that is steady in one frame of reference. However, we consider it also in another frame so that the flow is then unsteady. (The computations thus also illustrate the relationship between patterns in different frames, a matter mentioned in Section 3.1.) This is, of course, a rather special type of unsteady flow, but it serves the present purpose well. Illustration of the same ideas for a truly unsteady flow requires cine-film; Ref. [226] provides a very effective example.

The computations described in this section are presented more fully in Ref. [388]. As well as giving an extended discussion and more detailed results for the case described below, this paper contains corresponding results for a different value of the Reynolds number, 5.

Figure 6.2 shows the result of a numerical solution [140] of the Navier–Stokes and continuity equations for flow past a circular cylinder at a Reynolds number of 40. (In this and subsequent figures only half the flow is shown; the pattern is, of course, symmetrical about the centre plane PQ.) This type of flow was the topic of Chapter 3, where we saw that 40 is about the highest Reynolds number at which there is no unsteadiness due to wake instability. In Fig. 6.2 the cylinder is at rest and the fluid far from it moves from left to right at speed u_0. The figure shows streamlines, which, since the flow pattern is steady, are also particle paths and streaklines. The closed streamlines in the attached eddies behind the cylinder are thus regions where fluid particles are moving on closed paths; the eddies always consist of the same fluid.

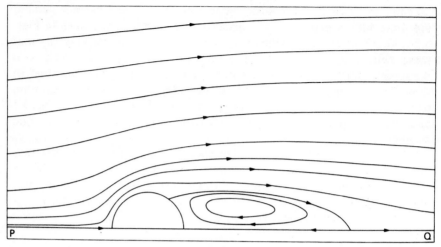

Fig. 6.2   Flow past a circular cylinder at Re = 40. Ref. [140].

Detailed experimental comparison with this computed pattern is difficult. In practice, a Reynolds number as low as 40 can be achieved only by working with a small length scale, at which observation of the flow details is not readily achievable. However, the overall features of the flow are in agreement with the observations described in Chapter 3. In addition, agreement between the value of the drag coefficient given by the computations and the observed value suggests that the details are in agreement.

Figure 6.3 shows the corresponding streamline pattern in the frame of

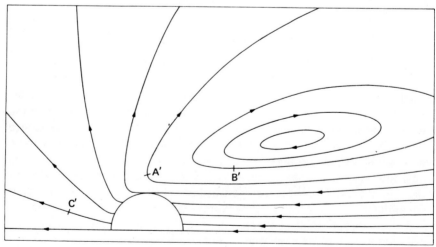

Fig. 6.3   Streamlines for cylinder moving through stationary ambient fluid.

reference in which the cylinder moves from right to left at speed u_0, the fluid far from it being a rest. It is thus related to Fig. 6.2 by the transformation of Fig. 3.1. The regions where the fluid is now (Fig. 6.3) moving from left to right correspond to regions where it was previously (Fig. 6.2) moving downstream faster than u_0; regions where it is now moving from right to left are those where it was previously moving downstream more slowly than u_0. The motion in the perpendicular direction is the same at corresponding points of the two figures.

The pattern of Fig. 6.3 is an instantaneous one; at any other time the same pattern is to be found displaced some distance to right or left.

The existence of the attached eddies is not apparent from this representation; the reason is that the speeds relative to the cylinder in the attached eddies are so small compared with u_0 that the superposition of u_0 swamps them. The fact that some of the fluid behind the cylinder moves towards it is implied in Fig. 6.3 by a region (apparent only on close examination) of fluid moving from right to left at speeds slightly above u_0.

Closed streamlines do appear in Fig. 6.3; indeed, on a figure of infinite extent, all streamlines would be closed (or begin and end on the cylinder). However, since the pattern is not steady, this does not imply that fluid particles are moving on closed paths nor indicate the existence of permanent eddies like the attached eddies.

Figure 6.4 shows particle paths in the frame of reference in which the cylinder travels from right to left. A fluid particle is initially at rest with the cylinder far to its right, is brought into motion as the cylinder approaches, follows some trajectory as the cylinder passes, and finally settles down to rest again as the cylinder moves away far to the left.

The single particle path in Fig. 6.4(a) is drawn to illustrate the relationship to the streamline pattern of fig. 6.3. At each instant the particle is moving with the velocity indicated by the appropriate point on the streamline pattern. For example, when the particle has reached point A, the cylinder is directly below it (in the position shown); the particle is thus at point A' of Fig. 6.3. The particle path at A is seen to be parallel to the streamline through A'. At a later instant, the particle is at point B. This corresponds to point B' of the streamline pattern, because, during the time in which the particle has moved from A to B, the cylinder and the whole streamline pattern have moved to the left a distance of $1.5d$ (d being the cylinder diameter). Similarly, when the cylinder was a distance $1.5d$ to the right of its illustrated position (in Fig. 6.4(a)), the particle was at position C; it was thus at position C' of the streamline pattern, and was moving accordingly.

Figure 6.4(b) shows a number of other particle paths for particles starting at different distances from the plane of symmetry. Different

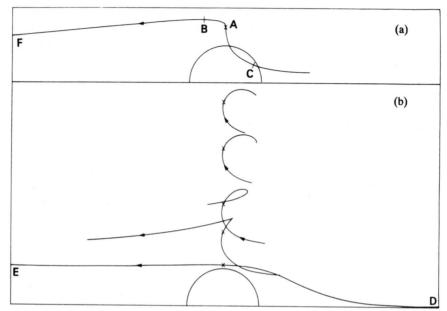

FIG. 6.4 Particle paths for cylinder moving through stationary ambient fluid. See text for significance of parts (a) and (b).

particles, initially the same distance from the plane of symmetry but separated in the direction of the cylinder's motion, follow paths of similar shapes but at different times and in different places. The paths in Fig. 6.4 have been drawn so that the particles concerned are all immediately above the cylinder at the same instant (which we may call $t = 0$); i.e. they are at the positions marked with crosses when the cylinder is in the position shown.

These particle paths are of finite length. However, computation of their complete length would require information extending outside the region for which the flow pattern has been computed. Hence, the paths shown correspond to tu_0/d (i.e. time in units of the time in which the cylinder moves one diameter) extending from -10 to 10. (For the cases where the edge of the diagram is reached first, points D, E, F correspond to values of tu_0/d of respectively -6.0, 3.7, and 7.3.) The motion before $tu_0/d = -10$ would only slightly extend the paths; but, because of wake formation, very slow motion continues for a very long time after $tu_0/d = 10$ so that complete particle paths would be significantly longer than those shown [388].

The above account covers the behaviour of all particles except those immediately to the right of the cylinder at any time. These are in the attached eddies and so are carried with the cylinder. They move from

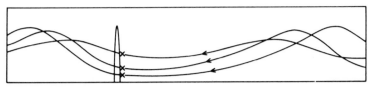

FIG. 6.5 Particle paths for fluid in attached eddies when cylinder moves through stationary ambient fluid. Scale in direction of movement contracted by 20:1 with respect to transverse scale. Cylinder is in position shown when particles are at crosses.

right to left with a slow up-and-down motion corresponding to the circulation in the attached eddy—so slow that one has to use different scales in different directions, as in Fig. 6.5, to present the computed paths. These paths are, of course, infinite ones. The existence of infinite paths shows the physical significance of the attached eddies even though they are not apparent in Fig. 6.3.

Particle paths can cross, as seen in Figs. 6.4 and 6.5. This means, of course, that the particles concerned were at the crossing position at different times.

Figure 6.6 exhibits two streaklines (the two solid lines). Dye is supposed to have been emitted continuously from a point S at a distance d from the plane of symmetry since $tu_0/d = -10$ ($t = 0$ being the instant at which the cylinder is in the position illustrated with its centre directly below the source). The lines show the computed distributions of that dye at $t = 0$ (curve SA) and at $tu_0/d = 10$ (curve SBC). Each streakline consists of particles that have passed through S at various times—the further along the streakline from S, the earlier the time.

To demonstrate the relationship between streaklines and particle paths, Fig. 6.6 also shows (broken lines) two trajectories of particles originating at the source. Line SAC originates at $tu_0/d = -10$, reaches A

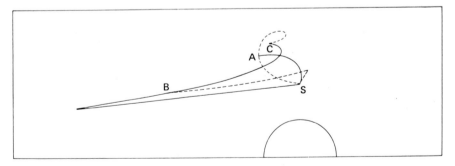

FIG. 6.6 Streaklines for cylinder moving through stationary ambient fluid (and related particle paths, shown as broken lines).

at $t = 0$ and C at $tu_0/d = 10$. Line SB originates at $t = 0$ and reaches B at $tu_0/d = 10$. These particle trajectories can be seen to be segments of paths like those in Fig. 6.4.

For reasons indicated above, these results are not of much value for direct comparison with observation. They do demonstrate, however, the complexity of the relationships between the various flow features in unsteady flow. They point out the need for care in interpreting flow visualization experiments and illustrate the type of reasoning that enters such interpretation.

6.3 The stream function

Since, for incompressible flow,

$$\text{div } \boldsymbol{u} = 0 \qquad (6.5)$$

one can put

$$\boldsymbol{u} = \text{curl } \boldsymbol{A} \qquad (6.6)$$

where $\boldsymbol{A}$ may be called a vector velocity potential. Since this replaces one vector variable, $\boldsymbol{u}$, by another vector variable, $\boldsymbol{A}$, this is not in general very useful. It is useful, however, in a two-dimensional or axisymmetric flow, where only one component of $\boldsymbol{A}$ is non-zero.

In a two-dimensional flow, for example, eqn (6.5) becomes

$$\frac{\partial u}{\partial x} + \frac{\partial v}{\partial y} = 0 \qquad (6.7)$$

and this can always be satisfied by introducing ψ such that

$$u = \frac{\partial \psi}{\partial y}, \quad v = -\frac{\partial \psi}{\partial x}. \qquad (6.8)$$

ψ is known as the stream function, because it is constant along a streamline. This can be verified as follows:

$$d\psi = \frac{\partial \psi}{\partial x} dx + \frac{\partial \psi}{\partial y} dy = -v \, dx + u \, dy. \qquad (6.9)$$

On a streamline

$$dx/u = dy/v \qquad (6.10)$$

and so

$$d\psi = 0. \qquad (6.11)$$

6.4 Vorticity [24]

The vorticity, $\boldsymbol{\omega}$, of a fluid motion is defined as

$$\boldsymbol{\omega} = \text{curl } \boldsymbol{u}. \tag{6.12}$$

It is thus a vector quantity defined at every point within the fluid. We shall see in later chapters that some flow configurations are most readily understood through a consideration of the vorticity.

The definition of curl by

$$\hat{\boldsymbol{n}} \cdot \text{curl } \boldsymbol{u} = \lim_{S \to 0} \frac{1}{S} \oint \boldsymbol{u} \cdot \text{d}\boldsymbol{l} \tag{6.13}$$

(where $\hat{\boldsymbol{n}}$ is the unit vector normal to the surface S, round the edge of which the line integral is taken) indicates that vorticity corresponds to rotation of the fluid; the line integral is non-zero only if fluid is going round the point under consideration. Flow without vorticity is called irrotational flow.

The physical meaning of 'rotation' and thus of vorticity when fluid particles can change shape (Section 5.6) requires some consideration. A convenient way of seeing how one may meaningfully talk about the rotation of a fluid particle in the presence of deformation is the following. Consider a flow, such as a two-dimensional flow, in which only the z-component of the vorticity is non-zero,

$$\omega_z = \frac{\partial v}{\partial x} - \frac{\partial u}{\partial y}, \tag{6.14}$$

and consider two short material lines instantaneously at right angles to one another in a plane perpendicular to z (Fig. 6.7). A short time later these lines of fluid may have changed their orientation as shown. (We ignore the effect of second velocity derivatives in making them curved, as we are concerned with the first-order effect at a point.) The rate at which the two lines are turning will not in general be the same, and it would not be sensible to define the rotation rate of a fluid particle in terms of the

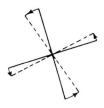

FIG. 6.7 Example of change of orientation (with translation subtracted out) in short time interval of two material lines instantaneously at right angles.

angular velocity of one such line. However, the average angular velocity of any two such lines at right angles is the same and is equal to $\omega_z/2$. This may be seen from the case in which the lines are parallel to the x- and y-axes; then the angular velocity of the former is $\partial v/\partial x$ and that of the latter is $-\partial u/\partial y$. Since the definition of curl is not related to any particular set of axes, the result is valid for other pairs of perpendicular lines in an xy-plane.

For further discussion of the physical significance of vorticity, it is convenient to refer to three examples:

1. The first is a fluid rotating as if it were a rigid body with angular velocity $\boldsymbol{\Omega}$. The velocity field is

$$\boldsymbol{u} = \boldsymbol{\Omega} \times \boldsymbol{r} \tag{6.15}$$

giving

$$\boldsymbol{\omega} = 2\boldsymbol{\Omega}. \tag{6.16}$$

The vorticity is the same at every point and equal to twice the angular velocity. For comparison with the next example we write this also in cylindrical polar coordinates:

$$u_\phi = \Omega r, \quad u_r = u_z = 0 \tag{6.17}$$

$$\omega_z = 2\Omega, \quad \omega_r = \omega_\phi = 0 \tag{6.18}$$

2. The second example again has every fluid particle moving on a circular path about the z-axis, but with a different radial distribution of azimuthal velocity:

$$u_\phi = \frac{K}{r}, \quad u_r = u_z = 0. \tag{6.19}$$

This gives

$$\omega_\phi = 0, \quad \omega_r = 0,$$

$$\omega_z = \frac{1}{r} \frac{\partial}{\partial r}(r u_\phi) = 0 \quad \text{for} \quad r \neq 0 \tag{6.20}$$

but leaves ω_z indeterminate on the axis. In fact, ω_z becomes infinite on the axis, as can be seen from Stokes's theorem:

$$\oint \boldsymbol{u} \cdot d\boldsymbol{l} = \int (\text{curl } \boldsymbol{u}) \cdot d\boldsymbol{S} = \int \boldsymbol{\omega} \cdot d\boldsymbol{S}. \tag{6.21}$$

Carrying out the line integration round a circular path at constant r makes the left-hand side equal to $2\pi K$ independently of r. Hence, there must be a singularity in ω_z such that the contribution $\omega_z \, dS_z$ from the element at $r = 0$ to the surface integral is $2\pi K$. Hence, this case corresponds to irrotational motion everywhere except on the axis. Obviously, exactly this flow cannot occur in practice; however, there are situations in which the vorticity is very high over a narrow linear region and practically zero elsewhere.

3. The third example is the simple shear-flow

$$u = u(y), \quad v = w = 0 \tag{6.22}$$

(in Cartesian coordinates) considered previously for different reasons (Sections 1.2 and 2.2). This has

$$\omega_x = \omega_y = 0, \quad \omega_z = -\partial u / \partial y. \tag{6.23}$$

Examples (2) and (3) illustrate an important distinction. Rotation, as specified by vorticity, corresponds to changing orientation in space of the fluid particle and *not* to motion of the particle on a closed path. In example (2) each fluid particle moves round a circular path, but its vorticity is zero. In example (3) each fluid particle moves in a straight line but it has vorticity. The concept of an isolated rigid particle moving on a circular path without rotating is readily illustrated (Fig. 6.8). It is here contrasted with rigid body rotation in which the particle both moves on a circular path and changes its orientation in space, so that the same side of it always faces the axis of rotation. A fluid particle in the flow of example (1) behaves like the particle in Fig. 6.8(a); its vorticity corresponds to its changing orientation. The behaviour of a fluid particle in example (2) is less closely analogous to that of the particle in Fig. 6.8(b), because it is also being deformed. Successive configurations of an initially rectangular fluid particle are shown in Fig. 6.9. For small displacements, one can see the meaning of the statements that the particle is not changing its orientation and that perpendicular sides are rotating in opposite senses to give zero vorticity.

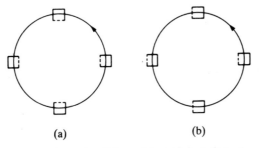

(a)   (b)

FIG. 6.8 Successive positions of solid particle in (a) rigid body rotation and (b) circulation without rotation. One side of particle is shown dotted to indicate its orientation.

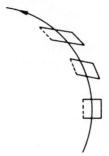

FIG. 6.9 Successive positions and shapes of initially square fluid particle circulating without vorticity (size of particle exaggerated relative to radius of circle).

Figure 6.10 illustrates the interpretation of example (3), where rotational flow in straight lines occurs. Subsequent configurations of an initially square fluid particle are parallelograms. Similar geometries may be reached by a non-rotational distortion, stretching along one diagonal and compressing along the other (corresponding mathematically to the principal axes of the rate-of-strain tensor). Rotation is then necessary, however, to give the actual orientation. Considering small displacements, we see that one of two instantaneously perpendicular sides is rotating, the other is not, thus the average is non-zero and there is vorticity.

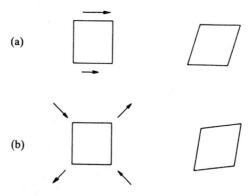

FIG. 6.10 Square in (a) shear flow, eqn (6.22), and (b) non-rotational distortion.

6.5 Circulation

The circulation round any arbitrary closed loop in the fluid is defined as

$$\Gamma = \oint \boldsymbol{u} \cdot d\boldsymbol{l} \qquad (6.24)$$

a quantity that we have already encountered in the discussion of example (2) in Section 6.4. Note that we are here concerned with a finite loop in contrast with the infinitesimal one involved in the definition of vorticity in equations (6.12) and (6.13). The circulation is related to the vorticity by Stokes's theorem, eqn (6.21). There must be vorticity within a loop round which circulation occurs.

The existence of closed streamlines in a flow pattern implies that there are loops for which $\Gamma \neq 0$ and thus that the flow is not irrotational everywhere. However, the converse does not apply: a flow without closed streamlines may (and frequently will) involve circulation. Example (3) of Section 6.4 is a case in point.

6.6 Vorticity equation

An equation for the vorticity in incompressible flow is obtained by applying the curl operation throughout the Navier–Stokes equation (5.22) (with $F = 0$). We have, as a vector identity,

$$\nabla \times (u \cdot \nabla u) = u \cdot \nabla(\nabla \times u) - (\nabla \times u) \cdot \nabla u + (\nabla \cdot u)(\nabla \times u). \quad (6.25)$$

The continuity equation (5.10) makes the last term zero, and so

$$\frac{\partial \omega}{\partial t} + u \cdot \nabla \omega - \omega \cdot \nabla u = \nu \nabla^2 \omega. \quad (6.26)$$

The pressure has disappeared because it occurs in a conservative term, but, to offset this simplification, the equation involves the velocity itself as well as the vorticity. Equation (6.25) may be rewritten

$$\frac{D\omega}{Dt} = \omega \cdot \nabla u + \nu \nabla^2 \omega, \quad (6.27)$$

giving an expression for the rate of change of the vorticity of a fluid particle.

It is useful to gain some physical understanding of the processes that tend to change the vorticity of a fluid particle; i.e. to 'interpret' each of the terms on the right-hand side of eqn (6.27). The significance of the second term on the right-hand side is readily seen through its mathematical similarity to the corresponding term in the Navier–Stokes equation. The action of viscosity produces diffusion of vorticity down a vorticity gradient, just as it produces diffusion of momentum down a momentum gradient; net changes to the vorticity of a fluid particle arise from spatial variations in the diffusion rate.

The first term of the right-hand side of eqn (6.27) represents the action of velocity variations on the vorticity. Its physical interpretation is

less straightforward, and it is convenient to develop it in the context of a
flow for which it is the only term, so that

$$D\boldsymbol{\omega}/Dt = \boldsymbol{\omega} \cdot \nabla \boldsymbol{u}. \tag{6.28}$$

This means that we are considering flow without viscous effects: inviscid
flow. We shall be putting this into its fluid dynamical context in Section
8.3 and Chapter 10.

In two-dimensional flow only one component of the vorticity is
non-zero,

$$\boldsymbol{\omega} = (0,\,0,\,\zeta). \tag{6.29}$$

(In Section 6.4, ζ was denoted by ω_z, but it is now convenient to have a
symbol without a suffix). Hence, (6.28) becomes

$$D\boldsymbol{\omega}/Dt = \zeta \partial \boldsymbol{u}/\partial z. \tag{6.30}$$

Since in two-dimensional flow also $\boldsymbol{u}$ does not vary with z, (6.30) in turn
becomes

$$D\zeta/Dt = 0. \tag{6.31}$$

The vorticity of a fluid particle is a conserved quantity in two-dimensional
inviscid flow.

In three-dimensional flow, in contrast, there are vorticity-changing
processes associated with the right-hand side of (6.28), and it is these that
we seek to understand. We may consider a fluid particle with its vorticity
instantaneously in the z-direction (just by so choosing the axes) and of
magnitude ζ. We may then again consider (6.28) in the form (6.30).
Writing this as three component equations (and denoting the components
of $\boldsymbol{\omega}$ by $(\xi,\,\eta,\,\zeta)$)

$$D\xi/Dt = \zeta \partial u/\partial z; \quad D\eta/Dt = \zeta \partial v/\partial z; \quad D\zeta/Dt = \zeta \partial w/\partial z. \tag{6.32}$$

Because $\partial u/\partial z$ and $\partial v/\partial z$ represent variations of a velocity component
in a perpendicular direction but $\partial w/\partial z$ represents variations of a velocity
component in its own direction, rather different interpretations are
needed for the first two equations of (6.32) on the one hand and the last
equation on the other. These interpretations are summarized respectively
by the phrases 'vortex twisting' and 'vortex stretching'.

Underlying these names is an important property of inviscid flow. Just
as one can draw streamlines everywhere in the direction of the local
velocity, so one can draw vortex lines everywhere in the direction of the
local vorticity. In inviscid flow a vortex line always consists of the same
fluid particles. We shall derive this result below, but complete the present
discussion first.

Consider now the first equation in (6.32). Non-zero $\partial u/\partial z$ implies that fluid particles instantaneously separated in the z-direction are acquiring a separation in the x-direction also. Hence, vorticity in the x-direction is being generated from vorticity originally in the z-direction by a process of vortex twisting. The second equation in (6.32) corresponds, of course, to just the same process.

The third equation in (6.32) says that the magnitude of ζ increases in time if $\partial w/\partial z$ is positive; i.e. if the fluid particle is elongating in the direction of its vorticity. It must then be contracting in the x- and/or y-direction to conserve mass (note that the continuity equation entered into the derivation of (6.27)). It thus rotates faster, as does a solid body of decreasing moment of inertia (e.g. a ballet dancer or skater who brings in his or her arms, or the Earth as its denser materials settle to the centre.) The vorticity increases through vortex stretching. Negative $\partial w/\partial z$ correspondingly leads to the vorticity reducing; such vortex contraction or 'negative stretching' is often supposed to be included in the term vortex stretching. (See also Section 10.2.)

This discussion has been developed in the context of inviscid flow, because only then does the above property of vortex lines apply. However, having used this context to gain some physical understanding of the processes involved in the term $\boldsymbol{\omega} \cdot \nabla \boldsymbol{u}$, we can apply this understanding to situations for which the full equation (6.27) is relevant. We can refer to vortex twisting and vortex stretching as processes occurring within a more general context. For example, a situation in which $D\boldsymbol{\omega}/Dt = 0$ but the individual terms on the right-hand side of (6.27) were non-zero might be described as arising from a balance between vortex twisting (say) and viscous diffusion of vorticity.

It remains to prove the result that a vortex line always consists of the same fluid particles in inviscid flow. We consider a short segment AB of length ε of a vortex line (Fig. 6.11). Again we choose axes so that this is instantaneously in the z-direction

$$\boldsymbol{\omega}_{t=0} = (0, 0, \zeta) \qquad (6.33)$$

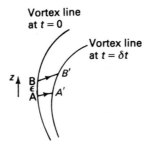

Vortex line
at $t = 0$

Vortex line
at $t = \delta t$

FIG. 6.11 Successive positions of vortex line and two points on it.

and its subsequent development is determined by equations (6.32). Then, after a small time δt, the vorticity of the same fluid is

$$\boldsymbol{\omega}_{t=\delta t} = \left(\zeta \frac{\partial u}{\partial z} \delta t, \ \zeta \frac{\partial v}{\partial z} \delta t, \ \zeta + \zeta \frac{\partial w}{\partial z} \delta t \right). \tag{6.34}$$

We now enquire about the orientation of A'B', A' being the position at $t = \delta t$ of the fluid that was at A at $t = 0$ and B' being similarly related to B. If the velocity components at A are (u, v, w), those at B are

$$\left(u + \frac{\partial u}{\partial z} \varepsilon, \ v + \frac{\partial v}{\partial z} \varepsilon, \ w + \frac{\partial w}{\partial z} \varepsilon \right).$$

Thus

$$\text{AA}' = (u\delta t, \ v\delta t, \ w\delta t) \tag{6.35}$$

$$\text{BB}' = \left[\left(u + \frac{\partial u}{\partial z} \varepsilon \right) \delta t, \ \left(v + \frac{\partial v}{\partial z} \varepsilon \right) \delta t, \ \left(w + \frac{\partial w}{\partial z} \varepsilon \right) \delta t \right] \tag{6.36}$$

and

$$\text{A}'\text{B}' = \left(\frac{\partial u}{\partial z} \varepsilon \delta t, \ \frac{\partial v}{\partial z} \varepsilon \delta t, \ \varepsilon + \frac{\partial w}{\partial z} \varepsilon \delta t \right). \tag{6.37}$$

The components of (6.37) are in the same ratios to one another as those of (6.34); the new orientation of the segment of vortex line is the same as that of the material line. Applying this result at each point along the vortex line, one concludes that it always consists of the same fluid particles.

7

DYNAMICAL SIMILARITY

7.1 Introduction [27, 28]

In only a minority of fluid dynamical situations can one determine the flow as an exact solution of the equations of motion. The necessary mathematical methods often do not exist. Even when an exact solution can be obtained it may not be unique and so may not correspond to what actually occurs. Hence, much of fluid mechanics concerns the development of both experimental and theoretical procedures for elucidating flows that cannot be rigorously calculated.

A useful starting point for this development is the following question: under what conditions do similar flow patterns occur in two geometrically similar pieces of apparatus? When these conditions are fulfilled, the two flows are said to be dynamically similar.

It is easier to think in terms of a specific system and we choose the example shown in Fig. 7.1. An elliptical obstacle of dimensions a and b as shown is placed in a channel of width c through which fluid moves. Other lengths may be necessary to specify the dimensions in the third direction. We consider two such pieces of apparatus, which are geometrically similar; i.e. one is just a scaled-up version of the other, so that

$$a_2/a_1 = b_2/b_1 = c_2/c_1 \tag{7.1}$$

(where the suffixes refer, of course, to apparatus 1 and apparatus 2). The two fluids in the two sets of apparatus may be different; ρ and μ are parameters of the problem. Given geometrical similarity, there is then the possibility that the two flow patterns will be similar; that is

$$u_{P2}/u_{P1} = v_{P2}/v_{P1} = w_{P2}/w_{P1} = u_{Q2}/u_{Q1} = v_{Q2}/v_{Q1} = w_{Q2}/w_{Q1} \tag{7.2}$$

for all pairs of geometrically similarly located points P1, Q1, P2, Q2 in the two pieces of apparatus. However, this dynamical similarity will occur only if further conditions are fulfilled, and we have to investigate what these are.

Although the discussion will refer to this example, the analysis to be given applies to all steady situations governed by the incompressible continuity equation and the Navier–Stokes equation without a body force. The same general method is applied with appropriate modifications to situations where other terms and/or other equations come in, as we shall be seeing in later chapters.

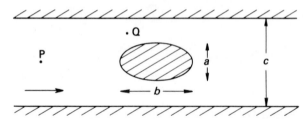

FIG. 7.1   Example of situation to which ideas in text can be applied.

There are several ways in which this question of dynamical similarity is important. It is evidently of practical importance for experiments with models. Tests of, say, an aeroplane or a ship using a small-scale model in a wind-tunnel or a towing tank will not be useful if the flow in the model experiment is quite different from that which occurs on the full scale. To take an extreme example, they will be useless if the flow with the model is laminar and that on the full scale is turbulent; less extreme differences, such as two laminar flows with different streamline patterns, are also undesirable. Dynamical similarity is likewise important in more fundamental experimental work. If one has observed a certain flow behaviour, one needs to know when this behaviour will occur and whether further experiments are necessary to investigate the full range of behaviour. Dynamical similarity also plays an important role in the theoretical development; it leads to useful approximations to the equations of motion (see Chapter 8).

7.2   Condition for dynamical similarity: Reynolds number

The derivation of the condition for dynamical similarity starts by introducing length and velocity scales, L and U. These are typical measures of the size of the apparatus and the rate at which the fluid is moving in it. In our example, a possible choice of L would be the width of the obstacle, $L = a$, and a possible choice of U would be the speed at the centre of the channel well upstream of the obstacle, $U = u_P$. However, the choices are arbitrary and the conclusion does not depend on the details of them. No pressure difference scale is introduced as the pressure differences cannot be varied independently of the velocity field. It would be possible to introduce a pressure difference scale instead of a velocity scale, but again this would alter only the form in which the conclusion is expressed, not the conclusion itself.

No time scale is introduced, because we are restricting attention to steady imposed conditions. The analysis will include time-variation of the flow, because, as we have seen in Chapters 2–4, unsteadiness can arise spontaneously. When this happens, the unsteadiness does not provide any time scale of its own; the rapidity of the fluctuations is related to the length and velocity scales. A flow that is inherently unsteady because of variations in the imposed conditions does involve an independent time scale, indicating the rapidity of these variations; this case is not covered by the following analysis.

There are two routes by which we may reach the conclusion. We will consider each in turn. The first is the quicker. The second perhaps shows better what underlies the conclusion and illustrates the method used in more complicated problems where the first route is less easy to apply.

Dynamical similarity pertains if the ratio of each of the velocity components at geometrically similar points of the flow to the velocity scale is the same in both pieces of apparatus; that is if

$$u' = u/U, \quad v' = v/U, \quad w' = w/U \qquad (7.3)$$

are the same when

$$x' = x/L, \quad y' = y/L, \quad z' = z/L \qquad (7.4)$$

are the same. We focus attention on u'; exactly parallel considerations apply to v' and w'.

u' is a dimensionless quantity. In any expression for it, dimensional quantities must appear in non-dimensional combinations. We can thus write

$$u' = f\left(\frac{b}{L}, \frac{c}{L}, \ldots, x', y', z', \frac{\rho U L}{\mu}\right). \qquad (7.5)$$

The quantities on the right-hand side are all the independent non-dimensional combinations of the physical parameters on which u' can depend. Any further non-dimensional combinations, for example $\rho U b/\mu$, are also combinations of the above and are thus implicitly included.

If all the quantities on the right-hand side of (7.5) are the same for two systems, then u' is the same. Equality of b/L, c/L, etc. in the two cases is a restatement of geometrical similarity. Equality of x', y', and z' indicates that geometrically similar points are being considered. Hence, equality of $\rho U L/\mu$ is the significant criterion.

Dynamical similarity of two geometrically similar systems of the type under consideration exists if they have the same value of the Reynolds number

$$\mathrm{Re} = \frac{\rho U L}{\mu} = \frac{U L}{\nu}. \qquad (7.6)$$

The second route to this result starts with the governing equations, in this case eqn (5.10) and eqn (5.22) with $F = 0$. The argument is clearer if we expand these in Cartesian coordinates,

$$\frac{\partial u}{\partial x} + \frac{\partial v}{\partial y} + \frac{\partial w}{\partial z} = 0 \tag{7.7}$$

$$\frac{\partial u}{\partial t} + u\frac{\partial u}{\partial x} + v\frac{\partial u}{\partial y} + w\frac{\partial u}{\partial z} = -\frac{1}{\rho}\frac{\partial p}{\partial x} + v\left(\frac{\partial^2 u}{\partial x^2} + \frac{\partial^2 u}{\partial y^2} + \frac{\partial^2 u}{\partial z^2}\right) \tag{7.8}$$

with similar equations for v and w. Length and velocity scales, L and U, are introduced as before and are used to give non-dimensional forms (denoted by primes) of the variables:

$$x' = x/L, \quad y' = y/L, \quad z' = z/L, \quad t' = tU/L,$$
$$u' = u/U, \quad v' = v/U, \quad w' = w/U, \quad (\Delta p)' = \Delta p/\rho U^2. \tag{7.9}$$

(Δp is the difference between the pressure and some reference pressure.) Separate time and pressure difference scales are not introduced for the reasons discussed earlier. This means that t and Δp are non-dimensionalized using appropriate combinations of the other scales. The ways chosen are not unique (e.g. $\mu U/L$ has the dimensions of pressure as well as ρU^2); however, different choices would make no difference to the final conclusion. Substituting (7.9) into (7.7) and (7.8) gives

$$\frac{U}{L}\left(\frac{\partial u'}{\partial x'} + \frac{\partial v'}{\partial y'} + \frac{\partial w'}{\partial z'}\right) = 0 \tag{7.10}$$

$$\frac{U^2}{L}\frac{\partial u'}{\partial t'} + \frac{U^2}{L}\left(u'\frac{\partial u'}{\partial x'} + v'\frac{\partial u'}{\partial y'} + w'\frac{\partial u'}{\partial z'}\right)$$
$$= -\frac{U^2}{L}\frac{\partial(\Delta p)'}{\partial x'} + \frac{vU}{L^2}\left(\frac{\partial^2 u'}{\partial x'^2} + \frac{\partial^2 u'}{\partial y'^2} + \frac{\partial^2 u'}{\partial z'^2}\right) \tag{7.11}$$

with similar equations for v' and w'. Hence,

$$\nabla' \cdot u' = 0 \tag{7.12}$$

$$\frac{\partial u'}{\partial t'} + u' \cdot \nabla' u' = -\nabla(\Delta p)' + \frac{1}{\text{Re}}\nabla'^2 u' \tag{7.13}$$

where

$$\nabla' = \hat{x}\frac{\partial}{\partial x'} + \hat{y}\frac{\partial}{\partial y'} + \hat{z}\frac{\partial}{\partial z'} \tag{7.14}$$

$$u' = u/U \tag{7.15}$$

and Re is the Reynolds number, UL/v. The boundary conditions can be similarly converted to conditions in the non-dimensional variables. We see that if the Reynolds number is the same for two geometrically similar situations, then the equations for the non-dimensional variables are the same. Hence, they have the same solutions and the same flow patterns occur. If the Reynolds number is different, the equations are different and there is no reason to expect the same flow behaviour. We have the same result as before that the condition for dynamical similarity is equality of the Reynolds number.

This conclusion greatly reduces the amount of work needed to obtain full information about a situation such as that in Fig. 7.1. Whereas it might seem that one would need to investigate separately the effect of varying each of L, U, ρ, and μ, in fact one need investigate only the variations with Reynolds number.

7.3 Dependent quantities

One may need to know the value of some quantity dependent on the flow. One would expect that when dynamical similarity pertains, then information about such quantities can be transferred between systems; for instance, model experiments can give information not only about the flow on full scale but also about quantities associated with it.

We may use, as an example of such a quantity, the force acting on an obstacle like that in Fig. 7.1. We denote this by D (treated as a scalar on the assumption that its direction is known from symmetry; if not the argument can apply to each component). The drag coefficient, defined as

$$C_D = D/\tfrac{1}{2}\rho U^2 L^2 \qquad (7.16)$$

is a non-dimensional form of this (the factor $\tfrac{1}{2}$ has, of course, no role in the non-dimensionalization, but is conventionally included because $\tfrac{1}{2}\rho U^2$ has a certain physical significance; see Section 10.7). It is not the only non-dimensional form, but a different choice would give an equivalent conclusion. The drag coefficient for a circular cylinder has already been introduced in Section 3.4; the appearance of L^2 in the denominator of (7.16) compared with d in (3.3) is because D is now the total force on the obstacle whereas, previously, it was the force per unit length.

In the general case, the drag coefficient depends only on the Reynolds number

$$C_D = f(\text{Re}). \qquad (7.17)$$

There are again two routes to this conclusion. The first is a direct dimensional argument: one says that the only dimensionally satisfactory expressions of

$$D = f(L, b, c, \ldots, U, \rho, \mu) \qquad (7.18)$$

are of the form

$$C_D = f\left(\frac{b}{L}, \frac{c}{L}, \ldots, \text{Re}\right) \qquad (7.19)$$

and when there is geometrical similarity (7.19) becomes (7.17).

The second route starts with the fact that forces acting on the obstacle are pressure forces of Δp per unit area and viscous forces of the form $\mu \, \partial u / \partial y$ per unit area. In the non-dimensional variables, these are $\rho U^2 (\Delta p)'$ per unit area and $(\rho U^2 / \text{Re}) \, \partial u' / \partial y'$ per unit area. When Re is the same in two cases the solutions for the primed variables are the same and the forces per unit area are proportional† to ρU^2. The total forces are thus proportional to $\rho U^2 L^2$. Hence, when Re is the same, $D / \rho U^2 L^2$ is the same; and one has eqn (7.17).

This relationship enables information about forces to be transferred from one apparatus to another geometrically similar apparatus. If Re is the same, C_D is the same; one can then infer the force in one apparatus from measurements in the other. We now have the full justification for the viewpoint adopted in Chapter 3 that Fig. 3.15 contains complete information about the drag on a circular cylinder in steady incompressible flow.

It is well known that dimensional analysis sometimes gives a more explicit expression for a dependent quantity than a relationship of the type of (7.17). This occurs when the number of dimensional quantities involved is smaller. For example, we shall see in Section 8.2 that when the Reynolds number is low, the equation of motion simplifies to a form (eqn (8.7)) not involving ρ. Thus ρ drops out of (7.18) and so out of the relationship between Re and C_D. This requires

$$C_D \propto 1/\text{Re}. \qquad (7.20)$$

When the other quantities are fixed this corresponds to the drag being directly proportional to the speed.

If μ drops out instead of ρ, similar reasoning gives

$$C_D = \text{const. w.r.t. Re} \qquad (7.21)$$

corresponding to the drag being proportional to the square of the speed. This result relates in a rather complex way to the behaviour at high Reynolds numbers (Sections 8.3 and 12.5).

Although this section has been almost entirely concerned with drag, the discussion has been intended to illustrate general principles applying to any dependent quantity. Another example is the frequency of a vortex street. We see now how the statement (eqn (3.2)) that the Strouhal

---

† As one goes from one case to the other: *not* as one varies U.

number is a function of the Reynolds number is a consequence of dynamical similarity.

7.4 Other governing non-dimensional parameters

The above discussion of the Reynolds number illustrates a general procedure. When other equations or other terms in the equations are applicable, other dimensionless combinations of the parameters may be formulated in addition to or instead of the Reynolds number. An early stage of the investigation of a new configuration is usually the formulation of the governing non-dimensional parameters, which determine the dynamical similarity. The use of the Rayleigh and Prandtl numbers to specify the convection problems in Chapter 4 arises from such considerations. This and other cases will be considered in detail in later chapters, but we look straightaway at a few simple examples to illustrate further the principles:

1. The first is a situation similar to that considered in Sections 7.1–7.3 but with a free surface on which waves can develop. In certain circumstances, important in the dynamics of ships, these waves are dominantly gravitational and in addition to U, L, ρ, and v, the acceleration due to gravity, g, is a parameter of the flow. There are two independent non-dimensional combinations of these,

$$\mathrm{Re} = UL/v \quad \text{and} \quad \mathrm{Fr} = U/(gL)^{1/2}, \tag{7.22}$$

the Reynolds number and the Froude number. Dynamical similarity requires that

$$\mathrm{Re}_1 = \mathrm{Re}_2 \quad \text{and} \quad \mathrm{Fr}_1 = \mathrm{Fr}_2 \tag{7.23}$$

(suffixes 1 and 2 referring to the two geometrically similar systems being compared). The drag coefficient, for example, has the dependence

$$C_D = f(\mathrm{Re}, \mathrm{Fr}). \tag{7.24}$$

2. Sometimes one is concerned primarily with effects of the gravitational waves and the viscous flow below the surface is unimportant. The Froude number is then the only relevant governing non-dimensional quantity, and

$$C_D = f(\mathrm{Fr}). \tag{7.25}$$

(This applies, more precisely, when nearly all the energy, generated by the work done by force D is carried away by the waves rather than dissipated by viscosity.)

3. For a situation again similar to our main example but with a speed high enough for the fluid to be compressible, dynamical similarity

requires equality between systems of both the Reynolds number and the Mach number, Ma $= U/a$ (a is the speed of sound; see Section 5.8). Thus

$$C_D = f(\text{Re}, \text{Ma}). \qquad (7.26)$$

4. When the flow is unsteady as a result of changes in the imposed conditions, these changes will have a time scale Ψ associated with them. In problems such as the above there is then the additional non-dimensional parameter $U\Psi/L$, and dynamical similarity throughout the development of the flow requires equality of this in addition to the Reynolds number.

It should be noted that, in the context of model testing, the above discussion of dynamical similarity is the statement of an ideal. It is often not possible in practice to make all the governing non-dimensional parameters the same as on the full scale. In ship model testing, for instance, a reduction in L requires an increase in U to keep the Reynolds number the same but a reduction in U to keep the Froude number the same (since there is little manoeuvreability of ρ, ν, and g). Hence, tests have to made without full dynamical similarity, and special attention must be given to the errors arising in the transfer of information to the full scale.

8

LOW AND HIGH REYNOLDS NUMBERS

8.1 Physical significance of the Reynolds number

The Reynolds number, introduced in the last chapter in the context of dynamical similarity, can be given a physical interpretation. This is useful in gaining an understanding of the dynamical processes that are important in different Reynolds number ranges, and in formulating corresponding approximations to the equations of motion.

To discuss this we need a name for each of the terms in the dynamical equation of steady incompressible flow:

$$\rho \mathbf{u} \cdot \nabla \mathbf{u} = -\nabla p + \mu \nabla^2 \mathbf{u}. \tag{8.1}$$

The second and third terms are given the obvious names pressure force and viscous force. The first term is called the inertia force. Physically, it is not a force, but it has the dimensions of force per unit volume and it is sometimes convenient to think of the dynamical equation in terms of a static balance between forces. The procedure is analogous to the more familiar use of the term centrifugal force to represent the acceleration involved in circular motion. No new idea is involved here, just a new name.

In the non-dimensional form of eqn (8.1),

$$\mathbf{u}' \cdot \nabla' \mathbf{u}' = -\nabla'(\Delta p)' + \frac{1}{\mathrm{Re}} \nabla'^2 \mathbf{u}' \tag{8.2}$$

(cf. eqn (7.13)), the primed quantities (possibly excepting $(\Delta p)'$) may be expected to be of order unity in magnitude. We shall see later that there are important qualifications to that statement. However, as a starting point it is justified so long as the length and velocity scales, U and L, have been chosen as typical quantities. Then a general speed will be of order U and $|\mathbf{u}'| \sim 1$; a general distance over which quantities vary significantly will be of order L and $\partial/\partial x'$, etc. will be of order unity.

Hence the ratio of the first term to the third in eqn (8.2) is of order Re. The corresponding terms in eqn (8.1) are in the same ratio. This indicates a physical interpretation of the Reynolds number as

$$\mathrm{Re} \sim \frac{\text{inertia forces}}{\text{viscous forces}}. \tag{8.3}$$

An alternative (entirely equivalent) formulation of this result, cited because we shall proceed in this way in subsequent chapters, is to write

$$|\boldsymbol{u} \cdot \nabla \boldsymbol{u}| \sim U^2/L, \quad |\nu \nabla^2 \boldsymbol{u}| \sim \nu U/L^2. \tag{8.4}$$

Hence

$$\frac{|\boldsymbol{u} \cdot \nabla \boldsymbol{u}|}{|\nu \nabla^2 \boldsymbol{u}|} \sim \frac{UL}{\nu} = \mathrm{Re}. \tag{8.5}$$

The Reynolds number thus indicates the relative importance of two dynamical processes. At a general point within the flow, the ratios of these two terms will not be exactly equal to the Reynolds number, but their characteristic magnitudes will be in this ratio.

8.2 Low Reynolds number

When the Reynolds number is much smaller than unity the viscous force dominates over the inertia force so much that the latter plays a negligible role in the flow dynamics. One may use an approximate form of the equation of motion with the inertia term dropped. Equation (8.2) becomes

$$0 = -\nabla'(\Delta p') + \frac{1}{\mathrm{Re}} \nabla'^2 \boldsymbol{u}', \tag{8.6}$$

these terms being of order 1/Re and the neglected term of order 1. The pressure term must be retained since it is necessary to match the number of variables to the number of equations (Section 5.6). Physically, this means that the size of the pressure term is always governed by the other dynamically important terms—in this case by the viscous term.

Reverting to the dimensional form, eqn (8.6) is

$$\nabla p = \mu \nabla^2 \boldsymbol{u}. \tag{8.7}$$

At every point in the fluid there is an effective balance between the local pressure and viscous forces. Equation (8.7) is known as the equation of creeping motion. It is evidently much simpler than the full Navier–Stokes equation, and solutions have been found for many cases for which the full equation has not yielded a solution. One case will be discussed in Section 9.4. Such solutions are found to agree well with the observed behaviour at low Reynolds number (see, e.g., Fig. 9.3), thus justifying the procedure leading to the approximation.

Two characteristic features of low Reynolds number flow are worth mentioning. Firstly, solutions of the equation of creeping motion are reversible; that is to say, if one has a solution, then there is another one with the same streamline pattern but with the flow everywhere in the opposite direction (with all pressure gradients reversed). Hence, for example, the flow from right to left past an obstacle is the exact reverse of that from left to right. By extension, if one has an obstacle of a shape having upstream–downstream symmetry (its rear half is the mirror image of its front half), then the whole flow pattern has this symmetry; the pressure distribution is antisymmetric. We shall not derive these results formally, but it is readily seen that the solution to be presented in Section 9.4 possesses the above properties. We shall also be seeing some flows with this symmetry in Figs. 12.1, 12.6, and 12.7.

The second characteristic feature of low Reynolds number flows is that viscous interactions extend over large distances. For example, particles sedimenting at low Reynolds number affect each other's motion even when their separation is large compared with their size. Figure 8.1 illustrates this long-range viscous action for flow past a circular cylinder. It re-presents the information of Fig. 3.2 to show the velocity distribution across the mid-plane at a Reynolds number of 0.1. In the next section we shall be looking at the corresponding figure for high Reynolds number flow; comparison of the two provides a good illustration of the way in which different dynamical processes dominate in different Reynolds number ranges.

In this book further discussion of low Reynolds number flows will be confined to Sections 9.4 and 9.5 and parts of Chapter 12. There are

FIG. 8.1   Velocity distribution on centre plane in flow past circular cylinder at Re = 0.1.

nevertheless many interesting low Reynolds number phenomena, as is particularly well illustrated by the film of Ref. [53].

8.3 High Reynolds number

Corresponding arguments for high Reynolds number flow indicate that the viscous force is so small compared with the inertia force that it can be neglected. Equation (8.2) then approximates to

$$\mathbf{u}' \cdot \nabla \mathbf{u}' = -\nabla'(\Delta p)' \tag{8.8}$$

or dimensionally

$$\rho \mathbf{u} \cdot \nabla \mathbf{u} = -\nabla p. \tag{8.9}$$

This is Euler's equation of inviscid motion. When it applies, the fluid at each point has an acceleration directly related to the pressure gradient.

The argument applies also to unsteady flow ($\mu \nabla^2 \mathbf{u}$ being negligible compared with $\rho \mathbf{u} \cdot \nabla \mathbf{u}$ regardless of the size of $\rho \, \partial \mathbf{u}/\partial t$), and a more general form of Euler's equation is

$$\rho \frac{D\mathbf{u}}{Dt} = -\nabla p. \tag{8.10}$$

The relationship of this equation to the actual behaviour at high Reynolds numbers is much more complex than the relationship of the creeping motion equation to low Reynolds number flow. Comparing Euler's equation with the Navier–Stokes equation, we see that the discarded term is the highest-order differential term: the only one involving second space derivatives. The approximation thus reduces the order of the differential equation. A corresponding reduction must be made in the number of boundary conditions.

We saw in Section 5.7 that the no-slip boundary condition is a consequence of the action of viscosity. One may thus expect that this is the condition that should be discarded for mathematical consistency with the inviscid equation. This is indeed the case; we shall see in Chapter 10 that solutions of Euler's equation are obtained by matching only to the impermeability condition (eqn (5.33)). Imposition of the no-slip condition also would result in no solution being obtainable.

We now have a paradoxical situation. The statement that Euler's equation applies at high Reynolds number means, more precisely, that as the Reynolds number is increased the viscous term becomes relatively smaller and smaller, although never absolutely zero; Euler's equation becomes a better and better approximation. A boundary condition, however, cannot be similarly relaxed as the Reynolds number increases.

It either applies or it does not apply; there is no meaning to the statement that the no-slip condition is present but to a negligible extent. On the other hand, one would not expect there to be some Reynolds number at which the no-slip condition suddenly 'switches off', and it is found experimentally that it continues to apply no matter how high the Reynolds number.

Consequently, the viscous term in the dynamical equation must always remain significant in the vicinity of a boundary, so that the equation remains of the order appropriate to the boundary conditions. The region in which this happens is known as the boundary layer. The reasoning (eqns (8.2)–(8.5)) that the viscous force should be negligible breaks down in the boundary layer because the flow develops an internal length scale much smaller than the imposed length scale, L. This is the boundary layer thickness, δ. We shall see below, and in more detail in Chapter 11, that the size of the viscous term can be determined by δ whilst the size of the inertia term is still determined by L:

$$|\boldsymbol{u} \cdot \nabla \boldsymbol{u}| \sim U^2/L, \quad |\nu\nabla^2\boldsymbol{u}| \sim \nu U/\delta^2. \tag{8.11}$$

Inertia and viscous forces can thus remain of comparable order of magnitude if

$$U^2/L \sim \nu U/\delta^2 \tag{8.12}$$

that is if

$$\frac{\delta}{L} \sim \left(\frac{UL}{\nu}\right)^{-1/2} = \mathrm{Re}^{-1/2}. \tag{8.13}$$

The difference between the two length scales must become more marked as the Reynolds number increases.

We consider first the simplest example of a boundary layer. Suppose that a flat plate of negligible thickness is placed in a uniform stream, speed u_0, parallel to it (Fig. 8.2). Then, for a theoretical situation governed by Euler's equation and without the no-slip condition, the flow could obviously continue as if the plate were not there (Fig. 8.2(a)). The speed would be u_0 everywhere. In a real fluid, at large Reynolds number, the speed remains very close to u_0 over most of the flow. Close to and behind the plate, however, there are regions in which a large change occurs (Fig. 8.2(b)). These are the boundary layers on either side of the plate and the wake behind it. We shall not consider the wake further at the moment, but it can be thought of as the extension of the boundary layers downstream.

The rapid variation of speed in the boundary layer gives rise to much larger values of $\partial^2 u/\partial y^2$ than would otherwise exist. This makes the viscous force much larger and so makes it appropriate to use (8.11), rather than (8.4), for the orders of magnitude.

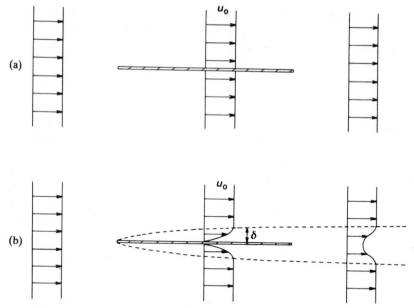

FIG. 8.2 Velocity profiles in flow past thin plate: (a) imagined inviscid flow; (b) real fluid at high Reynolds number. Dotted lines indicate edges of boundary layer and wake.

It is often useful to have a precise definition of the boundary layer thickness, δ (shown in Fig. 8.2(b)). There is, of course, no line beyond which the presence of the plate has absolutely no effect; the velocity still approaches asymptotically to u_0, but very rapidly. A common procedure is to choose δ such that

$$u = 0.99u_0 \quad \text{at} \quad y = \delta \qquad (8.14)$$

(y is distance from plate); the boundary layer is taken to be the region in which the velocity differs by more than 1 per cent from the free-stream velocity.

The longitudinal length scale L is provided in this example by the distance from the leading edge. We shall see in Section 11.4 (eqn (11.32)) that

$$\frac{\delta}{x} \propto \left(\frac{u_0 x}{\nu}\right)^{-1/2}, \qquad (8.15)$$

as expected from relationship (8.13).

The edge of the boundary layer is *not* a streamline. The only significance of the line, $y = \delta$, shown in Fig. 8.2(b) is that indicated by eqn (8.14). Fluid crosses this line. In the present example, fluid just outside the boundary layer at one value of x is inside it at larger x.

For any obstacle other than a flat plate parallel to the free-stream, the situation is more complicated. The fluid is diverted past the obstacle and the solution of Euler's equation is not just a uniform flow (which would not satisfy the impermeability boundary condition). It would thus be meaningless to define the boundary layer and wake as the regions in which the velocity departs significantly from u_0. They are defined instead as the regions in which the action of viscosity significantly affects the velocity distribution. Suppose one has found a solution of Euler's equation for the flow past such an obstacle (which we may call the inviscid flow solution). This will not satisfy the no-slip condition. Close to the wall of the obstacle, viscous action will modify the flow so that the no-slip condition is obeyed. The region in which this happens is the boundary layer.

We can thus generalize the specification of a boundary layer implied by eqn (8.14): the boundary layer is the region in which the velocity differs by more than 1 per cent from the inviscid flow solution.

Figure 8.3 shows an example (cf. Fig. 8.1 for the corresponding example at low Reynolds number). The inviscid flow solution for flow past a circular cylinder gives a velocity profile across the mid-plane as shown by the broken curve. The speed at the wall is twice the free-stream

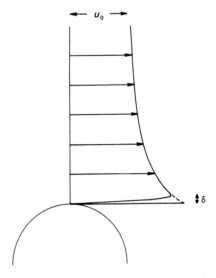

FIG. 8.3 Inviscid (broken line) and high Reynolds number (full line) velocity distributions across centre plane in flow past circular cylinder (neglecting effects of boundary layer separation).

speed. The full curve shows a high Reynolds number velocity profile, satisfying the no-slip condition. The boundary layer is the region in which the two profiles differ significantly.†

Chapter 10 will describe the elements of inviscid flow theory. Chapter 11 will discuss boundary layers. We can now see the role of these two aspects in the development of our understanding of flow at high Reynolds numbers. The first stage in tackling a new problem is usually a solution of the inviscid flow problem. This will, subject to qualifications made below, describe most of the flow. It does so, however, only because the region in which it applies is separated off from the no-slip condition by the boundary layer. One of the pieces of information obtained from the inviscid flow solution is the pressure distribution over the boundaries. This is part of the input to the second stage, a treatment of the boundary layer. If this indicates that the boundary layer undergoes separation (see Section 12.4), some modification to the inviscid solution will be required; in principle an iterative procedure is then needed, though in practice it may be difficult.

Historically, Euler's equation is older than the Navier–Stokes equation, and it was puzzling that it described some aspects of fluid behaviour very well whilst failing totally to describe others. We can now see the reason: those aspects described well were those unaffected by the presence of boundary layers. An important feature that does depend on boundary layers is the force on an obstacle, and we shall see that it was with this that failure of Euler's equation was particularly dramatic. The introduction of the concept of boundary layers, by Prandtl in 1904, was a landmark in the history of fluid dynamics; a very high proportion of subsequent developments stem directly from it.

We have seen in Chapters 2 and 3 that flow at high Reynolds number is prone to instability. Above we have been thinking mainly of laminar flow. The discussion is, however, by no means academic. Transition to turbulence occurs in regions such as boundary layers and wakes, whilst laminar flow continues in inviscid regions. Hence, the division into the two regions is still useful.

It is apparent that a much wider range of phenomena occur at high Reynolds number than at low. For this reason alone one would wish to give the former situation much more extensive consideration in a book of this sort. However, it is worth noting that this emphasis coincides to a large extent with the type of situation one meets most frequently in

---

† We know from Chapter 3 that Fig. 8.3 involves some oversimplification. In the first place, over a wide Reynolds number range, the separation of the flow from the wall leading to the formation of attached eddies occurs upstream of the station shown. Secondly, even when it occurs downstream, the existence of the separation will modify the inviscid flow solution—see Section 12.5.

practical situations. The values of the kinematic viscosity for water and air (at common values of the temperature and pressure) are respectively $1.0 \times 10^{-6} \, \mathrm{m^2 \, s^{-1}}$ and $1.5 \times 10^{-5} \, \mathrm{m^2 \, s^{-1}}$. In either fluid one needs only an object of a few centimetres in size moving at a speed of a few centimetres per second to reach a moderately high Reynolds number.

SOME SOLUTIONS OF THE VISCOUS FLOW EQUATIONS

9.1 Introduction

An obvious aim, once the equations of motion have been set up (in either the full form of Chapter 5 or in an approximate form such as those in Chapter 8) is to find solutions of them. Fluid dynamics thus constitutes an important branch of applied mathematics, although there are severe limitations to what can be learned by theory alone because of mathematical complexity, non-uniqueness, and instability. In this book, the mathematical aspects are somewhat underplayed to leave room for a full development of the experimental aspects. In this chapter we do look briefly at a few solutions of the Navier–Stokes equation and of the equation of creeping motion, both to illustrate the mathematical aspects of the subject and to provide information required elsewhere. For more systematic mathematical treatment see Refs. [11, 12, 14].

Incidentally, in the solutions considered in this chapter, the non-linear terms play either no role at all or a secondary role. They are unimportant either for geometrical reasons or, in the case of creeping motion, for the reasons discussed in Section 8.2. These solutions are, in this respect, untypical of much of fluid dynamics where non-linearity is responsible for both the mathematical difficulties and for the distinctive phenomena. The solutions of the boundary layer equations in Chapter 11 will provide examples of the mathematical handling of such problems.

9.2 Poiseuille flow

We have already looked in Chapter 2 at solutions for flow through a channel and through a pipe. For completeness we need to see that the equations set up there from first principles are indeed special cases of the full continuity and Navier–Stokes equations.

The assumption that there is only one non-zero component of the velocity reduces the continuity equation (5.10) to

$$\partial u / \partial x = 0 \tag{9.1}$$

which accords with the velocity profile being the same at all values of x. It also means that only one component of the Navier–Stokes equation is significant and this is

$$\rho u \frac{\partial u}{\partial x} = -\frac{\partial p}{\partial x} + \mu \nabla^2 u. \qquad (9.2)$$

The left-hand side is zero from (9.1), and the equation takes the form of either (2.6) or (2.14) for respectively Cartesian and cylindrical coordinates. Equations (2.8) and (2.17) are thus solutions of the equations of motion that we have now established.

Equation (9.2) with zero left-hand side is formally the same as the equation of creeping motion (8.7); both represent a direct balance between pressure and viscous forces. Yet, as we have seen in Chapter 2, the Poiseuille solution is not restricted to low Re. The reason is that the inertia term is absent, not as an approximation, but because the geometry makes it identically zero. When Poiseuille flow occurs with Re considerably above 1, the ideas of Section 8.3 relate to the fact that the entry length is then long (Section 2.5). The sequence of profiles in Fig. 2.5 may now be seen as being produced by an annular boundary layer on the pipe wall. This increases in thickness with distance down the pipe; when its thickness is comparable with the radius, viscous effects extend over the whole cross-section, and only then can the Poiseuille profile be approached. Equation (2.23) expresses the fact that, at high Re, the boundary layer is relatively thin; i.e. one has to go further down the pipe before its thickness becomes comparable with the radius.

9.3 Rotating Couette flow

Another simple case of some importance (as the principle of a device for measuring viscosity, amongst other applications) is rotating Couette flow. Fluid is contained in the annulus between two long concentric cylinders of radii a_1 and a_2 rotating about their common axis with angular velocities Ω_1 and Ω_2 (Fig. 9.1). We want to know the velocity distribution within the annulus (and, in viscosity measurements, the torque acting on the cylinders). A solution of the equations of motion may be obtained by assuming that the velocity is everywhere in the azimuthal (ϕ) direction and that the velocity and pressure are independent of ϕ and z (cylindrical polar coordinates shown in Fig. 9.1). These assumptions are the obvious ones suggested by the geometrical symmetry, but, as we shall consider in detail in Section 17.5, instability can produce a changeover to a flow with a more complex structure, to which the following theory does not apply.

The continuity equation $\partial u_\phi / \partial \phi = 0$ is automatically satisfied by these

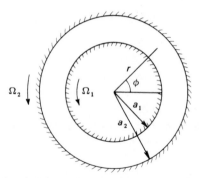

FIG. 9.1   Definition sketch for rotating Couette flow (z-axis is normal to paper).

assumptions, and the azimuthal and radial components of the Navier–Stokes equation become

$$0 = \mu\left(\frac{d^2 u_\phi}{dr^2} + \frac{1}{r}\frac{du_\phi}{dr} - \frac{u_\phi}{r^2}\right) \tag{9.3}$$

$$-\frac{\rho u_\phi^2}{r} = -\frac{dp}{dr} \tag{9.4}$$

with the boundary conditions

$$u_\phi = \Omega_1 a_1 \quad \text{at} \quad r = a_1; \quad u_\phi = \Omega_2 a_2 \quad \text{at} \quad r = a_2. \tag{9.5}$$

The first equation can be solved to give u_ϕ and this is then put into the second equation to give p. Thus, the distribution of the azimuthal velocity across the annulus is determined by the balance of viscous stresses, whilst the pressure distribution is determined by the balance between a radial pressure gradient and the centrifugal force associated with the circular motion.

The solution for u_ϕ (obtained by working in terms of the variable u_ϕ/r) is

$$u_\phi = Ar + B/r \tag{9.6}$$

where

$$A = (\Omega_2 a_2^2 - \Omega_1 a_1^2)/(a_2^2 - a_1^2), \quad B = (\Omega_1 - \Omega_2)a_1^2 a_2^2/(a_2^2 - a_1^2). \tag{9.7}$$

The torque Σ_1 acting on the inner cylinder (per unit length in the z-direction) is given by the viscous stress† $\mu[r\partial(u_\phi/r)/\partial r]_{r=a_1}$ multiplied by the area $2\pi a_1$ and by the radius a_1; i.e.

$$\Sigma_1 = 4\pi\mu a_1^2 a_2^2(\Omega_2 - \Omega_1)/(a_2^2 - a_1^2). \tag{9.8}$$

---

† That transformation to polar coordinates gives an expression of this form is to be expected from the fact that there will be no stress in rigid-body rotation, $u_\phi \propto r$.

Similarly, the torque on the outer cylinder

$$\Sigma_2 = -4\pi\mu a_1^2 a_2^2 (\Omega_2 - \Omega_1)/(a_2^2 - a_1^2). \tag{9.9}$$

We notice that Σ_1 and Σ_2 are equal and opposite, as they must be since the total angular momentum of the fluid is not changing.

9.4 Stokes flow past a sphere

The most famous solution of the equation of creeping motion (8.7) applies to low Reynolds number flow past a sphere. It leads to the relationship between the velocity and the drag used to determine viscosity in the familiar procedure of observing the rate of fall of a sphere through a viscous fluid. This is often known as Stokes flow.

In spherical polar coordinates with $\theta = 0$ in the flow direction in the frame of reference in which the sphere is at rest (Fig. 9.2) the equations are obtained from eqns (5.29)–(5.32) with the inertia terms (the left-hand sides of eqns (5.30)–(5.32)) and the body force terms put equal to zero, and with $u_\phi = 0$, $\partial/\partial\phi = 0$ by symmetry. The boundary conditions are

$$u_r = u_\theta = 0 \quad \text{at} \quad r = a \tag{9.10}$$

$$u_r \to u_0 \cos\theta, \quad u_\theta \to -u_0 \sin\theta \quad \text{as} \quad r \to \infty \tag{9.11}$$

where u_0 is the free-stream velocity and a is the radius of the sphere. The solution is

$$u_r = u_0 \cos\theta \left[1 - \frac{3a}{2r} + \frac{a^3}{2r^3} \right] \tag{9.12}$$

$$u_\theta = -u_0 \sin\theta \left[1 - \frac{3a}{4r} - \frac{a^3}{4r^3} \right] \tag{9.13}$$

$$p - p_0 = -\frac{3}{2}\frac{\mu u_0 a}{r^2} \cos\theta, \tag{9.14}$$

FIG. 9.2 Definition sketch for Stokes flow past a sphere (ϕ is azimuthal angle about $\theta = 0$ axis).

p_0 being the ambient pressure. This may readily be shown by substituting the solution into the equations; for the forward integration the reader is referred to other sources (e.g. Refs. [11, 12]).

The force per unit area in the flow direction acting at a point on the surface of the sphere is the sum of the appropriate components of the viscous and pressure forces; that is†

$$\sigma = -\mu\left(\frac{\partial u_\theta}{\partial r}\right)_{r=a} \sin\theta - (p - p_0)_{r=a} \cos\theta. \tag{9.15}$$

Substituting the solution into this gives

$$\sigma = 3\mu u_0/2a. \tag{9.16}$$

Since this happens to be independent of θ the total force on the sphere is just σ multiplied by the surface area of the sphere,

$$D = 4\pi a^2 \sigma = 6\pi\mu a u_0. \tag{9.17}$$

In terms of a drag coefficient defined by (7.16) with $L = 2a$, this is

$$C_D = 6\pi/\mathrm{Re} \tag{9.18}$$

(cf. (7.20)).

Equation (9.17) is the well-known result, due to Stokes, used in falling-sphere viscometry. Figure 9.3 shows a comparison of this result with experimental observations. The good agreement is important, not only for viscometry, but also because it demonstrates the validity of the reasoning leading to the equation of creeping motion and so encourages one to apply similar reasoning to other problems.

Figure 9.3 also shows the departures from Stokes's law that occur when the Reynolds number is too high for the creeping flow equation to apply. It is important to remember that eqn (9.17) can be used only when Re is less than about 0.5. (It is not sufficient, as is sometimes said, that the flow should be laminar.)

In applying this result it should also be remembered that, as remarked in Section 8.2, the viscous effects extend a long way at low Reynolds numbers. Distant boundaries may thus have a disturbingly large effect. In a falling-sphere viscometer, the container diameter must be more than one hundred times the sphere diameter for the error to be less than 2 per cent [192].

---

† The viscous stress in spherical polar co-ordinates reduces to this simple form for a boundary at rest.

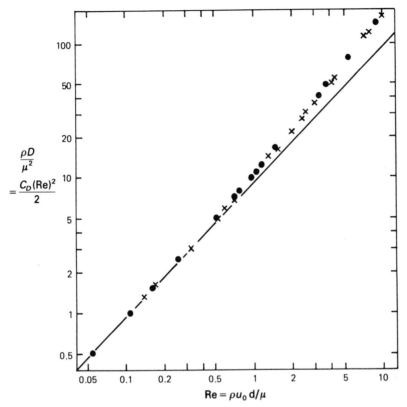

FIG. 9.3 Drag on a sphere at low Reynolds numbers. Experimental points from Refs. [248](×) and [336](●), both using the falling sphere method. The line represents eqn (9.17).

9.5 Low Reynolds number flow past a cylinder

Although the equation of creeping motion generally describes flow at low Reynolds number very satisfactorily, there is a complication that arises in two-dimensional flow, such as the flow past a circular cylinder. No solution of the equation can be found that matches to the boundary conditions both at the surface of the cylinder and at infinity (for mathematical details see, e.g. Ref. [11]). This fact is sometimes referred to as Stokes's paradox. Its resolution again involves the fact that the effect of a boundary on the velocity field extends to very large distances from that boundary. If we consider a moving cylinder in stationary ambient fluid, the fluid velocity remains comparable with the cylinder velocity to distances so large compared with the cylinder radius that the Reynolds number based on the flow length scale is of order unity. There is thus a region in which inertia forces are significant. The lower the Reynolds number the more remote this region is from the cylinder, but it always exists.

A low Reynolds number theory allowing for the existence of this region has been developed [6]. It gives, for the dependence of the drag coefficient on Reynolds number for a circular cylinder, the expression (in the notation of Section 3.4)

$$C_D = 8\pi/\mathrm{Re}(2.002 - \ln \mathrm{Re}). (9.19)$$

This provides the low Reynolds number end of the curve in Fig. 3.15 (the rest of which is compiled from experimental data). There is always a small departure from (7.20).

In real situations, of course, it may be other boundaries rather than a remote inertial region that should be considered in resolving Stokes's paradox.

10

INVISCID FLOW

10.1 Introduction

The relationship of inviscid flow theory to actual flow at high Reynolds number has been considered in Section 8.3. We now look at some of its features. Although Euler's equation is non-linear, we shall see that in many important cases it reduces to a linear equation. Consequently, it yields solutions much more readily than the full Navier–Stokes equation. This is one of the most mathematically developed branches of fluid mechanics in which experimental work plays a more minor role. It is very fully treated in many books, e.g. Refs. [6, 9, 13, 14, 276]. Consequently, we shall confine attention here to the important basic ideas, without developing the methods of application of these ideas to particular cases.

10.2 Kelvin circulation theorem

The circulation round a loop is defined by eqn (6.24). One may consider a loop which consists continuously of the same fluid particles; i.e. each element d*l*, is moving with the fluid. We call such a loop a material loop. The Kelvin circulation theorem states that, for any flow governed by Euler's equation, the circulation round a material loop is conserved. This may be written

$$\frac{D}{Dt} \oint \boldsymbol{u} \cdot \mathrm{d}\boldsymbol{l} = 0 \qquad (10.1)$$

(where the significance of the differential operator D/D*t* has been extended slightly from its application at a single point to indicate that each point of the loop is moving with the fluid).

The theorem is proved as follows:

$$\frac{D}{Dt} \oint \boldsymbol{u} \cdot \mathrm{d}\boldsymbol{l} = \oint \frac{D\boldsymbol{u}}{Dt} \cdot \mathrm{d}\boldsymbol{l} + \oint \boldsymbol{u} \cdot \frac{D(\mathrm{d}\boldsymbol{l})}{Dt}. \qquad (10.2)$$

Each of the two terms on the right-hand side of (10.2) may be shown to be zero. Substituting Euler's equation (8.10) in the first term gives

$$\oint \frac{D\boldsymbol{u}}{Dt} \cdot \mathrm{d}\boldsymbol{l} = -\frac{1}{\rho} \oint \nabla p \cdot \mathrm{d}\boldsymbol{l} = -\frac{1}{\rho} \oint \mathrm{d}p = 0 \qquad (10.3)$$

since ρ is a constant and the integration is round a closed loop. The second term in (10.2) is zero for essentially geometrical reasons. One has

$$D(d\boldsymbol{l})/Dt = d\boldsymbol{u} \qquad (10.4)$$

since one end of the spatial vector $d\boldsymbol{l}$ moves with velocity $\boldsymbol{u}$ and the other end with velocity $\boldsymbol{u} + d\boldsymbol{u}$. Thus

$$\oint \boldsymbol{u} \cdot D(d\boldsymbol{l})/Dt = \oint \boldsymbol{u} \cdot d\boldsymbol{u} = \oint (u\, du + v\, dv + w\, dw)$$

$$= \frac{1}{2} \oint d(u^2 + v^2 + w^2) = \frac{1}{2} \oint d(q^2) = 0 \qquad (10.5)$$

again because the integration is round a closed loop and so q^2 has the same value at the start and the end of it.

This proof depends, of course, on Euler's equation being applicable. The circulation round a material loop is not necessarily conserved in a viscous flow.

The circulation round a loop is related to the vorticity field by Stokes's theorem (6.21):

$$\Gamma = \int \boldsymbol{\omega} \cdot d\boldsymbol{S}. \qquad (10.6)$$

Hence, the vorticity flux through a material loop is conserved in inviscid flow. Changes in the area of a small material loop imply compensating changes in the magnitude of the vorticity component normal to the area. This is the process already discussed in Section 6.6 under the name vortex stretching.

10.3 Irrotational motion

A particularly important application of Kelvin's theorem concerns the case in which $\Gamma = 0$ at an initial instant. The circulation round the same material loop remains zero for all subsequent times. A flow in which $\Gamma = 0$ round each and every loop is necessarily irrotational; if $\boldsymbol{\omega}$ were non-zero in some region one could always find some loops for which eqn (10.6) made $\Gamma \neq 0$. Hence, one may infer that a flow that is initially irrotational throughout remains so at all subsequent times—a result known as the permanence of irrotational motion.

In two-dimensional flow the result may be inferred more directly from the conservation of fluid particle vorticity mentioned in Section 6.6. In three-dimensional flow it is obviously consistent with the inviscid vorticity equation (6.28), each side of which is individually zero in permanently irrotational motion.†

The physical explanation of the result is that, when Euler's equation applies, the only stresses acting on a fluid particle are the pressure stresses. These act normally to the particle surface and so cannot apply a couple to the particle to bring it into rotation. (Once it is rotating, its vorticity can change in three-dimensional flow through vortex twisting or stretching (Section 6.6) but these cannot produce the initial rotation.)

Many interesting flows are initially irrotational—for example, a uniform flow approaching an obstacle. Hence, the study of inviscid motion may for many purposes be reduced to the study of irrotational motion. This does not cover all solutions of Euler's equation, but in the others, the vorticity would have to be introduced initially by some other process.

In irrotational motion

$$\boldsymbol{\omega} = \operatorname{curl} \boldsymbol{u} = 0 \tag{10.7}$$

throughout the flow. Hence, one may introduce ϕ such that

$$\boldsymbol{u} = \operatorname{grad} \phi, \tag{10.8}$$

where ϕ is known as the velocity potential (by analogy with other potentials; it is not, however, associated with energy in the way electrostatic and gravitational potentials are).

From continuity

$$\operatorname{div} \boldsymbol{u} = 0, \tag{10.9}$$

and so

$$\nabla^2 \phi = 0; \tag{10.10}$$

the velocity potential obeys Laplace's equation.

Two parenthetic comments about this theory are worth making. Firstly, the boundary condition corresponding to the impermeability condition at a stationary wall is

$$\partial \phi / \partial n = 0 \tag{10.11}$$

where n is the direction normal to the boundary. We can now see that if we had also a condition on ϕ corresponding to the no-slip condition, the problem would be over-specified and no solution would be available.

Secondly, the analysis leading to eqn (10.10) applies to both steady and unsteady flow. Yet any time variation does not appear explicitly in

---

† But one cannot *rigorously* infer the principle from this equation. There are examples of equations giving $dX/dt = 0$ whenever $X = 0$ but having solutions with $X = 0$ at $t = 0$ and $X \neq 0$ at $t \neq 0$.

(10.10). This means that the instantaneous flow pattern depends only on the instantaneous boundary conditions and not on the history of the flow. Every point of the flow responds immediately to any change in the boundary conditions (a change, for example, in the speed at which an object is moving through a fluid). The physical mechanism by which this response is brought about is a change in the pressure distribution. Evidently there must be an upper limit to the rate at which a pressure change can actually transmit through the fluid, and so a limit to the applicability of the present theory. Small pressure changes transmit at the speed of sound; larger ones transmit at a speed of the same order of magnitude. Hence, the criterion for validity of the theory is closely related to the low Mach number criterion. If

$$L/a\Psi \ll 1 \qquad\qquad (10.12)$$

(where L is the length scale, Ψ is the time scale introduced in example (4) of Section 7.4, and a is the speed of sound), then pressure changes transmit a distance large compared with L in the time in which significant changes in the flow pattern occur. Hence, the response can be effectively instantaneous.

Techniques for solving Laplace's equation are highly developed as a result of its importance in a variety of other physical contexts. The problem of inviscid flow past an obstacle is directly analogous to that of the flow of electric current around a cavity in a conductor, or to that of the electric field distribution in a region of high dielectric constant around a cavity of low dielectric constant. However, we shall see in Section 12.5 that, because of the phenomenon of boundary layer separation, many simple solutions do not correspond closely to real flows. The most important solutions are for cases, such as the flow past an aerofoil, that have no physically significant counterpart in other branches of physics. Consequently, there has been a lot of work on solutions of Laplace's equation in fluid dynamical contexts. The special methods developed are described in the references mentioned above.

10.4 Bernoulli's equation

As well as the velocity distribution, derivable straightaway from a solution for ϕ, one may also want to know the pressure distribution. In deriving an equation for this, we will now confine attention to steady flow. However, we will derive it in the first place for any steady inviscid flow, irrotational or rotational. Then Bernoulli's equation relates the variation of speed and variation of pressure along a streamline. For steady flow we have a fixed streamline pattern. We consider one streamline and denote the distance along it by l. At any point on it the component of Euler's equation in the streamline direction is

$$\rho q \frac{\partial q}{\partial l} = -\frac{\partial p}{\partial l} \tag{10.13}$$

where q is the magnitude of the velocity, $q = |\boldsymbol{u}|$. Equation (10.13) follows from the facts that, by definition of a streamline, the velocity component along it is q and the velocity components normal to it are zero. Integrating eqn (10.13), and remembering that we are taking ρ to be constant,

$$\tfrac{1}{2}\rho q^2 + p = \text{constant along a streamline.} \tag{10.14}$$

This is called Bernoulli's equation.

At positions along the streamline where the velocity is high the pressure is low and vice versa. $\tfrac{1}{2}\rho q^2$ is the kinetic energy per unit volume, and the equation may be interpreted in the following way. When the pressure is increasing in the flow direction, a fluid particle is doing work against the pressure gradient and so loses kinetic energy. When the pressure is decreasing it gains kinetic energy.

When the flow is irrotational, Bernoulli's equation can be extended. We start with the vector identity

$$\boldsymbol{u} \cdot \nabla \boldsymbol{u} = \tfrac{1}{2}\nabla(\boldsymbol{u} \cdot \boldsymbol{u}) - \boldsymbol{u} \times (\nabla \times \boldsymbol{u}). \tag{10.15}$$

For irrotational flow, the second term is zero and

$$\boldsymbol{u} \cdot \nabla \boldsymbol{u} = \tfrac{1}{2}\nabla q^2. \tag{10.16}$$

Euler's equation is then

$$\nabla(\tfrac{1}{2}\rho q^2 + p) = 0 \tag{10.17}$$

and so

$$\tfrac{1}{2}\rho q^2 + p = \text{constant throughout the flow.} \tag{10.18}$$

The Bernoulli constant is the same for every streamline. The relationship between high velocity and low pressure and its converse apply throughout the flow.

We may see the physical significance of this result by referring to two simple examples in which q is constant. If, further, $\boldsymbol{u}$ is constant—the flow is entirely uniform—there can be no pressure variations between streamlines as there is nothing to balance the pressure gradient. The Bernoulli constant is constant throughout the flow. The other example is flow in circular streamlines with u_ϕ independent of r (a case intermediate between examples (1) and (2) in Section 6.4, but more difficult to generate in practice than either of those). This involves vorticity. The pressure varies with radius, the pressure gradient balancing the centrifugal force of the circular motion. The Bernoulli constant is thus a constant only on each streamline.

10.5   Drag in inviscid flow: d'Alembert's 'paradox'

Bernoulli's equation indicates the pressure distribution over the surface of an obstacle placed in a stream and thus the force that would act on that obstacle if the boundary layers could be ignored. We consider this for an obstacle that is symmetrical upstream and downstream, as in Fig. 10.1. If the flow were from right to left instead of from left to right, the streamline pattern would be the mirror image pattern about the plane of symmetry AB. However, eqn (10.10) has the property that, if $\phi = \phi_1$ is a solution, then so is $\phi = -\phi_1$. Thus the streamline pattern for the flow from right to left is the same as that for the flow from left to right; just the direction of flow along the streamlines is reversed.† Hence, considering now just the flow from left to right, the flow pattern downstream of the plane of symmetry is the mirror image of that upstream. The speed q is the same at symmetrically placed points such as C and D (Fig. 10.1).

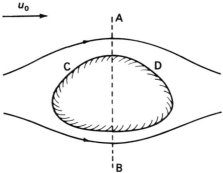

FIG. 10.1   Inviscid flow past a symmetric body, considered in discussion of d'Alembert's paradox.

From Bernoulli's equation the pressure is also the same. The pressure distribution around the rear half of the obstacle is the same as that round the front half, and there is thus no net force on the obstacle. The drag is zero.

† We have noted in Section 8.2 that low Reynolds number flow has a similar property. It is thus a property of both 'extreme cases': if one neglects either the inertia force or the viscous force, then this symmetry property arises. 'Intermediate cases' for which the flow pattern is influenced by both inertia and viscous forces do not give rise to such symmetry. The reason is apparent: reversing $\boldsymbol{u}$ everywhere reverses $\mu \nabla^2 \boldsymbol{u}$ but leaves $\boldsymbol{u} \cdot \nabla \boldsymbol{u}$ unchanged; one would therefore not have the same balance between them in a reversed flow.

This result is actually true whether or not the obstacle is symmetrical (although it is not demonstrated so quickly in the latter case [11, 14]). Irrotational flow calculations give zero drag on any obstacle placed in a fluid stream.

At one time it was thought that theory of this type ought to produce the correct drag. Hence, the above result became known as d'Alembert's paradox. This name lingers today, although the result ceases to be paradoxical once one knows that the region of applicability of the theory is separated from the surface of the obstacle by a boundary layer. (See also Section 12.5).

Indeed it would be more paradoxical if completely inviscid theory did give a drag. We can see this most readily by switching attention to the frame of reference in which the obstacle is moved through stationary fluid. Work would then be done against any drag and energy would be fed into the fluid. But, in the absence of viscous action, there is no process by which this energy could be dissipated, and no steady state would exist. (Arguments of this character require some caution. They do not apply when there are waves present which can carry energy away 'to infinity'.)

The result that inviscid motion produces no drag applies only for steady flow. In unsteady flow, although the velocity distribution is the same as that for the instantaneously corresponding steady flow, the pressure distribution is different. The work done against the drag as an obstacle is accelerated through a fluid provides the change in the kinetic energy of the fluid.

10.6 Applications of Bernoulli's equation

Although, in steady flow, inviscid theory predicts no force on an obstacle in the direction of relative motion between it and the fluid, it can predict a force at right angles to this direction. Such a force is called a lift. Lift generation is obviously of great practical importance; Bernoulli's equation can be used to understand in a general way why an aeroplane can fly. Consider an aerofoil such as that shown in Fig. 10.2. We use the frame of reference in which the aerofoil is fixed, so that the flow is steady. Above the convex upper surface the streamlines are pushed together and the fluid moves faster than the free-stream speed; the pressure is therefore reduced, as indicated in the figure. Similarly, the speed is reduced below the concave lower surface, and the pressure is increased. One sees that these pressure changes are such as to produce a net upward force on the aerofoil. In Section 11.2 we shall see that the pressure difference across the boundary layers is small, and so does not affect this argument, unless

FIG. 10.2 Pressure changes associated with Bernoulli's equation around an unsymmetrical aerofoil.

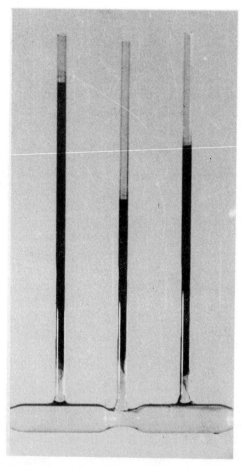

FIG. 10.3 Venturi demonstration. (Flow is from left to right; fluid in side-tubes is dyed.)

boundary layer separation (Section 12.4) occurs. However, a fuller understanding of the processes involved does require consideration of viscous effects, and we shall return to the topic of lift in Chapter 13.

Another system that is readily understood in terms of Bernoulli's equation is the Venturi tube. Indeed, this provides the simplest convenient demonstration of the relationship between speed and pressure. A pipe with a fluid passing through at high Reynolds number (thin boundary layers) has a constriction in it—a short length of reduced diameter. Because of continuity, the speed increases at the constriction and so the pressure is reduced. This is demonstrated by the apparatus shown in Fig. 10.3; the height the water rises up the vertical open side tubes is a measure of the pressure. The pressure is seen to drop markedly at the constriction and to rise again after it. According to Bernoulli's equation, the pressures on either side of the constriction should be the same, but, because of some viscous effects (notably those to be described in Section 12.4), the downstream pressure is a little lower.

The Venturi tube is of importance not merely as a demonstration device. For a carefully designed constriction, the pressure differences can be related quantitatively to the flow rate; this thus provides one method of measuring flow rates through pipelines. The Venturi is also a convenient simple way of obtaining pressures below atmospheric; for example, the suction pumps, fitted to taps and used by chemists to speed up filtration, operate on the Venturi principle.

Bernoulli's equation is also the basis of one of the most important instruments for measuring fluid velocities, the Pitot tube. This will be described in Section 25.2.

10.7 Some definitions

Since all the terms in Bernoulli's equation have the dimensions of pressure, the following nomenclature has developed. $\frac{1}{2}\rho q^2$ is known as the dynamic pressure and $(\frac{1}{2}\rho q^2 + p)$ as the total pressure or sometimes (for reasons indicated below) as the stagnation pressure. To distinguish it from these quantities, p is sometimes called the static pressure, although this is a somewhat misleading name (and it is quite different from the hydrostatic pressure to be introduced in Section 14.2). The static pressure is, of course, the pressure physically existing in the fluid.

The total pressure has the physical significance that it is the pressure at which the fluid comes to rest. Let us consider, for example, the flow past an obstacle placed in an otherwise uniform flow of velocity u_0 at pressure p_0 (Fig. 10.4). The streamlines pass on different sides of the obstacle. One streamline, in the middle, ends on the obstacle at point S and the

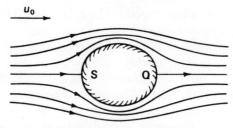

FIG. 10.4 Inviscid flow past an obstacle showing front and rear stagnation points, S and Q.

fluid at this point is at rest. According to inviscid flow theory, this and a corresponding point Q at the rear of the obstacle are the only points at which the velocity is zero. S and Q are known as the forward and rear stagnation points. Applying Bernoulli's equation to the streamline ending there shows that the pressure at S is

$$p_S = \tfrac{1}{2}\rho u_0^2 + p_0, \qquad\qquad (10.19)$$

i.e. it is the total pressure associated with the undisturbed flow. In theory, the same pressure exists at Q, but in practice the action of the boundary layer almost always greatly alters this. At the forward stagnation point, however, the effect of the boundary layer is slight, and the pressure p_S is actually observed. This is the highest pressure anywhere in the flow.

BOUNDARY LAYERS, WAKES, AND JETS

11.1 Viscous regions in high Reynolds number flow

The reason for the occurrence of boundary layers and their role in high Reynolds number flows have been considered in Section 8.3. Now that we have considered (in Chapter 10) the flow external to the boundary layers, we need to look in more detail at the boundary layers themselves. This is the first purpose of this chapter. However, wakes and jets—other regions where viscous forces are significant even at high Reynolds number—involve similar ideas and equations, and so we extend the discussion to include them.

The fact that flow outside the boundary layers is irrotational (Section 10.3) provides another way of viewing the process of boundary layer formation. Fluid particles can acquire vorticity only by viscous diffusion (i.e. through the action of the term $\nu\nabla^2\omega$ in eqn (6.27)). The action of viscosity comes in at the boundary through the need to satisfy the no-slip condition. As a result vorticity is introduced into the flow at the boundary, and then diffuses away from it. The boundary layer can be defined as the region of appreciable vorticity. The boundary layer is long and thin $(L \gg \delta)$ when the fluid travels a long distance downstream during the time that the vorticity diffuses only a small distance away from the boundary. This happens when the Reynolds number is large.

High Reynolds number wakes and jets are regions into which vorticity has been advected; the vorticity was introduced upstream where the fluid was close to boundaries—the walls of the obstacle producing the wake or of the orifice through which the jet emerges. These flow features are long and thin for just the same reason as a boundary layer is.

11.2 The boundary layer approximation [23]

Because of the difference in length scales in different directions, certain terms in the equations of motion play a negligible part in the dynamics of boundary layers. We now see in a systematic way how this can be used to formulate an appropriate approximation to the equations. This will provide further justification for the ideas introduced in Section 8.3. Also, the resulting equations can sometimes be solved when the exact equations cannot. We shall be looking (Sections 11.4 and 11.6) at a couple of solutions both for their own interest and as our examples of the mathematical methods used for fully non-linear problems.

From the outset we confine attention to steady, two-dimensional boundary layers—a severe restriction from a practical point of view, but one that still allows us to see the general principles involved. We suppose that the boundary layer is forming on a flat wall (with the x-coordinate in the flow direction and y normal to the wall). A free-stream velocity outside the boundary layer is prescribed as a function of x. This could be achieved by making the wall one side of a channel of variable width as in Fig. 11.1 (with the channel width always large compared to the boundary layer thickness). In fact, however, it makes negligible difference if the surface is curved, so long as there are no sharp corners—more precisely, so long as the radius of curvature of the surface is everywhere large compared to the boundary layer thickness. Thus, the prescribed free-steam velocity could be a solution of Euler's equation for flow past an obstacle (x then being a curvilinear coordinate in the surface).

We denote the free-stream velocity by u_0 and the pressure associated with it by p_0.

We take the boundary layer to have length scales L and δ in the x- and y-directions, as in Section 8.3. We may expect that the velocity scales will also be different in different directions and we denote the scales of u and v by U and V. Similarly the order of magnitude of the pressure differences across the boundary layer in the y-direction may not be the same as the order of magnitude of the imposed pressure differences outside the boundary layer; we denote the scale of the former by Λ and the scale of the latter by Π. We now consider each of the equations in turn, labelling the terms with their orders of magnitude.

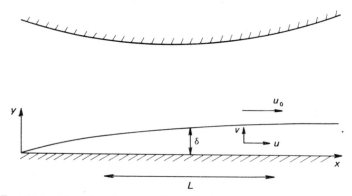

FIG. 11.1   Boundary layer on flat wall of channel: definition sketch.

The continuity equation is

$$\frac{\partial u}{\partial x} + \frac{\partial v}{\partial y} = 0 \qquad (11.1)$$

$$\frac{U}{L} \qquad \frac{V}{\delta}\,.$$

The two terms must be of the same order of magnitude; fluid entering or leaving the boundary layer at its outer edges must be associated with variations in the amount of fluid travelling downstream within the boundary layer. Hence,

$$V \sim U\delta/L; \qquad (11.2)$$

the velocity component normal to the wall is small compared with the rate of downstream flow when the boundary layer is thin.

The x-component of the Navier–Stokes equation is

$$u\frac{\partial u}{\partial x} + v\frac{\partial u}{\partial y} = -\frac{1}{\rho}\frac{\partial p}{\partial x} + v\frac{\partial^2 u}{\partial x^2} + v\frac{\partial^2 u}{\partial y^2} \qquad (11.3)$$

$$\frac{U^2}{L} \qquad \frac{VU}{\delta} \sim \frac{U^2}{L} \qquad \frac{\Pi}{\rho L} \qquad \frac{vU}{L^2} \qquad \frac{vU}{\delta^2}\,.$$

The second expression for the order of magnitude of $v\,\partial u/\partial y$ has been written using relationship (11.2). The two parts of the inertia term are comparable with one another, the smallness of V/U compensating for the more rapid variation of u with y than with x. The two parts of the viscous term are however of different sizes when δ/L is small, and $v\,\partial^2 u/\partial x^2$ may be neglected.

The y-component of the Navier–Stokes equation is

$$u\frac{\partial v}{\partial x} + v\frac{\partial v}{\partial y} = -\frac{1}{\rho}\frac{\partial p}{\partial y} + v\frac{\partial^2 v}{\partial x^2} + v\frac{\partial^2 v}{\partial y^2} \qquad (11.4)$$

$$\frac{UV}{L} \sim \frac{U^2\delta}{L^2} \qquad \frac{V^2}{\delta} \sim \frac{U^2\delta}{L^2} \qquad \frac{\Lambda}{\rho\delta} \qquad \frac{vV}{L^2} \sim \frac{vU\delta}{L^3} \qquad \frac{vV}{\delta^2} \sim \frac{vU}{L\delta}\,.$$

In both eqn (11.3) and eqn (11.4) the pressure term will be of the same order of magnitude as the largest of the other terms. Hence,

$$\Pi/\rho L \sim U^2/L \sim vU/\delta^2 \qquad (11.5)$$

$$\Lambda/\rho\delta \sim U^2\delta/L^2 \sim vU/L\delta \qquad (11.6)$$

and so

$$\Lambda/\Pi \sim \delta^2/L^2. \qquad (11.7)$$

The pressure differences across the boundary layer are much smaller than those in the x-direction. Hence, at any value of y the difference between $(1/\rho)\,\partial p/\partial x$ and $(1/\rho)\,dp_0/dx$ is much smaller than the significant terms in eqn (11.3) and we may replace the former by the latter, giving

$$u\frac{\partial u}{\partial x} + v\frac{\partial u}{\partial y} = -\frac{1}{\rho}\frac{dp_0}{dx} + v\frac{\partial^2 u}{\partial y^2}. \tag{11.8}$$

Outside the boundary layer there is no variation with y and

$$u_0\frac{du_0}{dx} = -\frac{1}{\rho}\frac{dp_0}{dx}, \tag{11.9}$$

a result which could also be obtained from Bernoulli's equation. Hence

$$u\frac{\partial u}{\partial x} + v\frac{\partial u}{\partial y} = u_0\frac{du_0}{dx} + v\frac{\partial^2 u}{\partial y^2}. \tag{11.10}$$

This equation together with

$$\frac{\partial u}{\partial x} + \frac{\partial v}{\partial y} = 0 \tag{11.11}$$

constitute the boundary layer equations—two equations in the two variables u and v.

11.3 Classification of boundary layers

Solution of (11.10) and (11.11) requires $u_0(x)$ to be specified, both to give the third term in (11.10) and as a boundary condition for integration with respect to y. This is why the solution of Euler's equation for the particular configuration is needed before the boundary layer can be analysed.

Obviously, many different distributions of $u_0(x)$ can arise. In the next section we shall consider the simplest case of all—when u_0 is constant. We shall not consider any other case quantitatively, but some general remarks may be made. A useful broad classification is given by the sign of du_0/dx or, equivalently through eqn (11.9), the sign of dp_0/dx. When

$$du_0/dx > 0; \quad dp_0/dx < 0 \tag{11.12}$$

(the external flow is accelerating as the pressure decreases) one talks of a boundary layer in a favourable pressure gradient. When

$$du_0/dx < 0; \quad dp_0/dx > 0 \tag{11.13}$$

(the external flow is decelerating as the pressure rises) one talks of an adverse pressure gradient. One can, of course, have regions of each type of pressure gradient within a given flow—indeed, this is usually the case for the boundary layer on an obstacle.

Boundary layers in favourable pressure gradients are relatively thin. In a region of strong enough pressure gradient the boundary layer thickness can actually decrease with distance downstream; the effect of the pressure gradient more than counteracts the viscous spreading process (explained qualitatively in Section 11.1 and to be seen quantitatively in Section 11.4). We shall also be noting in Section 18.2 that instability, leading to transition to turbulence, is delayed by a favourable pressure gradient. Such a pressure gradient does not, however, introduce flow phenomena qualitatively different from those occurring in boundary layers with zero pressure gradient.

The effects of an adverse pressure gradient are in the first place just the reverse of those just described. Much more significantly, however, a boundary layer in such a pressure gradient is prone to the phenomenon of separation. We shall discuss the nature and implications of this in Chapter 12 and particularly Sections 12.4 and 12.5. It should be noted here, however, that the effect of separation can be to modify the solution of Euler's equation for the region outside the boundary layer. Consequently, $u_0(x)$ may differ from the form that one initially assumes.

11.4 Zero pressure gradient solution

The simplest, and in a sense most fundamental, case is the one where the pressure gradient is zero. Equivalently, u_0 is constant; we consider the boundary layer beneath a uniform flow. Such a boundary layer is readily observed on a thin flat plate set up parallel to the free-stream; one wall of an empty wind-tunnel or water-channel is sometimes used.

The equations for this case are

$$u\frac{\partial u}{\partial x} + v\frac{\partial u}{\partial y} = v\frac{\partial^2 u}{\partial y^2} \tag{11.14}$$

$$\frac{\partial u}{\partial x} + \frac{\partial v}{\partial y} = 0 \tag{11.15}$$

with boundary conditions

$$u = v = 0 \quad \text{at} \quad y = 0$$

$$u \rightarrow u_0 \quad \text{as} \quad y \rightarrow \infty. \tag{11.16}$$

We look for a solution of the form

$$u = u_0 g(y/\Delta) \tag{11.17}$$

where Δ is a function of x. That the solution should be of this form is an assumption. It corresponds to the velocity profile having the same shape at all values of x, although with a different scale in the y-direction, and is thus physically plausible. Δ is directly proportional to the boundary layer thickness, but it is convenient to define it slightly differently from δ.

Equation (11.15) can be satisfied by introducing a stream function ψ such that

$$u = \partial\psi/\partial y, \quad v = -\partial\psi/\partial x \tag{11.18}$$

as in Section 6.3. If we take

$$\psi = u_0\Delta f(y/\Delta) \tag{11.19}$$

then (11.18) gives (11.17) as required with

$$g = f' \tag{11.20}$$

where the prime indicates differentiation with respect to

$$\eta = y/\Delta. \tag{11.21}$$

The second of equations (11.18) also gives

$$v = u_0(-f + yf'/\Delta)\,\mathrm{d}\Delta/\mathrm{d}x \tag{11.22}$$

and further differentiation leads to

$$\frac{\partial u}{\partial x} = -\frac{u_0\,yf''}{\Delta}\frac{\mathrm{d}\Delta}{\mathrm{d}x}; \quad \frac{\partial u}{\partial y} = \frac{u_0 f''}{\Delta}; \quad \frac{\partial^2 u}{\partial y^2} = \frac{u_0 f'''}{\Delta^2}. \tag{11.23}$$

Substitution into eqn (11.14) then gives

$$\frac{u_0^2}{\Delta}\frac{\mathrm{d}\Delta}{\mathrm{d}x}ff'' + \frac{\nu u_0}{\Delta^2}f''' = 0. \tag{11.24}$$

If the solution is of the assumed form this must reduce to a total differential equation in f as a function of η; i.e. the two coefficients must have the same dependence on x, so that this cancels out:

$$\frac{u_0^2}{\Delta}\frac{\mathrm{d}\Delta}{\mathrm{d}x} \propto \frac{\nu u_0}{\Delta^2} \tag{11.25}$$

and so

$$\Delta^2 \propto \nu x/u_0 + \text{const.} \tag{11.26}$$

It is convenient to choose the constant of proportionality and the origin of x so that

$$\Delta = (\nu x/u_0)^{1/2}. \tag{11.27}$$

It is found experimentally that this choice of the origin of x corresponds fairly closely to the leading edge of a flat plate set up in an otherwise unobstructed flow. Equation (11.27) is essentially the same result as eqn (8.13).

Equation (11.24) now becomes

$$ff'' + 2f''' = 0. \tag{11.28}$$

The boundary conditions transform to

$$f = f' = 0 \quad \text{at} \quad \eta = 0$$
$$f' \to 1 \quad \text{as} \quad \eta \to \infty. \tag{11.29}$$

The solution of this total differential equation has to be obtained numerically [22, 23]. The resulting variation of f' with η, and so the velocity profile is shown in Fig. 11.2. This curve is known as the Blasius profile.

It has the property that

$$f' = 0.99 \quad \text{when} \quad \eta = 4.99. \tag{11.30}$$

The boundary layer thickness as previously defined (Section 8.3) is thus

$$\delta = 4.99(vx/u_0)^{1/2}. \tag{11.31}$$

Other ways of writing this are

$$\delta/x = 4.99 \, \mathrm{Re}_x^{-1/2} \quad \text{and} \quad \mathrm{Re}_\delta = 4.99 \, \mathrm{Re}_x^{1/2} \tag{11.32}$$

($\mathrm{Re}_x = u_0 x/v$; $\mathrm{Re}_\delta = u_0 \delta/v$). The boundary layer thickness is small when the Reynolds number is large, as expected. This is, of course, a necessary condition for the theory to apply. Also Re_δ is large when Re_x is large; there is no ambiguity in talking about large Reynolds number.

Figure 11.2 includes experimental observations for several values of Re_x (from two separate experiments). The agreement with the theoretical profile is good, providing support for the various approximations and assumptions made in the course of the theory. The experimental results have been scaled to the coordinates $\eta (= y(u_0/vx)^{1/2})$ and $f' (= u/u_0)$. One sees the way in which the profile maintains its shape with distance downstream although the boundary layer thickness is changing—as assumed in eqn (11.17).

At higher values of the Reynolds number, the Blasius profile is unstable and the boundary layer becomes turbulent. The transition process will be described in Chapter 18, and the nature of the turbulent boundary layer in Chapter 21. The instability depends on Re_δ, which, as we see from eqn (11.32), increases with Re_x. Thus, any zero pressure gradient boundary layer undergoes transition if it extends far enough downstream. However, provided that the disturbance level is not too

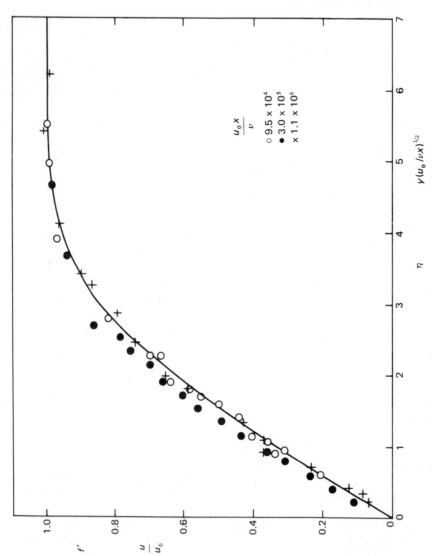

Fig. 11.2   Theoretical Blasius profile and experimental confirmation from Refs. [141] and [249].

high, the range in which the Reynolds number is high enough for boundary layer formation but low enough for laminar flow is significant.

11.5 Wakes

The fluid that has become rotational in a boundary layer retains this property as it travels downstream. High Reynolds number flow past an obstacle thus departs from irrotational theory not only around the obstacle but also behind it. The latter region of rotational flow is called the wake.

Another way of understanding wake formation has been mentioned in Section 3.4 (Fig. 3.14)—the fact that drag on the obstacle implies a reduction in momentum transport. It is not immediately obvious that the two explanations are equivalent; the link is the fact that an irrotational flow produces no drag (d'Alembert's paradox—Section 10.5).

The structure of a wake immediately behind an obstacle (the near-wake) depends very much on the obstacle geometry. General remarks are not very useful, except for a classification as to what determines the initial wake width. We shall see in Section 12.5 that on 'streamlined' obstacles (which include the limiting case of a flat plate parallel to the flow—Fig. 8.2) the boundary layers remain attached right up to the trailing edge. The initial wake width is then given by the thickness of these layers (Fig. 11.3). However, this obviously does not apply when one has flows like those in Fig. 3.13, and, in general, for 'bluff' obstacles (Section 12.5) the initial wake width is much more related to the width of the obstacle itself rather than to the boundary layer thicknesses.

Far enough downstream the wake structure becomes much less dependent on the particular obstacle. The far-wake region is considered

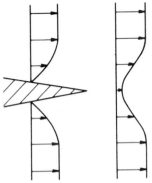

Fig. 11.3 Boundary layer and wake velocity profiles at trailing edge of streamlined body (schematic).

to be that where the velocity deficit at the centre of the wake profile is small compared with the velocity itself (i.e. as shown schematically in Fig. 3.14). The boundary layer equations for a laminar far-wake can be solved, but we shall not consider this; the procedure is similar to that for jets described below [23]. As for jets, comparison with experiment is inhibited by instability, giving rise to a vortex street (Section 17.8) or a turbulent wake (Sections 21.3, 21.4).

11.6 Jets

A jet is produced when fluid is ejected from an orifice. We are concerned here with the case when the ambient fluid into which the jet emerges is the same fluid as the jet itself. Because of their simplicity, the geometries that have been most investigated are a circular orifice giving an axisymmetric jet and a long thin slit giving a two-dimensional jet. The former is obviously the easier experimental configuration, but experiments with the latter have been performed; care is needed to make the jet nearly uniform along its length.

At low Reynolds numbers, the fluid from an orifice spreads out in all directions. At high Reynolds numbers, with which we are concerned here, a jet, like a wake, is long and thin; the equations of motion may be used in the form of the boundary layer approximation.

Jets become unstable at too low a Reynolds number for the laminar flow solution of these equations to be observed. We will, however, take a look at the solution for a two-dimensional jet for three reasons: it is interesting to see a soluton like the Blasius solution but not requiring numerical analysis; some results for turbulent jets (Section 21.1) may usefully be compared with a laminar jet; and in Section 11.7, we can illustrate certain basic properties of viscous flows in this context. Nor is the result wholly academic; it can be a starting point for investigations of the instability. What actually happens in high Reynolds number jet flows will be considered in Section 18.4.

The equations for a laminar two-dimensional jet are just as for the Blasius boundary layer,

$$\frac{\partial u}{\partial x} + \frac{\partial v}{\partial y} = 0 \tag{11.33}$$

$$u\frac{\partial u}{\partial x} + v\frac{\partial u}{\partial y} = v\frac{\partial^2 u}{\partial y^2} \tag{11.34}$$

(with x in the jet direction and y across the slit from which it emerges). There is no pressure gradient for reasons parallel to those considered in

Section 11.2; on the boundary layer approximation the effective pressure gradient is that imposed from outside and this is zero since the ambient fluid is at rest. The ejection of the fluid from the orifice will normally have been achieved by maintaining a higher pressure behind it. This will provide the fluid with its momentum, but the pressure variations do not extend significantly downstream of the orifice.

The difference from the boundary layer solution lies in the boundary conditions, which are

$$u \to 0 \quad \text{as} \quad y \to \infty$$
$$\partial u / \partial y = 0 \quad \text{and} \quad v = 0 \quad \text{at} \quad y = 0, \tag{11.35}$$

expressing the facts that the ambient fluid is at rest and the jet is symmetrical about its centre plane. (See also footnote to Section 11.7, p. 137.)

The procedure for solution closely follows that for the boundary layer. It is again assumed that the velocity profile is a similar shape at different distances downstream. However, the velocity scale, as well as the jet width, can now vary with x; the form of successive velocity profiles is shown schematically in Fig. 11.4. Hence, the velocity profile is taken to be of the form

$$u = u_{\max} g(y / \Delta) = u_{\max} g(\eta) \tag{11.36}$$

where both the maximum speed $u_{\max}$ and Δ (proportional to the jet-width δ) are functions of x. Drawing on our experience with the boundary layer solution we take them to be proportional to powers of x:

$$u_{\max} \propto x^m, \quad \Delta \propto x^n. \tag{11.37}$$

One relationship between m and n is obtained from the requirement that the inertia term and the viscous term must vary in the same way with

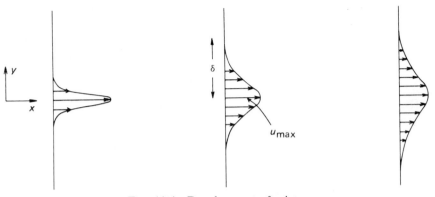

FIG. 11.4   Development of a jet.

x:

$$u\,\partial u/\partial x \propto x^{2m-1}; \quad v\,\partial^2 u/\partial y^2 \propto x^{m-2n} \tag{11.38}$$

and so

$$2m - 1 = m - 2n. \tag{11.39}$$

(This is entirely parallel to the derivation for the boundary layer that $\Delta \propto x^{1/2}$, but presented in a somewhat different way.)

A second relationship is obtained from the fact that the total momentum per unit time passing every cross-section of the jet must be the same. This is a statement of momentum conservation for the jet as a whole, which must apply since the jet does not exert any force on any external body. (Its relationship to eqn (11.34) for the momentum of individual fluid particles will be discussed in Section 11.7.) Since the jet is two-dimensional we consider the momentum per unit length in the z-direction. The momentum per unit volume is ρu and the rate at which it is being transported downstream is u. Hence, the total rate of momentum transport at any x is

$$M = \int_{-\infty}^{\infty} \rho u^2 \, \mathrm{d}y. \tag{11.40}$$

M is conveniently used as a measure of the strength of the jet. Conservation of momentum indicates that

$$\mathrm{d}M/\mathrm{d}x = 0. \tag{11.41}$$

Substituting (11.36) in (11.40),

$$M = \rho u_{\max}^2 \Delta \int_{-\infty}^{\infty} g^2 \, \mathrm{d}\eta \propto x^{2m+n}. \tag{11.42}$$

Hence,

$$2m + n = 0. \tag{11.43}$$

From eqns (11.39) and (11.43) together we have

$$m = -\tfrac{1}{3}, \quad n = \tfrac{2}{3}. \tag{11.44}$$

The ways in which the maximum velocity and the jet-width vary with x have been determined.

The parameters determining the velocity profile at any x are x itself, the jet momentum, and the properties of the fluid:

$$u_{\max} = u_{\max}(x, M, \rho, v); \quad \Delta = \Delta(x, M, \rho, v). \tag{11.45}$$

Dimensional analysis together with the power laws already determined indicates that

$$u_{max} = C(M^2/\rho^2 vx)^{1/3}; \quad \Delta = (\rho v^2 x^2/M)^{1/3} \quad (11.46)$$

where C is a numerical constant (there is no need for a similar numerical constant in the expression for Δ, as its definition was arbitrary to that extent).

Equation (11.33) is satisfied by deriving u and v from a stream function ψ. The form appropriate to eqns (11.36) and (11.46) is

$$\psi = C(Mvx/\rho)^{1/3}f \quad (11.47)$$

with

$$f' = g. \quad (11.48)$$

Substituting in eqn (11.34), with intermediate algebra similar to that around eqns (11.19)–(11.24), gives

$$\frac{3}{C}f''' + ff'' + f'^2 = 0 \quad (11.49)$$

with the boundary conditions

$$\begin{aligned} f = f'' = 0 \quad &\text{at } \eta = 0 \\ f' \to 0 \quad &\text{as } \eta \to \infty. \end{aligned} \quad (11.50)$$

This can be integrated directly to give

$$f = A\tanh(AC\eta/6) \quad (11.51)$$

and so

$$g = \tfrac{1}{6}A^2 C\operatorname{sech}^2(AC\eta/6). \quad (11.52)$$

A appears as an arbitrary constant, but substitution in eqn (11.40) indicates that

$$(AC)^3 = 9/2. \quad (11.53)$$

Hence the solution for the velocity profile is

$$u = \left(\frac{3M^2}{32\rho^2 vx}\right)^{1/3}\operatorname{sech}^2\left[y\left(\frac{M}{48\rho v^2 x^2}\right)^{1/3}\right]. \quad (11.54)$$

Although, as remarked before, this flow is not observable because of instability, it may be supposed that this result is to be interpreted in the following way. $u_{max} = \infty$ at $x = 0$, and the solution will not be valid in this region. A real jet starts off with a finite velocity but a velocity profile different from that given by the theory. As it travels downstream, the profile tends to the theoretical one, which will occur asymptotically with an origin of x displaced from the jet orifice (either upstream or downstream depending on the initial profile). The solution has the property that

$$\frac{d}{dx}\int_{-\infty}^{\infty} \rho u \, dy > 0. \tag{11.55}$$

Hence, the amount of fluid being transported downstream by the jet increases with distance downstream. The jet draws fluid into itself from the sides—a process known as entrainment. Far enough downstream, only a small proportion of the fluid in the jet is fluid that came out of the orifice. We shall see in Section 11.7 that this result follows from rather general considerations and is thus not a property of the particular solution.

It is for this reason that M, and not the rate of efflux from the orifice, is used as a measure of the jet strength.

The reader may have noticed that no boundary condition on v at large y was applied. Once one knows about the entrainment the reason is apparent; such a boundary condition would overconstrain the system.

11.7 Momentum and energy in viscous flow

The foregoing theory of a laminar jet provides a convenient context for a few further remarks about the basic properties of the equations of viscous flow.

Equation (11.41) was established by applying the principle of momentum conservation to the jet as a whole, a different type of procedure from those used so far. Although the equation follows directly from the laws of mechanics, one would expect it to be related to the equation for the momentum of a fluid particle and we now see how it is derived from this:

$$\frac{dM}{dx} = \frac{d}{dx}\int_{-\infty}^{\infty} \rho u^2 \, dy = 2\rho \int_{-\infty}^{\infty} u \frac{\partial u}{\partial x} \, dy \tag{11.56}$$

From eqn (11.34)

$$\int_{-\infty}^{\infty} u \frac{\partial u}{\partial x} \, dy + \int_{-\infty}^{\infty} v \frac{\partial u}{\partial y} \, dy = v \int_{-\infty}^{\infty} \frac{\partial^2 u}{\partial y^2} \, dy = v \left[\frac{\partial u}{\partial y}\right]_{-\infty}^{\infty} = 0 \tag{11.57}$$

since the velocity gradient is zero far outside the jet. Also

$$\int_{-\infty}^{\infty} v \frac{\partial u}{\partial y} \, dy = [uv]_{-\infty}^{\infty} - \int_{-\infty}^{\infty} u \frac{\partial v}{\partial y} \, dy = 0 + \int_{-\infty}^{\infty} u \frac{\partial u}{\partial x} \, dy \tag{11.58}$$

using continuity. The two terms on the left-hand side of eqn (11.57) are thus equal to one another, and so

$$dM/dx = 0. \tag{11.59}$$

This result is thus shown to be an integrated form of (an approximation to) the Navier–Stokes equation.† Such integrated forms are used in a variety of contexts. In general (e.g. when the boundary layer approximation does not apply or when there is a boundary on which the flow exerts a force), they are more complicated, but we shall not consider the general case in this book.

It is interesting to discuss kinetic energy similarly. An equation for the energy of a fluid particle is obtained in general by taking the scalar product of the Navier–Stokes equation with u. In a case such as the jet (where the two-dimensional boundary layer approximation applies and there is no pressure gradient), only one velocity component contributes significantly to the energy; the equation is obtained by multiplying eqn (11.34) by u:

$$\tfrac{1}{2}u\frac{\partial(u^2)}{\partial x} + \tfrac{1}{2}v\frac{\partial(u^2)}{\partial y} = vu\frac{\partial^2 u}{\partial y^2} = -v\left(\frac{\partial u}{\partial y}\right)^2 + \tfrac{1}{2}v\frac{\partial^2(u^2)}{\partial y^2}. \tag{11.60}$$

The left-hand side is the rate of change of kinetic energy (per unit mass) following a fluid particle. The right-hand side has two parts: one is essentially negative and so represents dissipation of energy; the other can have either sign and has the nature of a diffusion of energy down the energy gradient.

We can learn more from the integrated form. The rate of transport of kinetic energy in the x-direction by the jet is

$$E = \tfrac{1}{2}\rho \int_{-\infty}^{\infty} u^3 \, dy. \tag{11.61}$$

By a procedure similar to that for dM/dx, it can be shown that

$$\frac{dE}{dx} = -\mu \int_{-\infty}^{\infty} \left(\frac{\partial u}{\partial y}\right)^2 dy \tag{11.62}$$

and thus is negative, as could have been anticipated.

Alternatively the argument can be turned round to show that since dE/dx must be negative, μ must be positive. A fluid with a negative viscosity coefficient could gain mechanical energy from heat in a way that would violate the second law of thermodynamics.

† It may be thought surprising that the laminar jet solution uses both eqn (11.34) and an integrated form of it. The extra information involved in the latter is the boundary condition $\partial u/\partial y \to 0$ as $y \to \infty$. The solution, eqn (11.54), satisfies this, although it was not used explicitly. Equation (11.41) could be omitted if this condition were used explicitly, but the procedure adopted much simplifies the determination of the exponents m and n.

The discussion has been formulated in terms of a two-dimensional jet, but, from their physical significance, we expect the results to apply generally. For a jet of arbitrary cross-section, the momentum and energy transports are

$$M = \int \rho u^2 \, dS, \quad E = \frac{1}{2} \int \rho u^3 \, dS, \tag{11.63}$$

where dS is an element of cross-sectional area perpendicular to the main flow direction, and the integrations are carried out over the whole cross-section. M and E are respectively constant and decreasing with distance downstream.

These results enable us to make some general remarks about the action of viscosity on momentum and energy. The jet illustrates these points well because the complicating effects of boundaries and of a pressure gradient are absent. Viscous action tends towards equalizing the momentum of different fluid particles. This is a redistribution of the momentum which conserves the total. The effect on the kinetic energy consists partly of a dissipation that reduces the energy at every point and partly of a redistribution, similar to the momentum redistribution, with no net effect. Some parts of the fluid can be gaining kinetic energy by viscous action, the redistribution more than counteracting the dissipation, but on average there is a loss. The situation is analogous to the collision of inelastic bodies: conservation of total momentum together with a redistribution such that the momentum of different bodies is more nearly equal results in a loss of total kinetic energy.

From the facts that

$$\frac{d}{dx} \int u^2 \, dS = 0 \tag{11.64}$$

and

$$\frac{d}{dx} \int u^3 \, dS < 0 \tag{11.65}$$

one may infer that

$$\frac{d}{dx} \int u \, dS > 0. \tag{11.66}$$

This is a rigorous argument if the profiles are similar at different x (as in eqn (11.36)), a plausibility argument otherwise. Hence, the fact that a jet entrains ambient fluid is a consequence of the combination of momentum conservation and energy dissipation.

12

SEPARATION AND ATTACHMENT

12.1 The phenomenon of separation

Figures 12.1 and 12.2 illustrate the phenomena with which this chapter is concerned. Each shows effectively two-dimensional flow past a square block on a wall. The flows are at low and high Reynolds number respectively.

Each picture actually shows two examples of the phenomenon of separation and also illustrates the phenomenon of reattachment to be discussed towards the end of this chapter. However, for the moment we

FIG. 12.1 Two-dimensional flow (shown by streaks produced by suspended aluminium powder) past a block on a wall at Re = 0.02. From Ref. [366].

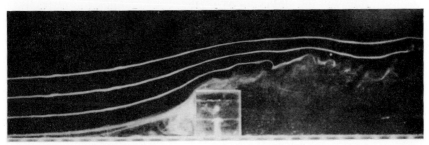

FIG. 12.2 Two-dimensional flow (shown by smoke) past a block on a wall at Re = 3300.

focus attention on the left-hand region of each flow, upstream of the block. Fluid initially close to the wall and moving in the main flow direction moves away from the boundary; further along the boundary the flow is in the opposite direction, as part of a recirculating region. The streamline pattern is thus locally as sketched in Fig. 12.3. The flow is said to separate, and the point S is called the separation point. Because of the no-slip condition, the fluid right at the wall is, of course, at rest. Remarks about the direction of flow thus apply to fluid close to the wall; the distinction between points on the wall to the left and right of S is in the sign of $(\partial u/\partial y)_{y=0}$. S is thus defined as the point at which this quantity is zero.

There is an evident similarity between the upstream regions of Figs. 12.1 and 12.2, despite the very different values of the Reynolds number. The differences in the dominant dynamical processes of low and high Reynolds number flow (Chapter 8) make such similarity remarkable. Separation is evidently a phenomenon that can occur under widely varying conditions.

For several reasons separation at high Reynolds number, i.e. boundary layer separation, has received far more attention than low Re separation. Firstly, high Re flows are encountered much more widely. Secondly, there are some flow configurations for which separation occurs at high Re but not at low (compare, for example, Figs. 3.2, 3.3, and 3.4), whereas the converse is probably not true. Thirdly, and most importantly, high Re separation can have profound consequences for the whole flow structure in a way that relates to questions such as what determines the drag on a bluff body and how a low-drag configuration is designed. As a result, explanations of separation are commonly (as in the first edition of this book) presented in a way that assumes that it is a boundary layer that is separating. The fact, however, that separation is not confined to boundary layers suggests the desirability of a more general approach.

Hence, Section 12.2 considers, in a way applicable to all Reynolds numbers, how and when separation may occur. The two subsequent sections consider the implications of these generalities for respectively low and high Reynolds number flow; the discussion of the latter is

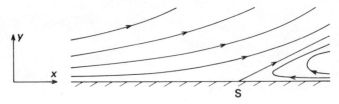

FIG. 12.3 Typical streamline pattern in vicinity of separation point.

extended, in Section 12.5, to cover the important consequences mentioned above.

Of course, with separation occurring at both low and high Re, it is also to be observed at intermediate Re. We shall not discuss such cases specifically, but the generalities obviously cover them. It was mentioned above that flows such as that past a circular cylinder (which is typical of flow past many bluff bodies) exhibit separation only at higher Re values. These are therefore naturally discussed in terms of boundary layer separation. However, the lowest Reynolds number at which separation occurs may well be such that it is scarcely a boundary layer that is separating; e.g. Fig. 3.3.

12.2 Conditions for separation

In discussing when and how separation occurs we confine attention to steady two-dimensional flow. This is sufficient to illustrate the principles involved, although the details for three-dimensional and/or unsteady flow are considerably more complicated. The restriction to two-dimensional flow is simplifying because the argument involves vorticity; since only one component of the vorticity is non-zero, we can discuss it as a scalar. Further, $\omega \cdot \nabla u = 0$ in the vorticity equation (6.27); the argument is not complicated by the processes contained in this term (Section 6.6). Only advection ($u \cdot \nabla \omega$) and viscous diffusion ($\nu \nabla^2 \omega$) of vorticity are involved.

We take coordinates as shown in Fig. 12.3. The non-zero vorticity component is thus the z-component, ζ, governed by the equation

$$u \cdot \nabla \zeta = \nu \nabla^2 \zeta. \tag{12.1}$$

By definition,

$$\zeta = \frac{\partial v}{\partial x} - \frac{\partial u}{\partial y} \tag{12.2}$$

and consequently

$$\frac{\partial \zeta}{\partial y} = \frac{\partial^2 v}{\partial x \, \partial y} - \frac{\partial^2 u}{\partial y^2}. \tag{12.3}$$

At the wall, $y = 0$, one has $u = 0$ and $v = 0$ for all x, giving

$$(\partial v / \partial x)_{y=0} = 0. \tag{12.4}$$

Additionally the continuity equation

$$\frac{\partial u}{\partial x} + \frac{\partial v}{\partial y} = 0 \tag{12.5}$$

implies that

$$(\partial v/\partial y)_{y=0} = 0 \tag{12.6}$$

and thus

$$(\partial^2 v/\partial x\, \partial y)_{y=0} = 0. \tag{12.7}$$

Hence, at the wall

$$\zeta = -\partial u/\partial y \tag{12.8}$$

and

$$\frac{\partial \zeta}{\partial y} = -\frac{\partial^2 u}{\partial y^2}. \tag{12.9}$$

We saw in the previous section that the difference between $x < x_S$ and $x > x_S$ is in the sign of $(\partial u/\partial y)_{y=0}$. From eqn (12.8), this means that the vorticity changes sign. Separation necessarily involves the existence of a region in which the vorticity has opposite sign from that associated with the flow as a whole. The key to understanding when separation may occur is to understand how this reversed vorticity is introduced into the flow.

Note incidentally that, with the conventions of Fig. 12.3 that the main flow is in the positive x-direction and the distance from the wall is $+y$, ζ is negative in the oncoming flow. Hence, the region of reversed vorticity is one of positive ζ.

It is also worth noting parenthetically that, in a flow such as the upstream part of Fig. 12.1 or 12.2, the region of reversed vorticity is much smaller than the region of recirculating motion. Evidently the sense of the circulation is such that the vorticity at its centre has the same sign as the vorticity of the upstream flow. Nevertheless there must be reversed vorticity in a region as sketched in Fig. 12.4.

Consider now how the x-component of the Navier–Stokes equation

$$u\frac{\partial u}{\partial x} + v\frac{\partial u}{\partial y} = -\frac{1}{\rho}\frac{\partial p}{\partial x} + v\frac{\partial^2 u}{\partial x^2} + v\frac{\partial^2 u}{\partial y^2} \tag{12.10}$$

simplifies close to the boundary. From considerations similar to those above about the effects of the boundary conditions, this becomes

$$-\frac{1}{\rho}\frac{\partial p}{\partial x} + v\frac{\partial^2 u}{\partial y^2} = 0 \tag{12.11}$$

at $y = 0$. Equation (12.9) then gives

$$\frac{1}{\rho}\frac{\partial p}{\partial x} + v\frac{\partial \zeta}{\partial y} = 0 \tag{12.12}$$

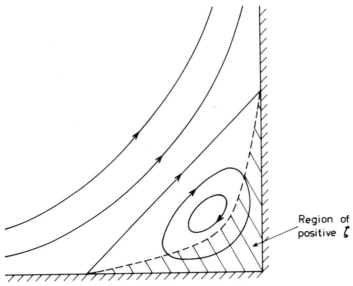

FIG. 12.4   Region of reversed vorticity: see text.

at $y = 0$. The pressure gradient along the wall and the vorticity gradient normal to the wall are thus related. It should be noted that eqn (12.12) is not a general relationship applying throughout a flow but one that applies specifically at a wall as a result of the boundary conditions. On the other hand, it does have generality in the sense that no approximation applying only over a certain Reynolds number range has been introduced.

As discussed in Section 6.6, the action of viscosity on vorticity is essentially one of diffusion down the vorticity gradient. The sign of the vorticity gradient at the wall thus determines the sign of the vorticity being introduced into the flow at the wall. Introduction of positive ζ into a flow of predominantly negative ζ thus requires a region on the wall over which $\partial \zeta / \partial y$ is negative.

Equation (12.12) tells us that this corresponds to there being a region with

$$\partial p / \partial x > 0. \tag{12.13}$$

The flow before separation has been taken to be in the positive x-direction, and (12.13) therefore means that the pressure must be rising in the flow direction. Extending the use of the nomenclature introduced in Section 11.3, the pressure gradient must be adverse (but note that it is an extension because $\partial p / \partial x$ at the wall is effectively the same as dp_0/dx of Section 11.3 only when the boundary layer approximation applies).

The relation (12.13) is a necessary but not sufficient condition for separation. Whether diffusion of positive vorticity into the flow produces a region of positive vorticity depends on whether this diffusion more than counterbalances diffusion and/or advection from the regions of negative vorticity; i.e. on the whole vorticity balance of eqn (12.1). The important point is that, without this diffusion, there is no mechanism for a region of positive vorticity to arise.

The question of whether separation will occur is thus a question firstly of whether the flow will develop a region of adverse pressure gradient on the wall and secondly of whether this is large enough. The answer to the first question is often, although not always, obvious; that to the second almost always difficult, and beyond the scope of this book. Both answers, however, must be given in terms of the specific flow. In particular, one cannot carry the discussion any further in a way that applies simultaneously to all ranges of Reynolds number. This is apparent from the different relationships between the velocity and pressure fields characteristic of, on the one hand, flows with a balance between viscous and pressure forces (as exemplified by Poiseuille flow) and, on the other hand, flows to which Bernoulli's equation is relevant.

Figure 12.5 summarizes the variations in velocity profile that underlie the above discussion. It shows u and $-\partial u/\partial y$ distributions characteristic of favourable (a), zero (b) and adverse (c and d) pressure gradients. The vorticity distribution has similarities to the $-\partial u/\partial y$ distribution; the two are identical at $y = 0$ (eqn (12.8)) and thus close to one another for low y, but may become significantly different at larger y. The form of the curves is governed by the requirement on the sign of $\partial^2 u/\partial y^2$ close to the wall given by eqn (12.11); again this is exactly true at the wall, approximately true close enough to the wall, but inapplicable further from the wall. In

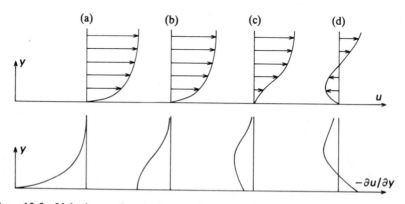

FIG. 12.5 Velocity and velocity gradient profiles: see text. (a) Favourable pressure gradient; (b) zero pressure gradient; (c) and (d) adverse pressure gradient.

an adverse pressure gradient, for example, the equation expresses the requirement that, in order to overcome the retarding effect of the pressure gradient on fluid close to the wall, external fluid must exert a larger forward drag than the backward drag exerted by the wall. Cases (c) and (d) correspond to the pressure gradient being respectively insufficient and sufficient to produce separation. One can see that a large enough pressure gradient is bound to produce separation; the curve for u in case (c) can reverse its curvature only sufficiently far from the wall that eqn (12.11) has become a poor approximation; the total velocity change over this distance may be too large to match to the external flow. A curve of the form shown in (d) resolves this problem.

12.3 Low Reynolds number separation

We have seen an example of low Reynolds number flow with separation in Fig. 12.1, and Figs. 12.6 and 12.7 show two further examples. In this section we discuss briefly various aspects of such flow with the aid of these examples.

In Section 8.2 the reversibility of low Reynolds number flows was mentioned, with its implication that flow past a symmetrical body must be symmetrical. This symmetry is apparent in our examples. In particular the recirculating region in front of the block in Fig. 12.1, as already discussed, is matched by a similar one behind it, produced by a second

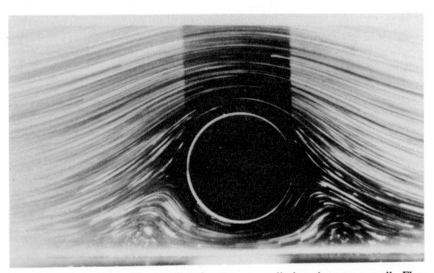

FIG. 12.6 Low Reynolds number flow past a cylinder close to a wall. Flow visualization as in Fig. 12.1. From Ref. [366].

FIG. 12.7 Low Reynolds number flow past a cavity in a wall. Flow visualization as in Fig. 12.1. From Ref. [366].

separation. The contrast with the high Reynolds number flow of Fig. 12.2 is apparent; in that there is again a second separation but the overall pattern is quite unsymmetrical.

How do the ideas of the previous section apply to low Reynolds number flow? As stated there, detailed prediction of whether and where separation will occur requires a much fuller analysis than we can give here. However, one can often see that the regions of adverse pressure gradient are likely to arise and thus gain a general understanding of the broad features of a flow. The one shown in Fig. 12.6 is conveniently used to show how such arguments are developed. Fluid flows at low Reynolds number past a circular cylinder close to a wall. Regions of separated flow are to be seen at either end of the gap between the cylinder and the wall. The pressure must be higher at the upstream end of the gap than at the downstream end in order to force the fluid through it against viscous action; there is qualitative similarity to channel flow (Section 2.2). Moreover, if the gap is narrow enough, this pressure difference may be larger than the typical pressure difference between the front and rear of the cylinder associated with flow round it. The pressure along the wall must then rise towards the high near the beginning of the gap and/or from the low near the end. The upstream–downstream symmetry mentioned in the last paragraph requires that the 'and/or' is 'and'.

Hence, one gets a pair of symmetrical separation zones as seen in the photograph.

Figure 12.7 shows flow past a square cavity in a wall. Separation occurs close to the edge of the cavity, generating a circulating region within it. The main comment to be made on this flow is that the behaviour is actually less obvious than it may appear at first sight. This is a rather negative comment but still worth making—it relates to how one gains an understanding of various types of flow. Intuition may mislead when one is dealing with low Reynolds number flows, mainly because they are seldom encountered in 'everyday' situations. One tends to react to Fig. 12.7 by saying that of course the fluid will 'shoot past' the edge of the cavity rather than turning into it. But 'shooting past' is an inertial effect, and this is a flow in which such effects are negligible. It is much less apparent that a balance of viscous and pressure forces will give rise to the observed behaviour.

12.4 Boundary layer separation

Flows of practical importance involving separation mostly occur at high Reynolds number. The flow can then be divided into inviscid and boundary layer regions, and separation is associated with the latter. (One might say that an inviscid flow pattern such as that in Fig. 10.4 is exhibiting separation at Q. However, the ideas in this chapter, and specifically the analysis in Section 12.2, are concerned with understanding separation when it occurs at a position that is *not* a stagnation point of the inviscid flow solution.) In Sections 11.2 and 11.3 we saw that the longitudinal pressure gradient in a boundary layer is that imposed by the inviscid flow and thus given by Bernoulli's equation. Hence, separation may occur when (11.13) applies, for example on the walls of a diverging channel or on the rear part of an obstacle placed in a flow. If a wall has a sharp corner, separation will normally occur there (e.g. the second separation in Fig. 12.2). Otherwise, it occurs at a position without special characteristics.

Both laminar and turbulent boundary layers can separate. (In the latter case, Fig. 12.3 must be interpreted as a mean-flow streamline pattern.) Laminar layers usually (depending on the initial velocity profile) require only a relatively short region of adverse pressure gradient to produce separation. Turbulent layers separate less readily.

The overall flow pattern, when separation occurs, depends greatly on the particular flow. The upstream flow behind the separation point is normally fed by recirculation of some of the separating fluid, but little can be said in general beyond this. Sometimes the effect is quite localized

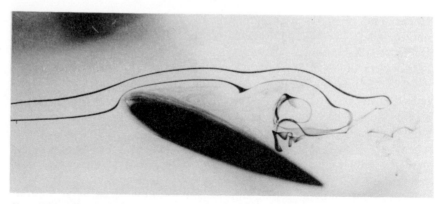

FIG. 12.8   Flow over an aerofoil at large angle of attack (view is slightly oblique from above and front).

as we shall be seeing in Section 12.6. More often it is not: we have had examples of flow patterns in which boundary layer separation was a major feature in Figs. 3.4, 3.13 and 12.2. Figure 12.8 shows another example: the flow past an aerofoil (see Chapter 13) sharply inclined to the flow direction. Separation occurs quickly on the upper side. The consequent pattern is affected by the fact that the separated flow becomes turbulent and so there is a highly fluctuating recirculating motion over the whole of the top of the aerofoil.

The occurrence of an extended region of separated flow has far-reaching consequences for the dynamics of the flow as a whole. We look at these in the next section, and we shall also return to Fig. 12.8 in Section 13.2.

12.5   Bluff and streamlined bodies at high Reynolds number

Knowing about separation, we can consider aspects of high Reynolds number flow more fully than was possible in Section 8.3.

In flows where separation is a major feature rotational motion is not confined to thin layers next to boundaries plus a thin wake. Vorticity introduced in the boundary layer is carried by the separated flow into regions that one would suppose to be irrotational in obtaining the inviscid flow solution. The presence of separation thus significantly modifies the inviscid flow, and the assumption that the inviscid flow can be analysed without reference to the boundary layers (except as a reason for ignoring the no-slip condition) breaks down. As an example, consider the observation that, over a wide Reynolds number range, separation on a circular cylinder occurs upstream of the maximum width (Fig. 3.13). On

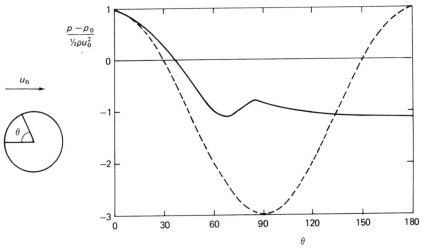

FIG. 12.9 Pressure distribution on a circular cylinder. Broken line: inviscid flow solution. Solid line: distribution measured at $\mathrm{Re} = 1.9 \times 10^5$ (data obtained by Flachsbart, given in Ref. [7]).

the inviscid flow solution for flow past a circular cylinder this is a region of favourable pressure gradient (Fig. 12.9). Inviscid flow is still occurring outside the boundary layer and wake, but not that particular inviscid flow solution.

Figure 12.9 includes a measured pressure distribution over the surface of a cylinder. It can be seen that, as a result of the boundary layer separation, the rear region (around $\theta = 180°$) is at a much lower pressure than indicated by the inviscid flow solution. The difference between this pressure and that at the front (around $\theta = 0°$) produces a drag on the cylinder. At high Reynolds numbers, this drag is much larger than the one produced by the viscous stress over the surface. Thus for bluff bodies, such as a circular cylinder, the inapplicability of d'Alembert's 'paradox' arises primarily from the indirect effect of viscous action on the pressure distribution and only secondarily from direct viscous stresses.

Over a wide range of high Reynolds number, the drag coefficient for a bluff body varies little. One example is Fig. 3.15, and it is a useful general rule in a variety of situations that the drag is roughly proportional to the square of the speed. We have seen in Section 7.3 (eqn (7.21)) that constancy of the drag coefficient is predicted by dimensional considerations if one assumes that the viscosity is not a relevant parameter. At first sight, d'Alembert's paradox renders this result useless. However, we can now see how it relates to observation. Over the Reynolds number range concerned, the separation point moves little. The pressure distribution,

FIG. 12.10   Streamlined body.

like that shown in Fig. 12.9, is thus always much the same. The pressure differences are proportional to the square of the speed, and, since these generate most of the drag, that has the same proportionality. The drag coefficient is not completely constant because of small changes in the pressure distribution and because of the small contribution of the viscous stress. When the separation point moves substantially, as when the boundary layer becomes turbulent, the pressure distribution is changed markedly and eqn (7.21) breaks down completely as described at the end of Section 3.4.

The fact that the drag on an obstacle at high Reynolds number is primarily due to the redistribution of the pressure from the inviscid flow distribution, as a result of separation, has important practical implications. If separation can be delayed till near the rear, or eliminated altogether except right at the rear where boundary layers from opposite sides meet, the drag will be very low; the pressure forces on the front and rear will almost balance (d'Alembert's 'paradox' is seen to have highly useful implications). This can be achieved with two- and three-dimensional bodies of cross-section like that in Fig. 12.10. Such bodies are said to be streamlined. The drag on a two-dimensional streamlined body can be as low as 1/15 of that on the cylinder of the same thickness despite the greatly increased surface area. The most important feature is the slowly tapering tail. It is essential to their performance that the wings and the fuselage of an aeroplane and the parts of a submarine should have streamlined profiles. It is also apparent that Fig. 12.10 resembles the shape of many marine creatures such as fish, dolphins, etc.

12.6   Attachment and reattachment

Attachment is, obviously enough, the opposite of separation. Reverse the arrows on Fig. 12.3 and one has the local streamline pattern in the vicinity of an attachment point. Like separation, attachment may be a feature of a flow at any Reynolds number.

Figures 12.1, 12.6, and 12.7 all contain examples of attachment in low Reynolds number flows. Indeed, because of the reversibility of such flows (Sections 8.2, 12.3), once we have understood separation we have also understood attachment. External flow from right to left instead of from left to right in any of these pictures would leave them unchanged (and would still do so even if the imposed geometry were not symmetrical); separation points would become attachment ones and vice versa.

This argument does not apply except at low Re. At higher Re, the statement that the arrows on Fig. 12.3 can be reversed to give an attachment pattern is true only in so far as the diagram is schematic, not a precise flow pattern. Let us in fact pass straight to the high Re case—again the one of most practical importance—and look at an example. Suppose a two-dimensional jet emits into a region bounded by a side-wall, as shown in Fig. 12.11. It is found that the jet is pulled down onto the wall giving the type of flow shown in the figure [96, 334]. This behaviour may be understood by remembering (Sections 11.6, 11.7, particularly eqns (11.55), (11.66)) that, in the case of an unconfined jet, entrainment produces an inflow of fluid from the sides. In the presence of a wall, the jet cannot draw fluid into itself in this way. Instead it is drawn down onto the wall.

Two parallel jets will similarly be drawn together and so merge into a single jet. Thermal plumes behave similarly and will provide our main illustration of attachment when we discuss these—see Fig. 14.7.

This tendency for flows to attach to walls or to one another is known as the Coanda effect. This is used as a rather general name for wall attachment occurring under a variety of conditions, and it is not certain that the same dynamical mechanism is acting in all cases. However, the explanation based on entrainment is frequently relevant.

For reasons to be considered in Section 21.2, a turbulent flow has a much higher entrainment rate than the corresponding laminar one. Consequently, the Coanda effect occurs much more strongly; for example a turbulent jet will attach to a more distant wall than a laminar one. Similarly, two-dimensional flows exhibit the effect more strongly than three-dimensional; in the latter case fluid for entrainment can enter from the sides.

The process producing attachment acts on a separated boundary layer. There is thus a tendency for separation to be followed by reattachment

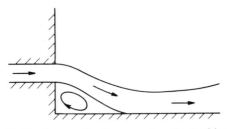

FIG. 12.11 Schematic diagram of wall-attaching jet.

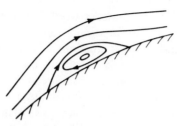

FIG. 12.12   Streamline pattern for separation followed by local reattachment.

unless the adverse pressure gradient continues long enough to prevent it. The result is a localized separation 'bubble' as sketched in Fig. 12.12. This can occur with the flow remaining laminar throughout. However, flows away from boundaries are much more prone to become turbulent than ones alongside a boundary (Section 19.4). Consequently, the more usual sequence is that transition to turbulence occurs soon after separation and this then promotes reattachment. Figure 12.12 then represents the *mean* streamline pattern. Downstream the boundary layer is turbulent. We saw in Chapter 3 (Fig. 3.13) that there is a Reynolds number range, 3×10^5–3×10^6, in which this sequence of events occurs on a circular cylinder. Of course, we have also seen in Section 12.4 that there are many situations in which local reattachment does not occur; Figs. 12.2 and 12.8 are examples.

13

LIFT

13.1 Introduction [13]

We have seen that the drag on an obstacle placed in a flow arises essentially from viscous action (Sections 10.5 and 12.5), whereas the generation of lift on a suitably shaped obstacle can be understood, at least in a general sense, from inviscid theory (Section 10.6). In fact viscous action also enters into the process of lift generation in an essential, although more subtle, way. We now look at the reasons for and consequences of this.

Lift can be obtained either by appropriate asymmetrical shaping as discussed in Section 10.6, or by having a symmetrical body inclined to the flow direction, or, of course, by a combination of asymmetry and inclination (Fig. 13.1).

The curvature of the centre line in Fig. 13.1(a) and (d) is called the camber. The inclination angle α is called the angle of attack. The simplest geometry giving lift is an inclined flat plate (Fig. 13.1(b)). However, boundary layer separation at the sharp leading edge results in a 'stalled' flow configuration (see below) except for very small values of α and this does not provide a good context for a discussion of the important ideas. We turn instead to those shapes (known as aerofoils) that have been developed specially to produce high lift; our understanding of the subject has been strongly influenced by the obvious application to aircraft wings. An aerofoil is a slender body with a rounded leading edge and a sharp trailing edge (Fig. 13.1(a), (c) and (d)); we shall see that these features help to provide a good lifting behaviour as well as helping to reduce drag (Section 12.5). A well-designed aerofoil can produce a much larger lift than drag, as is illustrated by Fig. 13.2.

It will be convenient in the following discussion to use the standard names for the different dimensions of an aerofoil, as shown in Fig. 13.3.

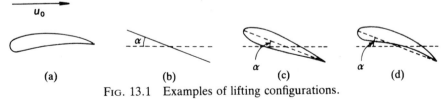

FIG. 13.1 Examples of lifting configurations.

It will also be convenient to talk of the top and bottom of the aerofoil with the lift thought of as upwards, although the direction of the vertical does not enter into the dynamics; in general, a lift is any force perpendicular to the direction of relative motion between a body and a fluid.

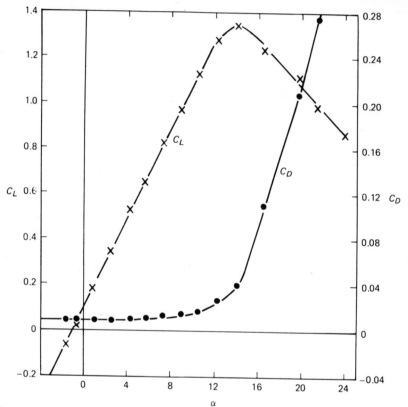

FIG. 13.2   Example of variation with angle of attack of lift coefficient and drag coefficient for an aerofoil. Note different scales for C_D and C_L. (Aerofoil type RAF 34 at Re based on chord of 4.5×10^6; measurements by Relf, Jones, and Bell, given in Ref. [7]).

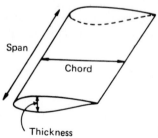

FIG. 13.3   Aerofoil terminology.

13.2 Two-dimensional aerofoils

The first part of the following discussion applies to strictly two-dimensional flow. This should obviously be relevant to the motion around a body, such as a wing, of large span. However, there are complications that arise in *any* three-dimensional system, which we must consider later.

The explanation of lift generation in terms of Bernoulli's equation, given in Section 10.6, implies that the circulation

$$\Gamma = \oint \boldsymbol{u} \cdot d\boldsymbol{l} \tag{13.1}$$

around the aerofoil is non-zero. To see this, we may consider a thin (thickness $\ll$ chord) aerofoil inclined at a small angle to the flow direction, so that nearly every point of the surface is nearly parallel to the flow direction (Fig. 13.4). The upward force (per unit length in the spanwise direction) on the element dx is $(p_B - p_T) \, dx$, where p_B and p_T are the pressures below and above the aerofoil at this station along the chord. Bernoulli's equation gives

$$p_B - p_T = \tfrac{1}{2}\rho(u_T^2 - u_B^2) = \tfrac{1}{2}\rho(u_T + u_B)(u_T - u_B). \tag{13.2}$$

For a thin aerofoil, the variations of the velocity from the free-stream velocity u_0 will be small, and we may approximate this by

$$p_B - p_T = \rho u_0(u_T - u_B). \tag{13.3}$$

Hence the total lift per unit span,

$$L = \rho u_0 \int_0^c (u_T - u_B) \, dx. \tag{13.4}$$

The circulation round a contour following the surface of the aerofoil (physically round a contour just outside the boundary layer) may be

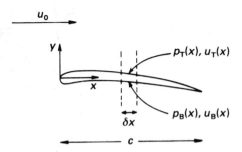

FIG. 13.4 Definition sketch for discussion of lift and circulation of a thin aerofoil.

approximated by

$$\Gamma = \int_0^c u_B \, dx + \int_c^0 u_T \, dx = - \int_0^c (u_T - u_B) \, dx. \qquad (13.5)$$

Thus

$$L = -\rho u_0 \Gamma, \qquad (13.6)$$

where Γ is the circulation round any loop enclosing the aerofoil (since the circulation $\oint \boldsymbol{u} \cdot d\boldsymbol{l} = \int \boldsymbol{\omega} \cdot d\boldsymbol{S}$ round any loop not enclosing it is zero for irrotational flow).

This is an example of a general exact result of inviscid irrotational flow theory, the Kutta–Zhukovskii (or Joukovski) theorem. The force per unit length on any two-dimensional body in relative motion to the ambient fluid with velocity u_0 and with circulation Γ round it is

$$L = -\rho u_0 \Gamma \qquad (13.7)$$

acting at right angles to the direction of u_0. (The sign corresponds to the convention that u_0 is positive when the fluid moves past a stationary obstacle from left to right or the obstacle moves through stationary ambient fluid from right to left; Γ is positive when the circulation is anticlockwise; and L is positive upwards.)

This much can be said without reference to viscosity. However, the generation of the circulation in the first place requires viscous action. This is apparent from the Kelvin circulation theorem (Section 10.2) if we consider an aerofoil started from rest in stationary fluid, so that there is no initial circulation. There is always a solution of the inviscid flow equations with no circulation. For a cambered and/or inclined aerofoil that would normally generate lift, this solution has the rear stagnation point some distance ahead of the trailing edge on the upper surface (Fig. 13.5(a)). This is the flow pattern† which actually occurs immediately after motion starts and which would necessarily occur at all subsequent times in a truly inviscid fluid. However, boundary layer processes change it. The boundary layer on the lower side separates at the trailing edge; the eddy so produced interacts with the inviscid flow to make its stagnation point coincide with the separation point (Fig. 13.5(b)). The new inviscid flow, outside the boundary layers, now involves circulaton, of an amount determined by the geometry of the aerofoil in conjunction with the requirement that the rear stagnation point is at the trailing edge.

Any force on an obstacle must be balanced by a rate of momentum change integrated over the fluid as a whole. We can see from the figure

---

† The pattern is drawn for the frame of reference in which the aerofoil is at rest, although the simplest way of producing the effects described is to start the aerofoil moving.

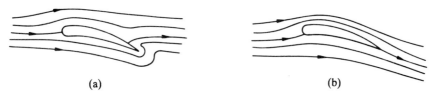

(a)   (b)

FIG. 13.5 Inviscid flow around aerofoil (schematic): (a) flow without circulation; (b) flow with circulation, in which rear stagnation point is located by boundary layer separation.

that the movement of the stagnation point does correspond to a deflection of the flow in the appropriate direction.

The starting process by which the circulation is generated involves discharge of vorticity into the fluid. One can see that this must happen by applying the Kelvin circulation theorem to a circuit enclosing both the initial position and the present position of the aerofoil; this circulation must remain zero. An eddy, of opposite sense to the circulation round the aerofoil, is left at the place from which it started. In principle, if one ignores further viscous action, this eddy remains there for ever.

Figure 13.6 shows a demonstration of this effect. Instead of starting the motion, the aerofoil is fixed in a steady flow but its angle of attack is suddenly changed (actually from one sign to the other to produce a large effect). The resulting change in lift corresponds to a change in the circulation and a 'starting eddy' must be generated as described above. The photograph, taken when the disturbance produced by the rotation of the aerofoil has travelled some distance downstream, shows that the pattern, although somewhat complicated, is dominated by an eddy circulating in the opposite sense to the change in circulation round the aerofoil. (The Reynolds number was comparatively low to avoid complications due to wake instability.)

We return now to considering an aerofoil in steady motion. When the angle of attack becomes large, boundary layer separation occurs on the upper surface close to the leading edge, giving a flow pattern like that

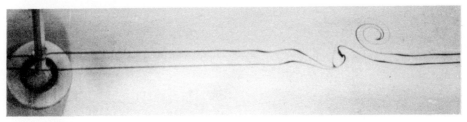

FIG. 13.6 Eddy produced by changing angle of attack of aerofoil. Flow visualization: two dye lines originating far upstream.

shown in Fig. 12.8. The flow is said to have stalled. The pressure in the separated region is less than that below the aerofoil (around the front stagnation point). There is thus still some lift, but much less than before stalling; there is also a sharp increase in the drag (Fig. 13.2).

13.3 Three-dimensional aerofoils

The relationship between lift and circulation gives rise to further complications in three-dimensional situations, i.e. for an aerofoil of finite span. The existence of circulation in combination with Stokes' theorem,

$$\oint \boldsymbol{u} \cdot \mathrm{d}\boldsymbol{l} = \int \boldsymbol{\omega} \cdot \mathrm{d}\boldsymbol{S} \qquad (13.8)$$

implies that there must be vorticity generated in the vicinity of the aerofoil (but outside the boundary layer); the surface over which the right-hand integral is carried out can extend over an end of the aerofoil as shown in Fig. 13.7. In other words, the concept of circulation in an entirely irrotational motion is meaningful only for a multiply-connected geometry.

The vorticity is generated near the ends of the aerofoil. Because of the pressure difference, $p_B - p_T$, fluid flows around each end from bottom to top. Vorticity is produced in the boundary layers associated with this motion and is then carried into the wake by the main-stream, giving the pattern shown in Fig. 13.8.

The fact that div $\boldsymbol{\omega} \equiv 0$ is in principle satisfied by the longitudinal vortices from the two ends of an aerofoil extending back to the position where the aerofoil started and there joining up through the starting vortex, as shown in Fig. 13.8. However, as they go back a long distance, the vortices become so diffuse and weak, through the action of viscosity, that there is no appreciable motion associated with them.

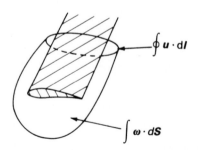

Fig. 13.7 Loop and surface integrals at end of aerofoil.

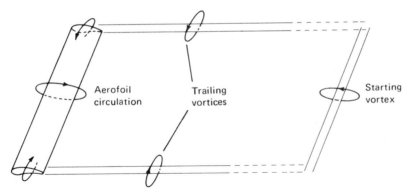

Fig. 13.8 Schematic diagram of circulation, trailing vortices, and starting vortex.

Usually, an aerofoil tapers towards its ends and the circulation drops gradually to zero; the longitudinal vorticity in the wake is then more spread out in the spanwise direction. Lift on a body of which the span is not long compared with the thickness cannot, of course, be discussed in these terms at all. The detailed dynamics of the boundary layer must then be considered, the generation of a wake with longitudinal vorticity again being an important part of the process.

13.4 Spinning bodies

It is well known that a spinning ball is deflected sideways. Studies have been made of the lift force generated on rotating cylinders and on rotating spheres. The usual name for such force generation is the Magnus effect, although sometimes this name is restricted to the force on cylinders with that on spheres being called the Robins effect [78].

A cylinder rotating about its axis and moving through a fluid in a direction perpendicular to its axis experiences a force perpendicular to both the direction of motion and the axis. Like the lift on an aerofoil, this phenomenon can be understood through Bernoulli's equation—by working in a frame of reference in which the axis is at rest and so the flow is steady. On the side on which the cylinder moves in the same direction as the flow, the fluid velocity is increased and so the pressure is reduced; on the other side the velocity is decreased and the pressure increased. There is thus a force acting on the cylinder from the latter to the former. But, again as with the aerofoil, it is necessary to invoke viscous action to understand why the rotation of the cylinder should produce a corresponding circulation of fluid. Indeed, but for the no-slip condition the

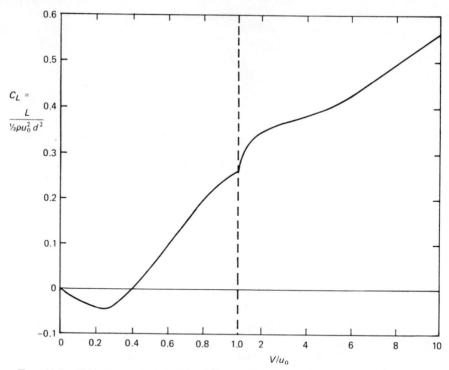

FIG. 13.9 Experimental variation of lift coefficient with relative rotation rate for spinning sphere. Note change in abscissa scale by factor of 8 at $V/u_0 = 1$. Curve is average of data in Reynolds number range 1.5×10^3–1.1×10^5 from Refs. [78, 135, 261].

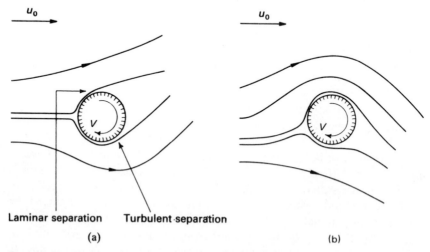

Laminar separation   Turbulent separation

(a)   (b)

FIG. 13.10 Sketch of flow pattern (a) produced by asymmetrical boundary layer separation, contrasted with flow pattern (b) associated with the 'conventional' Magnus effect.

flow would make no distinction between a rotating cylinder and a stationary one. At relatively low rotation rates, complications, like those described below for a sphere, can occur.

A rotating sphere also experiences a force perpendicular to both the rotation axis and its motion. As for an aerofoil, the three-dimensional problem is inherently more complicated than the two-dimensional. Figure 13.9 shows observations of the variation of the lift coefficient with the ratio of the circumferential speed of the sphere (V) to the flow speed (u_0), although combining data from widely varying Re into a single curve is undoubtedly an oversimplification. At large V/u_0, the behaviour can be understood qualitatively in the way described for a rotating cylinder, but any quantitative information must come from experiment.

At low V/u_0, the force reverses in direction (although not if the Reynolds number is too low [394]). This effect is probably the result of transition to turbulence occurring in the boundary layer on the side where the relative velocity between the sphere and the fluid is larger, but not on the other side [135]. Since transition delays separation (Section 12.4), this will produce a flow of the general form shown in Fig. 13.10(a). the resulting upward (in the geometry of the figure) deflection of the wake produces a downward force on the sphere; an effect of opposite sign from the 'conventional' Magnus effect (Fig. 13.10(b)).

14

CONVECTION

14.1 Introduction

This chapter returns to fundamentals: the equations of motion, dynamical similarity, and so forth. We need to do so because so far (apart from Chapter 4) the discussion has been centred on flow of fluids of uniform properties and without significant body forces. How do we handle a situation in which the temperature varies from place to place, perhaps giving rise to a gravitational body force; or one in which there is a variable concentration of some substance mixed with or dissolved in the fluid?

The practical importance of such flows needs little illustration. The use of moving fluids to transport or remove heat is well known, ranging from the circulation of coolant through a nuclear reactor to the mounting of a power transistor on a block with cooling fins. Equally apparent is the importance of thermal processes in meteorology and other branches of geophysics; indeed, the ultimate source of energy for almost all atmospheric motions and much of the oceanic circulation is solar radiation.

The variable salinity (salt concentration) of the oceans can be dynamically important. Air and water pollution studies require understanding of the processes determining pollutant concentration. Chemical engineering processes often involve the mixing of different substances.

These two types of topic are in fact closely analogous—sufficiently so that, as we shall see particularly in Chapter 15, one can sometimes use concentration variations to investigate the effects of temperature variations, or vice versa. It is thus appropriate to discuss them alongside one another. We shall in fact develop the theory for flows with temperature variations. This is the more complicated case, and flows with concentration variations can then be quickly covered by analogy (in Section 14.3 and, in passing, elsewhere). We shall be looking briefly at one or two particular flows in this chapter, but, for the most part, we are concerned with the background to Chapters 4, 15, and 22 and Sections 17.3 and 21.7. Some distinctive phenomena arise in the presence of both temperature and concentration variations, and these form the topic of Chapter 23.

The name convection is given to the general category of flows with externally introduced temperature variations. The same name is sometimes used for more specific processes occurring within such flows;

however, we shall use it in this book only in the general sense. Another name for the general topic is heat transfer, used particularly of course when the effect of the flow in transporting heat is of importance. Analogously the topic of flows with concentration variations is also known by the name mass transfer.

14.2 Equations of convection [7, 37]

In the practical examples mentioned above, and in the more fundamental situations considered in this book, the temperature variations are introduced by some process independent of the flow dynamics. We are not concerned here with flow-induced temperature variations, arising from adiabatic expansion or compression or from viscous dissipation. (The conditions in which such effects are negligible are considered in Section 5.8 and in the appendix to this chapter.) Most commonly, the temperature variations are introduced through temperature differences between boundaries, or between a boundary and ambient fluid. Occasionally, they are produced by internal heat generation, which can arise from a variety of causes such as radioactivity, absorption of thermal radiation, and release of latent heat as water vapour condenses. Heating and cooling either at the boundaries or within the fluid results in the temperature being a continuous function of position.

The temperature variations give rise to variations in the properties of the fluid, in the density and viscosity for example. An analysis including the full effects of these is so complicated that some approximation becomes essential. The equations are commonly used in a form known as the Boussinesq approximation. Here, we shall make this approximation without considering just when it is valid or demonstrating the fact that its different parts are internally self-consistent. These matters will be considered in some detail for the particular case of free convection in the appendix to this chapter; this will illustrate the procedure. In part, the Boussinesq approximation is the counterpart for convection of the incompressibility approximation for the flows considered hitherto. As discussed at the end of Section 5.8, the latter implies that the flow is a 'thermodynamic perturbation', and we shall see in Section 14.7 that this is a characteristic of Boussinesq convection also.

In the Boussinesq approximation, variations of all fluid properties other than the density are ignored completely. Variations of the density are ignored except insofar as they give rise to a gravitational force. Thus the continuity equation is used in its constant density form

$$\nabla \cdot \boldsymbol{u} = 0. \tag{14.1}$$

Similarly, $\rho Du/Dt$ is replaced by $\rho_0 Du/Dt$, with ρ_0 constant—chosen, for example, as the density at one typical position. However, a body force term is included to allow for the effect of gravity; that is in eqn (5.21) one puts

$$F = \rho g. \tag{14.2}$$

We may anticipate that the density variations will be important here and write

$$\rho = \rho_0 + \Delta\rho. \tag{14.3}$$

Also the gravitational acceleration is derivable from a potential

$$g = -\nabla\Phi. \tag{14.4}$$

(On the laboratory scale where g is uniform $\Phi = gz$, where z is taken vertically upwards.) Hence

$$F = -(\rho_0 + \Delta\rho)\nabla\Phi = -\nabla(\rho_0\Phi) + \Delta\rho g. \tag{14.5}$$

If we introduce

$$P = p + \rho_0\Phi \tag{14.6}$$

the Navier–Stokes equation (5.21) becomes

$$\rho_0 \frac{Du}{Dt} = -\nabla P + \mu\nabla^2 u + \Delta\rho g. \tag{14.7}$$

If $\Delta\rho = 0$, this is the same as the Navier–Stokes equation without a body force except that P has replaced p. Provided that the pressure does not appear explicitly in the boundary conditions, this change makes no difference. We have thus verified the conclusion considered more physically in Section 5.6, that the gravitational force is unimportant if the density is uniform. One merely has to correct for the hydrostatic pressure (the difference between p and P) if the actual pressure is required.

We are now concerned with cases in which $\Delta\rho \neq 0$, because of the temperature variations. The essential reason why it is consistent to treat ρ as a constant except in the one place is that the last term of eqn (14.7) can produce very significant effects even when $\Delta\rho/\rho_0 \ll 1$. This implies, of course, that all accelerations involved in the flow are small compared with $|g|$ (i.e. $|Du/Dt| \ll |g|$).

In these circumstances one can linearize the dependence of ρ on T,

$$\Delta\rho = -\alpha\rho_0\Delta T, \tag{14.8}$$

where α is the coefficient of expansion of the fluid. Hence, the Boussinesq dynamical equation is

$$\frac{Du}{Dt} = -\frac{1}{\rho}\nabla p + \nu\nabla^2 u - g\alpha\Delta T \tag{14.9}$$

(where we have reverted to writing ρ and p for the quantities temporarily denoted by ρ_0 and P).

One requires in addition an equation for the temperature. In the spirit of the Boussinesq approximation, it is supposed that the fluid has a constant heat capacity per unit volume,† ρC_p; then $\rho C_p DT/Dt$ is equal to the rate of heating per unit volume of a fluid particle. (The choice of C_p, the specific heat at constant pressure, is physically sensible, since the pressure is not free to respond directly to the heating process; for a proper justification, however, one needs to look at the detailed nature of the Boussinesq approximation: see the appendix.) This heating is brought about by transfer of heat from neighbouring fluid particles by thermal conduction and sometimes also by internal heat generation. The corresponding terms in the thermal equation are analogous respectively to the viscous term and the body force term in the dynamical equation. The conductive heat flux

$$\boldsymbol{H} = -k \operatorname{grad} T \tag{14.10}$$

where k is the thermal conductivity of the fluid. Thus

$$\rho C_p DT/Dt = -\operatorname{div} \boldsymbol{H} + J \tag{14.11}$$

where J is the rate of internal heat generation per unit volume. Taking k to be constant, eqn (14.11) may be rewritten

$$\frac{\partial T}{\partial t} + \boldsymbol{u} \cdot \nabla T = \kappa \nabla^2 T + \frac{J}{\rho C_p} \tag{14.12}$$

where

$$\kappa = k/\rho C_p. \tag{14.13}$$

κ is known as the thermal diffusivity or sometimes as the thermometric conductivity.

Equations (14.1), (14.9), and (14.12) constitute the basic equations of convection in the Boussinesq approximation. They are one vector and two scalar equations for the one vector and two scalar variables $\boldsymbol{u}$, p and ΔT. (Since T appears differentiated throughout eqn (14.12) this is effectively an equation in ΔT, thus matching up with eqn (14.9).)

It is useful to have names for the new terms. The additional term in the dynamical equation, $-g\alpha\Delta T$, is known as the buoyancy force‡ (even when ΔT is negative and the term represents the tendency for heavy fluid to sink). The two terms on the right-hand side of eqn (14.12) have

---

† See comment about notation for extensive thermodynamic quantities in Notation section at the end of the book (p. 468).
‡ For a perfect gas $\alpha = 1/T$, and in texts concerned primarily with gases, the buoyancy term is often written $-g\Delta T/T$.

obvious names: the conduction term and the heat generation term. $u \cdot \nabla T$, which represents the transport of heat by the motion, may be called the advection term. (It is often called the convection term, but it is convenient to have a separate name from the processes represented by the equations as a whole.)

Equation (14.12) requires boundary conditions for the temperature field. The commonest type specifies the temperature of a boundary wall; the fluid right next to the wall must then also be at that temperature. There are, however, other types. As an example, we may mention the case in which the heat transfer through a wall is specified; the temperature gradient in the fluid is then fixed at the wall.

It is important to note that thermal conduction plays an integral role in convection. For example, when heat is introduced into a fluid by heating a boundary wall, there is no advection of heat through the boundary; the fluid first gains heat entirely by conduction, although further from the wall advection may be the principal process. The traditional division of heat transfer processes into radiation, conduction, and convection is not completely sharp. Whenever convection occurs, it and conduction become parts of a single process.

Evidently a wide range of dynamical behaviours is to be expected, depending on the importance of the buoyancy force relative to the other terms in eqn (14.9). The two extremes—when the buoyancy force is negligible, and when it is the sole cause of motion—are called forced convection and free convection respectively. They will be considered further in Sections 14.4 and 14.5. Intermediate cases are known as mixed convection; we shall be considering these only in the important special case when the imposed flow is predominantly horizontal so that the buoyancy force acts perpendicularly to it: this is known as stratified flow (Chapter 15 and Section 21.7).

14.3 Flows with concentration variations

Equations governing flows with concentration variations may be quickly formulated by analogy with the equations in Section 14.2. The amount of a substance carried by a fluid may be expressed by a concentration, c—say, the mass of the substance per unit volume. c is a continuously variable function of position. The presence of the substance increases the density by an amount $\Delta \rho$ above its value ρ_0 corresponding to $c = 0$. One may take a linear relationship between concentration and density

$$\Delta \rho = \rho_0 \alpha_c c \tag{14.14}$$

where α_c is a coefficient. ($\rho_0 \alpha_c = 1$ if the substance is absorbed into the fluid without increase in volume, but that is not necessarily the case. α_c can be negative, as in the case of a lighter gas mixing in the flow of a heavier gas.) Thus the dynamical equation may be taken as

$$\frac{D\boldsymbol{u}}{Dt} = -\frac{1}{\rho}\nabla p + \nu\nabla^2\boldsymbol{u} + \boldsymbol{g}\alpha_c c \qquad (14.15)$$

cf. eqn (14.9).

The distribution of c is determined by its advection by moving fluid particles and by its diffusion between fluid particles. Hence,

$$\frac{Dc}{Dt} = \kappa_c\nabla^2 c, \qquad (14.16)$$

where κ_c is a diffusion coefficient depending both on the fluid and on the diffusing substance. Equation (14.16) is to be compared with (14.12); in the absence of chemical reactions, there is no counterpart of the internal heat generation term, but that is often absent also in thermal problems.

Equations (14.15) and (14.16) have, of course, to be accompanied by the continuity equation, again normally taken to be

$$\nabla \cdot \boldsymbol{u} = 0 \qquad (14.17)$$

(cf. the discussion around eqn (14.1)).

The point at which the analogy is most likely to break down is in the boundary conditions. Ways of introducing concentration variations are usually not the counterparts of the commonest ways of introducing temperature variations as considered in Section 14.2. In particular, it is difficult to maintain the concentration at a fixed value at a boundary, the counterpart of the common fixed-temperature boundary condition. An example of a more likely way of introducing concentration variations, one that is more readily achieved in the laboratory, is for two meeting streams to have different concentrations.

One further point about the equations may now be made, partly to relate the present approach with that sometimes adopted elsewhere. It is sometimes permissible to ignore mass diffusion and so to approximate eqn (14.16) by

$$Dc/Dt = 0. \qquad (14.18)$$

This has more relevance to mass transfer than to heat transfer: in the latter, even when order-of-magnitude considerations suggest that diffusion is negligible (see Sections 14.4, 14.5), it usually remains important in boundary layers; but concentration differences are often introduced in a way that does not give the diffusivity an essential role at boundaries. When eqn (14.18) applies, each fluid particle always consists of the same material and thus conserves its density; i.e.

$$D\rho/Dt = 0. \tag{14.19}$$

Substituting this into the general form of the continuity equation (5.9) reduces that to (14.17) without further approximation. Hence, (14.17) remains applicable even when c is sufficiently large for (14.14) to produce large fractional density variations. This result is of limited use, however, because one will usually also wish to approximate ρ by ρ_0 in (14.15) and that does require $\Delta\rho/\rho_0 \ll 1$.

14.4 Forced convection

As indicated in Section 14.2, forced convection occurs when the buoyancy term in eqn (14.9) is negligible. For example, at high Reynolds number, it occurs if the buoyancy force is small compared with the inertia force; i.e. if

$$g\alpha\Theta L/U^2 \ll 1 \tag{14.20}$$

where L is a length scale, Θ a temperature difference scale, and U a velocity scale introduced independently of Θ. (Anticipating the nomenclature to be introduced in Section 15.1, relationship (14.20) says that the Richardson number must be low, or the internal Froude number high, for forced convection.)

Forced convection is frequently of importance in heat transfer applications. However, no new flow phenomena arise; the velocity field is unaffected by the temperature field and can be determined in the same ways as before. Hence, the only matter needing consideration is the temperature distribution resulting from the flow; once the velocity field is known this is given by (14.12). But, of course, it is frequently necessary to turn to experiment to get the required results. The transfer of information from one system to another requires not only dynamical similarity, i.e. that the flow patterns should be the same, but also thermal similarity, i.e. that the temperature fields should have the same patterns. The former is guaranteed by equality of the Reynolds number, but the latter gives rise to an extra condition.

We will confine attention to steady convection without internal heat generation. Equation (14.12) becomes

$$\boldsymbol{u} \cdot \nabla T = \kappa\nabla^2 T. \tag{14.21}$$

Analysis of this equation along the lines of the analysis of the Navier–Stokes equation in Section 7.2 shows that there will be thermal similarity when the two systems have the same value of

$$Pe = UL/\kappa, \tag{14.22}$$

where Pe is the Péclet number. This number may be interpreted as a measure of the relative importance of the two terms in eqn (14.21), analogously with the interpretation of the Reynolds number as the ratio of inertia forces to viscous forces:†

$$\text{Pe} \sim \frac{\text{advection of heat}}{\text{conduction of heat}}. \tag{14.23}$$

When Pe is small, eqn (14.21) approximates to

$$\kappa \nabla^2 T = 0 \tag{14.24}$$

the flow having negligible effect on the temperature distribution, which is determined by the same equation as in a material at rest. At high Pe, eqn (14.21) approximates at first to

$$\boldsymbol{u} \cdot \nabla T = 0 \tag{14.25}$$

but now conduction can be important in thermal boundary layers for reasons analogous to the origin of viscous boundary layers.

Full similarity of forced convection situations requires equality of both $\text{Re} = UL/\nu$ and $\text{Pe} = UL/\kappa$. An equivalent statement is that it requires the equality of both the Reynolds number and

$$\text{Pr} = \nu/\kappa. \tag{14.26}$$

This non-dimensional parameter, called the Prandtl number, is a property of the fluid, not of the particular flow. Hence, there is a restriction on the transfer of information from experiments with one fluid to those with another.

The Prandtl number is the ratio of two diffusivities, ν being the diffusivity of momentum and vorticity and κ that of heat. The meaning of this can be illustrated by considering irrotational fluid at a uniform temperature entering a pipe of which the walls are maintained at a higher temperature. Both the Reynolds number and the Péclet number are taken to be significantly above 1. We know from Section 11.1 that

† Curiously, when applied to equations with only two terms this type of argument can give rise to a conceptual difficulty that does not arise when there are more terms. There is an apparent contradiction, which sometimes worries people, in saying simultaneously that 'advection equals conduction' (eqn (14.21)) and that Pe measures the relative sizes of advection and conduction (so that, for example, if Pe ≫ 1, 'advection is much greater than conduction'). The point is the following. Equation (14.21) is an exact equation, and there is thus a sense in which it is true that advection is in exact balance with conduction at every point in the flow. One may nevertheless make an order-of-magnitude estimate of the two terms and find (for example) that advection comes out much the larger. The implication is that advection is the dominant process determining the temperature field, little modified by conduction. This leads one to say as an approximation 'advection equals zero' ((14.25)). Since this is then the equation determining the temperature field, 'advection equals zero' is just the opposite of 'advection is negligible'.

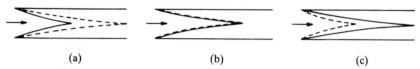

FIG. 14.1   Relative position of edges of velocity boundary layer (solid lines) and thermal boundary layer (broken lines) in pipe flow entry: (a) Pr significantly above 1; (b) Pr ~ 1; (c) Pr significantly below 1.

vorticity is introduced into the flow in a viscous boundary layer that increases in thickness with distance downstream. Similarly, heat spreads into the flow in a thermal boundary layer of increasing thickness, leaving a diminishing core of fluid that is still effectively at its initial temperature. Figure 14.1 shows schematically the relative thicknesses of the two boundary layers for values of the Prandtl number significantly greater than 1, around 1, and significantly less than 1. The relatively more rapid diffusion of heat than vorticity as the Prandtl number is decreased can be seen.

Gases have values of the Prandtl number around (but a little less than) 1. Most liquids have values greater than 1 but by widely varying amounts. Water is typical of the low end of the range with Prandtl number (at room temperature) around 6. In general, the kinematic viscosity varies much more widely than the thermal diffusivity, so the high Prandtl number liquids are very viscous ones. The important exception to these statements is liquid metals where high thermal diffusivities are produced by the free electrons, giving low Prandtl numbers; for example mercury has a value around 0.025.

The non-dimensional parameters given in (14.22) and (14.26) have their mass transfer counterparts. UL/κ_c does not have a special name. However, ν/κ_c is called the Schmidt number, Sc. One is usually concerned either with $Sc \approx 1$ (the case for one gas mixed in another) or with $Sc \gg 1$ (most other cases, such as aqueous solutions of electrolytes or small solid particles carried by a liquid or gas flow).

A quantity of frequent practical importance in convection problems is the heat transfer through a surface into or out of the fluid. If we denote the rate of transfer per unit area by H, a non-dimensional form of this is

$$Nu = HL/k\Theta \tag{14.27}$$

where k is the thermal conductivity and L and Θ are length and temperature difference scales (e.g. Θ might be the temperature difference between the surface under consideration and the ambient fluid). Nu is the Nusselt number. The Nusselt number may have both local and overall meanings depending on whether H is the local heat transfer or an average value over the whole surface. See Section 14.8 for an example. For the moment we are concerned principally with the latter meaning.

For forced convection, dimensional considerations indicate that

$$\text{Nu} = f(\text{Re}, \text{Pr}). \qquad (14.28)$$

Neither Re nor Pr involves Θ. Hence, when all other quantities are being held constant,

$$H \propto \Theta \qquad (14.29)$$

the heat transfer is proportional to the imposed temperature excess. This result, which comes essentially from the linearity in ΔT of eqn (14.21), is Newton's law of cooling. We shall see in Sections 14.5 and 14.8 that it does not apply to free convection; nor, in general, does it apply to mixed convection. The familiar statement that Newton's law of cooling applies in a strong draught can be made more precise: it applies when the forced convection approximation is valid.

14.5 Free convection: basic concepts [37]

A free convection flow is one produced by buoyancy forces. Temperature differences are introduced, for example through boundaries maintained at different temperatures, and the consequent density differences induce the motion; hot fluid tends to rise, cold to fall.

Figure 14.2 shows one example of free convection, the flow produced by a horizontal heated cylinder in a large expanse of cooler fluid, illustrating the complexity of the phenomena that can arise from the simple fact that hot fluid rises. One sees that the cylinder is surrounded by a boundary layer of hot fluid; from this a relatively narrow column (or 'plume') of hot fluid emerges, becoming unstable and ultimately turbulent as it rises.

If we confine attention to steady convection without internal heat generation the governing equations on the Boussinesq approximation (14.1), (14.9) and (14.12) are

$$\nabla \cdot \boldsymbol{u} = 0 \qquad (14.30)$$

$$\boldsymbol{u} \cdot \nabla \boldsymbol{u} = -\frac{1}{\rho}\nabla p + \nu \nabla^2 \boldsymbol{u} - g\alpha \Delta T \qquad (14.31)$$

$$\boldsymbol{u} \cdot \nabla T = \kappa \nabla^2 T. \qquad (14.32)$$

(An appendix to this chapter considers the conditions in which the Boussinesq approximation is valid for free convection and Section 14.6 describes a method of extending its applicability, important in geophysical contexts.) Equations (14.31) and (14.32) must now be considered simultaneously as both involve both $\boldsymbol{u}$ and T. The velocity distribution is

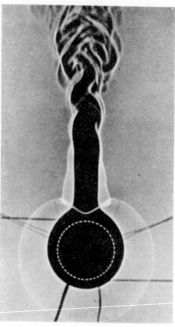

FIG. 14.2 Shadowgraph of convection around a heated horizontal cylinder (in position indicated by dotted line); approximately, dark regions are those of hot fluid. From E. Schmidt, Forschung auf dem Gebiete des Ingenieurwesens, Band 3, Seite 181 (1932), by kind permission of VDI-Verlag GmbH, Dusseldorf.

governed by the temperature distribution, but the temperature distribution depends through the advection of heat on the velocity distribution. There is no possibility of determining one independently of the other, as could be done in forced convection. For this reason, free convection is hard to treat theoretically and much of our information about it comes from experimental investigations.

In such experimental investigations one needs to know when dynamical similarity pertains. No velocity scale is provided by the specification of a free convection situation. Hence, one cannot define a Reynolds number using only parameters involved in setting the problem up. Instead, a temperature difference scale, Θ, will govern how vigorous the motion is. If eqns (14.30) to (14.32) are analysed by the methods introduced in Chapter 7, it is found that the dynamical similarity depends on two non-dimensional parameters:

$$Gr = g\alpha\Theta L^3/v^2 \qquad (14.33)$$

$$Pr = v/\kappa. \qquad (14.34)$$

These are respectively the Grashof number and the Prandtl number. Only when both of these are the same in two geometrically similar situations can the same flow patterns be expected. As already mentioned in the context of forced convection (Section 14.4), Pr is a property of the fluid and the transfer of information from one fluid to another is restricted. A full investigation of a given situation requires ranging of the Grashof number for fluids of a variety of values of the Prandtl number and thus is inherently longer than an investigation of a situation governed by a single non-dimensional parameter such as the Reynolds number.

For a given fluid, the Grashof number indicates the type of flow to be expected—which dynamical processes are dominant, whether the flow is laminar or turbulent, and so on—as the Reynolds number does for forced flow. The Grashof number cannot be given a general simple interpretation as the ratio of two dynamical processes. Nevertheless it indicates the relative importance of inertia and viscous forces, as the following discussion shows.

Either the inertia force or the viscous force, or both, must be of the same order of magnitude as the buoyancy force. The motion can reach a steady state only when other terms balance the buoyancy force; on the other hand, since the buoyancy force is the cause of the motion, these other terms cannot become large compared with it. We suppose in the first place that the inertia force is of the same order of magnitude as the buoyancy force:

$$|\boldsymbol{u} \cdot \nabla \boldsymbol{u}| \sim |g\alpha \Delta T| \tag{14.35}$$

that is,

$$U^2/L \sim g\alpha\Theta. \tag{14.36}$$

This enables a velocity scale to be written down indicating typically how fast the fluid will move as a result of the temperature variations:

$$U \sim (g\alpha L\Theta)^{1/2}. \tag{14.37}$$

We can now compare the orders of magnitude of the inertia and viscous forces:

$$\frac{|\boldsymbol{u} \cdot \nabla \boldsymbol{u}|}{|\nu\nabla^2\boldsymbol{u}|} \sim \frac{UL}{\nu} \sim \left(\frac{g\alpha\Theta L^3}{\nu^2}\right)^{1/2} = \mathrm{Gr}^{1/2}. \tag{14.38}$$

This tells us that, when the Grashof number is large, the viscous force is negligible compared with the buoyancy and inertia forces (subject to qualifications discussed below). On the other hand, it tells us nothing in the case of small Grashof number as the apparent prediction that the inertia force is small is in contradiction with the original assumption that the inertia force is comparible with the buoyancy force.

To deal with that case, we start with the alternative assumption that the viscous force is comparable with the buoyancy force

$$|\nu\nabla^2 \boldsymbol{u}| \sim |g\alpha\Delta T|. \tag{14.39}$$

The corresponding procedure gives

$$U \sim g\alpha\Theta L^2/\nu \tag{14.40}$$

and so

$$\frac{|\boldsymbol{u}\cdot\nabla u|}{|\nu\nabla^2 \boldsymbol{u}|} \sim \mathrm{Gr}. \tag{14.41}$$

This analysis indicates that small Grashof number implies negligible inertia forces, but is irrelevant to the case of high Grashof number.

Hence, in general, the Grashof number is a measure of the relative importance of viscous and inertial effects; but, because of the different powers to which Gr appears in relationships (14.38) and (14.41), one cannot write a general expression for Gr as a ratio of effects.

In convection problems, one needs to know not only which dynamical processes are important but also which processes are important in determining the temperature distribution.†

$$\frac{\text{Advection}}{\text{Conduction}} \sim \frac{|\boldsymbol{u}\cdot\nabla T|}{|\kappa\nabla^2 T|} \sim \frac{UL}{\kappa} \tag{14.42}$$

and so, when Gr is large,

$$|\boldsymbol{u}\cdot\nabla T|/|\kappa\nabla^2 T| \sim \mathrm{Gr}^{1/2}\mathrm{Pr} \tag{14.43}$$

and, when Gr is small,

$$|\boldsymbol{u}\cdot\nabla T|/|\kappa\nabla^2 T| \sim \mathrm{GrPr}. \tag{14.44}$$

When the Prandtl number is around 1 (as it is for gases and for some liquids, see Section 14.4), dominance of advection over conduction always occurs simultaneously with dominance of inertial forces over viscous forces. When the Prandtl number is small (as in liquid metals) or large (as in viscous oils and many other liquids), this correspondence does not apply.

The quantity

$$\mathrm{Ra} = \mathrm{GrPr} = g\alpha\Theta L^3/\nu\kappa \tag{14.45}$$

appearing in relationship (14.44) plays a special role in studies of convection in horizontal layers (see Chapters 4 and 22) and is called the Rayleigh number.

† See footnote on p. 169.

The quantities UL/ν and UL/κ appearing in (14.38) and (14.42) can, of course, be identified as a Reynolds number and a Péclet number. Their role, however, is different from that in forced flows, since they involve U which is a dependent scale, not an independent one. Hence, Re (and similarly Pe) is a dependent non-dimensional parameter and one can write

$$\text{Re} = f(\text{Gr}, \text{Pr}). \tag{14.46}$$

Another important dependent parameter is the Nusselt number, indicating the heat transfer as for forced convection (Section 14.4). From dimensional analysis

$$\text{Nu} = f(\text{Gr}, \text{Pr}). \tag{14.47}$$

In this relationship Θ appears in both Nu and Gr. Hence, the argument applying to forced convection that the heat transfer is proportional to the temperature difference ((14.29), Newton's law of cooling) does not apply to free convection. As one would expect, Nu always (or almost always) increases with increasing Gr, implying that, when all else is held constant, the heat transfer increases more rapidly than direct proportionality to the temperature difference. This is because of the dual role of the temperature difference: the larger it is the greater is the rate of heat transfer by a particular flow rate; but also the larger it is, the greater is the flow rate.

Large values of the Grashof number occur much more frequently than small ones. For example, the comparatively small temperature difference and length scales of 1°C and 10^{-2} m give Gr $\sim 10^3$ in water and $\sim 10^2$ in air. This is related to the fact that quite vigorous convection currents often arise as a result of stray temperature differences in any large volume of fluid left standing (and can be a nuisance if one is trying to investigate something else!).

When Gr is large (and Pr is not too small), relationships (14.38) and (14.43) imply the dominance of inertia forces over viscous and of advection over conduction. However, this is based on the assumption that the only length scale is the imposed one, L. This assumption will be invalidated by boundary layer formation for reasons similar to those explained in Section 8.3. (Notice that the conduction term is the highest-order differential term in eqn (14.32) just as the viscous term is in the Navier–Stokes equation.) In the commonest situation where the flow is produced by maintaining the temperature differences between impermeable boundaries, thermal conduction is responsible for the introduction of temperature differences into the fluid. Without thermal conduction the fluid next to a wall could remain at a different temperature from the wall (a state of affairs analogous to the theoretical possibility of a velocity difference between fluid and wall in the absence of viscosity). Omitting the action of conductivity would thus remove the convection problem altogether. The correct inference when $Gr^{1/2}$ and $Gr^{1/2}Pr$ are large is that the flow will have a boundary layer character. A flow considered in Section 14.8 will illustrate more explicitly the nature of such boundary layers.

Free convective flows are normally rotational. Buoyancy forces directly generate vorticity. Applying the curl operation throughout (14.9), we see how eqn (6.27) is modified by the addition of the buoyancy force:

$$\frac{D\boldsymbol{\omega}}{Dt} = \boldsymbol{\omega} \cdot \nabla \boldsymbol{u} + \nu \nabla^2 \boldsymbol{\omega} + \alpha \boldsymbol{g} \times \nabla(\Delta T). \tag{14.48}$$

Horizontal components of the temperature gradient $\nabla(\Delta T)$ contribute to the last term; the vorticity so generated is also horizontal but perpendicular to the temperature gradient. This result is physically quite obvious; the torque associated with the tendency for hotter fluid to rise and colder to fall, when these are horizontally separated, is producing rotation about a horizontal axis.

The formal justification for statements in Sections 4.3 and 4.4 is closely related to this. We saw that a situation with horizontal temperature gradients (as in a vertical slot) did not allow equilibrium with the fluid at rest, whereas one with only a vertical temperature gradient (as in a horizontal layer) could have such an equilibrium solution (which might be stable or unstable). One can again readily see physically why this should be. However, more formally, a rest configuration must be given by eqn (14.9) without the terms involving the velocity

$$-\frac{1}{\rho} \nabla p = \boldsymbol{g} \alpha \Delta T \tag{14.49}$$

i.e. by a balance between pressure and buoyancy forces. Since the curl of the pressure gradient is identically zero, this requires

$$\nabla \times (\boldsymbol{g} \alpha \Delta T) = -\alpha \boldsymbol{g} \times \nabla(\Delta T) = 0. \tag{14.50}$$

Only when $\nabla(\Delta T)$ is vertical is this satisfied.

14.6 The adiabatic temperature gradient

We digress here to consider a matter of little significance for laboratory fluid dynamics but of importance in the application of laboratory results to almost any geophysical situation. The matter concerns the effect of the hydrostatic pressure on the convection: as fluid rises, it expands and so cools; as fluid falls, it is compressed and so warms. This effect is ignored in the Boussinesq approximation, but it is both useful and possible to make allowance for it. (The point is discussed more systematically in the appendix to this chapter.)

The matter is not confined to free convection, but is conveniently considered in this context. Indeed, we consider it first in the context of the particular type of free convection, already introduced in Section 4.4, in which the imposed temperature differences are vertical. Considering such a situation on the basis of the Boussinesq approximation, a first requirement for instability to bring the fluid into motion is that the temperature should decrease with height. This is inferred by considering a fluid particle displaced a small distance vertically—upwards, say. Ignoring interactions with other fluid particles, the particle will be lighter than its new environment when the temperature decreases upwards and heavier when it increases upwards. In the former case, it will tend to move further upwards and the displacement is amplified; in the latter case, it tends to be restored to its original level. (When interactions, such as heat conduction between particles, are taken into account, additional requirements for instability come in.)

This argument is valid when the effect of changes in hydrostatic pressure can be neglected. If it cannot, then the criterion for instability must be modified to allow for the cooling of an upward moving fluid particle. Only if the particle's new temperature exceeds that of its new environment will it continue to be displaced upwards. There is a certain vertical temperature distribution for which a fluid particle moving vertically always has just the local temperature. For instability, the temperature must decrease with height more rapidly than this. We denote this distribution by $T_a(z)$.

To determine it, we again consider a fluid particle displaced vertically in an ideal inviscid non-heat-conducting fluid. Thermodynamically irreversible processes and heat transfer between particles involve the action of viscosity and conductivity. Hence, the displacement occurs at constant entropy, and

$$\frac{dT_a}{dz} = \left(\frac{\partial T}{\partial p}\right)_s \frac{dp}{dz}. \qquad (14.51)$$

The vertical pressure gradient is produced by the weight of the fluid. Hence,

$$\frac{dp}{dz} = -g\rho. \qquad (14.52)$$

Also, in conventional thermodynamic notation (with $\rho = 1/V$)

$$\left(\frac{\partial T}{\partial p}\right)_s = \left(\frac{\partial V}{\partial S}\right)_p = \left(\frac{\partial T}{\partial S}\right)_p \left(\frac{\partial V}{\partial T}\right)_p = \frac{\alpha T}{\rho C_p} \qquad (14.53)$$

and so

$$\frac{dT_a}{dz} = -\frac{g\alpha T}{C_p}.$$ (14.54)

The temperature gradient given by (14.54) is known as the adiabatic gradient, or, in geophysical usage, the adiabatic lapse rate.

The criterion for instability, $-\partial T/\partial z > 0$, is thus more correctly written

$$-\frac{\partial T}{\partial z} > \frac{g\alpha T}{C_p}.$$ (14.55)

The distinction is frequently important in geophysical situations, although rarely in the laboratory. We may illustrate this by considering the adiabatic temperature gradient for air, which, at $T \sim 300\,\text{K}$ is approximately $10^{-2}\,\text{K}\,\text{m}^{-1}$. A Rayleigh number of 10^4 (see Section 4.4) in a layer of $3 \times 10^{-2}\,\text{m}$ (typical in the laboratory) corresponds to a temperature gradient of around $10^2\,\text{K}\,\text{m}^{-1}$ whereas over a depth of 10 m (the small end of atmospheric scales) it corresponds to $10^{-8}\,\text{K}\,\text{m}^{-1}$. In natural situations it is thus usually the temperature gradient criterion (14.55) that governs whether convection occurs. The Rayleigh number criterion (Sections 4.4, 22.2) is rarely relevant.

We shall see in Section 22.7 that, for situations in which the adiabatic gradient is negligible, high Rayleigh number convection has a negligible temperature gradient except in thin boundary layers (Fig. 22.11). Correspondingly, vigorous convection in natural situations reduces the temperature gradient to the adiabatic gradient. For this reason, regions occur in which the temperature distribution is governed by an equation of the form of (14.54) with $T = T_a$. This is of particular importance in the theories of stellar and planetary interiors. Also the temperature difference between the bottom and top of a mountain can often be estimated by assuming that it is just due to the adiabatic gradient of $10^{-2}\,\text{K}\,\text{m}^{-1}$.

The foregoing considerations suggest that it would be useful to introduce θ such that

$$\frac{\partial \theta}{\partial z} = \frac{\partial T}{\partial z} - \frac{dT_a}{dz}$$ (14.56)

and so

$$\theta = T - (T_a - T_0)$$ (14.57)

where T_0 is a reference temperature. (One usually chooses to put both $T_a = T_0$ and $\theta = T_0$ at the level at which $T = T_0$; there are adjustable constants in the integrations of both (14.54) and (14.56)). θ is known as the potential temperature. This procedure is indeed useful, not only for the above type of stability consideration but quite generally. When temperature changes due to motion in the hydrostatic pressure field are taken into account, the equations become the same as those derived ignoring such changes, provided that T is replaced by θ. (The justification for this statement and the conditions in which it is true are both indicated in the appendix to this chapter.) The main body of our discussion of free convection is being developed in terms of laboratory situations for which the distinction between T and θ is insignificant. However, the applicability of these ideas is greatly extended by the concept of potential temperature.

Because of the adiabatic temperature gradient, it is not always satisfactory to describe flows without buoyancy forces as 'isothermal'. The alternative term is 'barotropic'. Flows with buoyancy forces are conversely described as 'baroclinic'. In general, a barotropic situation is one in which surfaces of constant pressure and surfaces of constant density coincide; a baroclinic situation is one in which they intersect. When the principal cause of pressure variation is the hydrostatic effect, barotropy implies that constant density surfaces are horizontal and thus that buoyancy forces are absent.

14.7 Free convection as a heat engine

A free convective flow is thermodynamically a heat engine. Heat enters the fluid at hot boundaries, is transported by the flow, and leaves it at colder boundaries; during the transport, forces are generated which feed kinetic energy into the flow. The fact that this energy is usually dissipated again within the fluid, rather than becoming available as work done externally, complicates the overall thermodynamics but does not change the character of the kinetic energy generation process. (One can do a 'thought experiment' in which the action of viscosity is simulated by a large number of tiny propellers which do deliver external work.) One might expect therefore that considerations involving the laws of thermodynamics would enter more explicitly into the analysis of free convection. The aim of this section is to clarify this point.

From eqns (14.30)–(14.32) one can derive expressions for the rate of kinetic energy generation and for the rate of heat transfer. In order of magnitude the ratio of these quantities is†

$$W/Q_1 \sim g\alpha L/C_p \qquad (14.58)$$

† The derivation of this result involves methods outside the scope of this book. They are somewhat similar to those mentioned in Section 11.7. The result is also closely related to eqn (14.111).

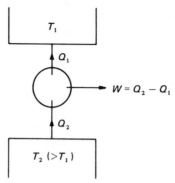

F<small>IG</small>. 14.3 Definition diagram for heat engine, drawing heat Q_2 from thermal reservoir at temperature T_2 and delivering heat Q_1 to thermal reservoir at temperature T_1 and doing external work W.

where L is the length scale of the convecting system and the notation of the left-hand side corresponds to conventional thermodynamic usage as in Fig. 14.3. This ratio is always very small in any situation to which the Boussinesq approximation applies (see appendix); that is, such situations have very low 'efficiency':

$$W \ll Q_1, \quad Q_2 \approx Q_1. \tag{14.59}$$

This fact has two consequences. Firstly, the kinetic energy generation is a negligible perturbation in the first law of thermodynamics. This therefore reduces to a heat conservation law (strictly an enthalpy conservation law, corresponding to the appearance of C_p, not C_V, in the equations). Such a law is expressed by an integrated form of eqn (14.32). (Note that this does not mean that mechanical effects on the heat transfer are in every way negligible; the heat transfer is usually much larger than it would be if the fluid remained at rest.)

Secondly, the efficiency is so low compared with the Carnot efficiency, $(T_2 - T_1)/T_1$, that the constraints imposed by the second law of thermodynamics are automatically fulfilled and do not need explicit consideration. (It might be remarked that the efficiency in (14.58) is, in order of magnitude, independent of $(T_2 - T_1)/T_1$, and that one could thus envisage a very weak convection produced by very small temperature differences in which the actual efficiency did apparently exceed the Carnot efficiency. The resolution of this apparent paradox lies in the fact that such temperature differences would be smaller than those associated with the adiabatic temperature gradient—which, from eqn (14.54), have order of magnitude $g\alpha TL/C_p$. The existence of the latter provides a Carnot efficiency at least as large as the actual efficiency.)

Situations to which the Boussinesq approximation does not apply do require much more explicit thermodynamic formulation than the situations being considered here.

14.8 Free convection: boundary layer type flows

For the most part this chapter has developed the basic background to topics considered elsewhere in the book. However, we now look briefly at a couple of other configurations to illustrate how their broad features may be inferred by applying ideas with which we are now familiar. At the end of Section 14.5 we noted the importance of high Grashof number situations and thus of 'boundary layer' type flows (using this name in the general sense of flows with much smaller transverse than longitudinal length scale). It is thus interesting to look at the simplest free-convection counterparts of the flows considered in Chapter 11. These are convection from a heated vertical plate and the plume over a localized heat source, shown schematically in Fig. 14.4, and the counterparts of respectively the Blasius boundary layer (Section 11.4) and a jet (Section 11.6).

In the first case, the plate, maintained at a constant uniform temperature T_1, is on one side of a large expanse of fluid of which the temperature is T_0 ($<T_1$) far from the plate. The horizontal dimension of the plate is supposed to be large enough for the motion to be considered as two-dimensional. Just as the distance from the leading edge provided the longitudinal length scale for the Blasius boundary layer, so the distance z from the lower edge provides it here; the presence of the upper edge affects the flow alongside the plate only very near the top.

The fluid next to the plate is heated by thermal conduction. As a result it rises up the plate. When the Grashof number, now defined as

$$Gr_z = g\alpha(T_1 - T_0)z^3/\nu^2 \tag{14.60}$$

is large enough, the speed generated in this way is sufficiently high that the heat is advected in the z-direction before it has penetrated far in the y-direction (Fig. 14.4(a)). The convection occurs entirely in a thin boundary layer (Fig. 14.5). Outside this the fluid remains almost at rest at temperature T_0. It does drift slowly towards the plate, as the boundary layer entrains fluid, the amount of fluid moving up the plate increasing with z.

The temperature is T_1 at the wall and T_0 outside the boundary layer. The vertical velocity is zero both at the wall and outside the boundary layer. Their profiles must therefore be of the general forms shown in Fig. 14.4(a). Both the boundary layer thickness and the maximum velocity vary with z, whilst the temperature difference across the boundary layer is, of course, fixed at $(T_1 - T_0)$.

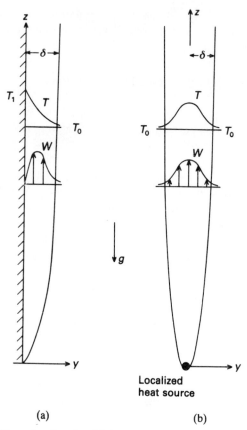

FIG. 14.4 Definition sketches for (a) free convection boundary layer on a heated vertical plate, (b) thermal plume above a heat source.

In general, as for forced convection (Fig. 14.1), there are actually two boundary layer thicknesses to be considered, that of the velocity profile and that of the temperature profile. One expects these to be about the same when the Prandtl number is around unity. When $\mathrm{Pr} \gg 1$, the velocity boundary layer is thick compared with the temperature one (as in forced convection). The heat diffuses only slowly away from the wall, but the buoyancy of the thin hot layer drags much more fluid into motion through the action of viscosity. On the other hand, when $\mathrm{Pr} \ll 1$, the velocity and temperature boundary layers remain of comparable thickness (in contrast to the behaviour in forced convection). Any fluid that is heated by thermal conduction moves under the action of buoyancy, even if viscous effects have not spread that far from the wall [271, 294].

Order-of-magnitude scaling analysis, similar to that in Section 8.3 for

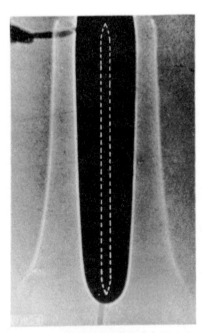

FIG. 14.5 Shadowgraph of convection around a heated vertical plate. Dotted lines show the position of the shadow of the plate in the absence of thermal effects. Inner bright lines are produced by closest negligibly refracted light and indicate edge of boundary layers; outer bright lines are produced by strongly refracted light at plate surface and indicate temperature gradient there. From E. Schmidt, Forschung auf dem Gebiete des Ingenieurwesens, Band 3, Seite 181 (1932), by kind permission of VDI-Verlag GmbH, Düsseldorf.

forced flows, can be used to indicate how the flow parameters vary. For simplicity we take $\mathrm{Pr} \sim 1$. Comparability of the first, third, and fourth terms of eqn (14.31), with the length scales used in different terms chosen in the way explained in Sections 8.3 and 11.2, implies that

$$w_{\max}^2/z \sim \nu w_{\max}/\delta^2 \sim g\alpha(T_1 - T_0) \qquad (14.61)$$

(with the notation indicated in Fig. 14.4(a)), giving

$$w_{\max} \sim [g\alpha(T_1 - T_0)z]^{1/2} \qquad (14.62)$$

$$\delta \sim [\nu^2 z/g\alpha(T_1 - T_0)]^{1/4}. \qquad (14.63)$$

Equation (14.32) has also to be considered, of course, but for $\mathrm{Pr} \sim 1$, it just gives a result consistent with (14.61); for other Pr, it relates to the question of the relative thicknesses of the two boundary layers, mentioned above. Relations (14.62) and (14.63) show that, in a given flow,

$$w_{\max} \propto z^{1/2}; \quad \delta \propto z^{1/4}. \qquad (14.64)$$

Writing the above results in non-dimensional form gives

$$\delta/z \sim \mathrm{Gr}_z^{-1/4}; \quad \mathrm{Re}_z = w_{max}z/v \sim \mathrm{Gr}_z^{1/2} \tag{14.65}$$

which may also be written

$$\mathrm{Gr}_\delta = g\alpha(T_1 - T_0)\delta^3/v^2 \sim \mathrm{Gr}_z^{1/4}; \quad \mathrm{Re}_\delta = w_{max}\delta/v \sim \mathrm{Gr}_z^{1/4}. \tag{14.66}$$

Thus the broad features of the flow may be inferred without detailed analysis. The details of the profiles are given by full solution of the boundary layer equations [7, 23, 294] with satisfactory agreement with measurements for those values of Pr at which experiments have been done.

It is interesting to consider the heat transfer from the plate, partly because it is a quantity of practical importance but particularly to illustrate the way in which departures from Newton's law of cooling occur in free convection. One may consider both the local heat transfer per unit area, H, at some z and the total heat transfer per unit area, H_T, from a plate of height l; i.e.

$$H_T = \int_0^l H \, dz/l. \tag{14.67}$$

The two are non-dimensionalized as Nusselt numbers,

$$\mathrm{Nu}_z = Hz/k(T_1 - T_0); \quad \mathrm{Nu}_T = H_T l/k(T_1 - T_0). \tag{14.68}$$

Heat transfer right at the wall is by conduction, so

$$H = -k(\partial T/\partial y)_{y=0} \sim k(T_1 - T_0)/\delta. \tag{14.69}$$

Since the boundary layer thickness decreases as the temperature elevation of the plate is increased and increases with distance up the plate in the ways indicated by (14.63), this implies that

$$\mathrm{Nu}_z = C \, \mathrm{Gr}_z^{1/4} \tag{14.70}$$

where C is a constant depending on the Prandtl number. Substitution in (14.67) leads to

$$\mathrm{Nu}_T = \tfrac{4}{3}C \, \mathrm{Gr}_l^{1/4} \quad (\mathrm{Gr}_l = g\alpha(T_1 - T_0)l^3/v^2). \tag{14.71}$$

This result applies provided that Gr_l is sufficiently large that the boundary layer approximation applies over the whole plate but not so large that the flow becomes turbulent (see below). This is a significant range. If one considers a given plate in a given fluid, then (14.71) gives the dependence of the heat transfer on the temperature elevation as

$$H_T \propto (T_1 - T_0)^{5/4}. \tag{14.72}$$

Although this result has been derived in the context of a particular configuration it is typical of many laminar free convection flows.

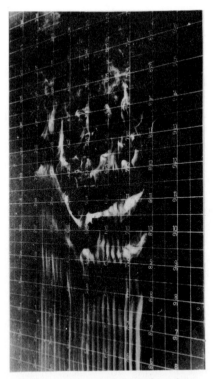

FIG. 14.6 Transition in free convection boundary layer, exhibited by smoke introduced through a row of holes close to the lower edge. From Čolak-Antić, P. 'Dreidimensionale Instabilitätserscheinungen des laminarturbulenten Umschlags bei freier Konvektion längs einer vertikalen geheizten Platte', *Sitzungsberichte der Heidelberger Akademie der Wissenschaften, Mathem.-naturw. Klasse. Jahrg. 1962/64, 6. Abhandlung*. Springer-Verlag, Berlin, 1964.

All the foregoing supposes that the flow remains laminar. In fact, free-convection boundary layers undergo transition (Fig. 14.6) and become turbulent at sufficiently high values of the Grashof number [127, 362]. We shall not discuss the dynamics of the transition process or the structure of the turbulent motion. The types of idea that go into understanding them are illustrated by other flows in later chapters. It is, however, worth noting that the counterpart of (14.72) if the flow is turbulent over almost all the plate is

$$H_T \propto (T_1 - T_0)^{4/3}. \tag{14.73}$$

The departure from Newton's law is somewhat stronger than for laminar flow. Unlike (14.72), (14.73) is not a theoretical result, but one used to summarize experimental data [121, 404]. It may be only an approximate representation of a more complicated dependence. Again, however, it is typical—for turbulent free convection in various configurations.

Figure 14.4(b) is a sketch of a thermal plume, a column of hot fluid that rises above a localized heat source—a line source producing a two-dimensional plume or a point source producing an axisymmetric one. The top part of Fig. 14.2 is of course, a plume and we shall be seeing another experimentally observed plume in Fig. 14.7(a).

We shall discuss this flow much more briefly than that of Fig. 14.4(a) [37]. A combination of the procedures explained in Section 11.6 for a jet and above for a thermal boundary layer yields the principal features of a laminar plume. Also, like jets, plumes become turbulent much more quickly than boundary layers, so that the range of Grashof number for which both the boundary layer approximation and laminar flow theory apply is much narrower.

The key to the analysis of a laminar plume is the identification of the conserved quantity that is the counterpart of M in jet theory in Section 11.6. (Actually this remark applies also to a turbulent plume; cf. the role of M in turbulent jets, Section 21.1.) This quantity is the heat being transported vertically by the plume. Consequently,

$$\frac{d}{dz} \int w(T - T_0)\, dS = 0 \tag{14.74}$$

(z being vertical and dS being an element of horizontal area). Using (14.74) along with a scaling analysis like that earlier in this section leads to the width, maximum velocity, and maximum temperature contrast in a two-dimensional laminar plume varying with height in the forms:

$$\delta \propto z^{2/5}; \quad w_{max} \propto z^{1/5}; \quad (T_{max} - T_0) \propto z^{-3/5}. \tag{14.75}$$

The vertical fluxes of mass and momentum (cf. eqns (11.55) and (11.40) respectively) are

$$\int_{-\infty}^{\infty} \rho w\, dy \propto w_{max}\delta \propto z^{3/5} \tag{14.76}$$

$$\int_{-\infty}^{\infty} \rho w^2\, dy \propto w_{max}^2\delta \propto z^{4/5} \tag{14.77}$$

(ρ being taken as a constant for this purpose, consistently with the Boussinesq approximation). Relation (14.76) indicates that, like a jet, a plume entrains ambient fluid. The increasing momentum of the fluid with height ((14.77)) is due to acceleration by the buoyancy force.

Again, we omit discussion of the corresponding transitional and turbulent flows, allowing other examples in later chapters to convey the important ideas. However, as anticipated in Section 12.6, we illustrate

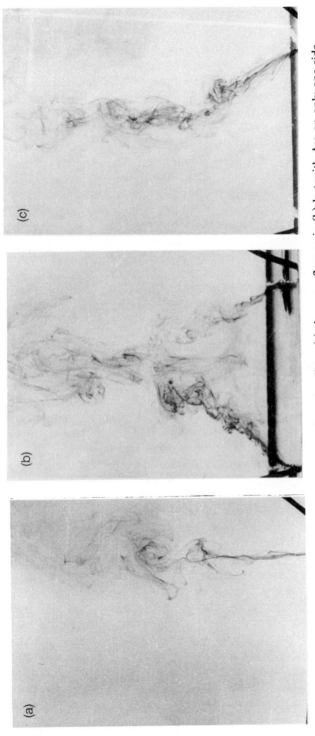

FIG. 14.7 (a) Single plume; (b) plumes showing Coanda effect; (c) the same flow as in (b) but with dye on only one side.

the Coanda effect by looking at plumes attaching to one another. The photographs of Fig. 14.7 were obtained by having two long thin parallel heating elements close to the floor in a tank of water. Dye could be introduced into the inflow to the plumes from two sources, also close to the floor, a little way to the side of each element remote from the other element. In Fig. 14.7(a) only the right-hand element is switched on and the single plume can be seen. In the other two pictures both elements are being heated. Each plume would normally entrain fluid from the space between them. However, since there is no way in which this fluid can be replaced, this entrainment is not possible. Instead, the two plumes are drawn together into a single plume rising above the line midway between the two elements. Figure 14.7(b) shows this flow. Figure 14.7(c) shows the same situation, but only the right-hand dye source is present; the change in direction of motion of fluid from the right as it meets fluid from the left can be seen. The plumes in Fig. 14.7 were, evidently, turbulent. As discussed in Section 12.6, this makes the Coanda effect occur more readily. Nevertheless similar observations have been made with laminar plumes [299].

Appendix: The Boussinesq approximation in free convection

In this appendix we examine systematically the conditions for the Boussinesq approximation to be a good approximation in the specific context of steady free convection. This is not the only context in which the matter is important (indeed, some of the points, such as the effect of the hydrostatic pressure on the density, should in principle be considered even for barotropic flows) but a general discussion becomes too complicated. The following analysis may serve as the principal demonstration in this book of the methods used in such considerations. Some of the discussion that follows the analysis has wider application. On the other hand, particular examples of free convection may involve complexities not covered by the general discussion—for example, through having more than a single length scale.

We have not made extensive use of the Boussinesq equations in this book. However, it should be remembered that the Boussinesq approximation underlies the statement that dynamical similarity of free convective flows depends on the Grashof and Prandtl numbers. When the approximation fails, other non-dimensional parameters come in. Thus, the considerations of this section are important not only for the mathematical formalism but also for the whole question of the transfer of results from one situation to another. This is of particular relevance to attempts to model geophysical situations in the laboratory.

We suppose that the flow is produced by the introduction of temperature differences of order Θ into a system of length scale L. It is important for some of the following considerations that the vertical dimension of the system should be of order L, but we shall not usually make a distinction between scales in different directions.

The principal criteria for the applicability of the Boussinesq approximation are

$$A = \alpha \Theta \ll 1 \tag{14.78}$$

$$B = g\rho\beta L \ll 1 \tag{14.79}$$

$$C = g\alpha L/C_p \ll 1 \tag{14.80}$$

(β = isothermal compressibility; other symbols as before). There are some supplementary criteria connected with the constancy of fluid properties.

If in addition to the above,

$$D = g\alpha L T_0/C_p\Theta \ll 1 \tag{14.81}$$

(where T_0 is the absolute temperature), the Boussinesq equations apply directly. If (14.78)–(14.80) are fulfilled but not (14.81), they apply with the variable T replaced by the potential temperature θ (eqn (14.57)). In this case Θ is the scale of the imposed variations in θ.

We shall consider the significance of relationships (14.78)–(14.81) after we have examined their roles in the Boussinesq approximation. We may remark, however, that relationships (14.79), (14.80), and (14.81) are likely to be violated only in large-scale geophysical situations and it is for these that the following considerations are most important.

We choose the two independent thermodynamic variables, in terms of which other thermodynamic quantities may be expressed, as the pressure and temperature, and write

$$p = p_0 + p_h + p' \tag{14.82}$$

$$T = T_0 + T'. \tag{14.83}$$

Here p_0 and T_0 are the absolute pressure and temperature at some arbitrarily chosen reference point, p_h is the hydrostatic variation of the pressure and is thus a function only of the vertical coordinate, p' and T' are the variations associated with the motion, and are thus general functions of position. T' encompasses all variations in T, whether associated with the boundary conditions or arising from internal effects such as viscous dissipation or adiabatic volume change.

The density may now be written

$$\rho = \rho_0 + \rho_h + \rho' \tag{14.84}$$

where ρ_0 is the density at p_0 and T_0, ρ_h is the departure from this in hydrostatic conditions, and ρ' is the further departure associated with the flow.

We now consider how the more general equations of motion reduce to the Boussinesq approximation, taking the continuity, dynamical and temperature equations in turn.

For steady flow the continuity equation is

$$\boldsymbol{u} \cdot \nabla\rho + \rho\nabla \cdot \boldsymbol{u} = 0. \tag{14.85}$$

The orders of magnitude of the two terms are $(\rho_h + \rho')U/L$ and $(\rho_0 + \rho_h + \rho')U/L$. To write the equation in its incompressible form (14.30) we require

$$\rho_h/\rho_0 \ll 1, \quad \rho'/\rho_0 \ll 1. \tag{14.86}$$

When these are true (we may assume them and look for self-consistency), the equation of state may be linearized to give

$$\rho_h = \rho_0\beta p_h, \quad \rho' = \rho_0(-\alpha T' + \beta p') \tag{14.87}$$

and (14.86) is equivalent to

$$\beta p_h \ll 1. \tag{14.88}$$

$$\alpha T' \ll 1. \tag{14.89}$$

$$\beta p' \ll 1. \tag{14.90}$$

Taking these in turn, condition (14.88) depends on the size of p_h, which may be estimated from the hydrostatic balance

$$\nabla p_h = (\rho_0 + \rho_h)\boldsymbol{g} \approx \rho_0\boldsymbol{g} \tag{14.91}$$

giving

$$p_h \sim \rho_0 g L. \tag{14.92}$$

Hence, (14.88) is fulfilled when B is small.

The principal requirement for condition (14.89) is evidently that A should be small, i.e. that the imposed temperature differences should not directly produce excessive density differences. Flow-induced temperature variations must also be subject to this restriction. This means that the requirements discussed below for the temperature equation to take the Boussinesq form are also implicitly requirements for (14.89), as this then guarantees that

$$T' \sim \Theta. \tag{14.93}$$

We shall see below that condition (14.90) is a much weaker condition than (14.88) and (14.89) and so does not need separate consideration.

Turning to the dynamical equation, the conditions in (14.86) enable us to replace $\rho u \cdot \nabla u$ by $\rho_0 u \cdot \nabla u$ and when eqn (14.30) applies the viscous term may be reduced to $\mu \nabla^2 u$ (for constant μ) as considered in Section 5.6. Hence, for steady flow

$$\rho_0 u \cdot \nabla u = -\nabla p' + \rho' g + \mu \nabla^2 u \tag{14.94}$$

where the hydrostatic balance of eqn (14.91) has been subtracted out as in Section 14.2.

To put eqn (14.94) into the form used in the Boussinesq approximation (eqn (14.31)) one also writes

$$\rho' = -\rho_0 \alpha T' \tag{14.95}$$

which will be valid if

$$\beta p' \ll \alpha T'. \tag{14.96}$$

The order of magnitude of p' is indicated by eqn (14.94); the second term in this will be of the same order of magnitude as the largest of the other terms. For free convection,

$$p'/L \sim g \rho_0 \alpha T'. \tag{14.97}$$

Hence, criterion (14.96) is equivalent to (14.79) (small B).

In circumstances in which (14.78) and (14.96) are fulfilled, (14.90), of which we postponed consideration above, will be much more strongly fulfilled.

The starting point for the development of the temperature equation (14.32) is the first law of thermodynamics in the form

$$dE = dQ - p \, dV \tag{14.98}$$

where E is the internal energy, dQ is the heat supplied, and V is the volume. If the thermodynamic quantities are taken to apply to unit mass,

$$V = 1/\rho. \tag{14.99}$$

Applying (14.98) to a fluid particle,

$$\frac{DE}{Dt} = \frac{DQ}{Dt} - p\frac{DV}{Dt}. \tag{14.100}$$

From the considerations in Section 14.2, the heat supplied to a fluid particle per unit mass is

$$\frac{DQ}{Dt} = \frac{1}{\rho}(k\nabla^2 T + J + \Phi) \tag{14.101}$$

where J is the internal heat generation per unit volume due to external causes, and Φ is the heating due to viscous dissipation of mechanical energy.

Using relationships derived by standard thermodynamic procedures we may write

$$\frac{DE}{Dt} = \left(\frac{\partial E}{\partial T}\right)_p \frac{DT}{Dt} + \left(\frac{\partial E}{\partial p}\right)_T \frac{Dp}{Dt} = \left(C_p - \frac{\alpha p}{\rho}\right)\frac{DT}{Dt} + \frac{\beta p - \alpha T}{\rho}\frac{Dp}{Dt} \qquad (14.102)$$

$$\frac{DV}{Dt} = \left(\frac{\partial V}{\partial T}\right)_p \frac{DT}{Dt} + \left(\frac{\partial V}{\partial p}\right)_T \frac{Dp}{Dt} = \frac{\alpha}{\rho}\frac{DT}{Dt} - \frac{\beta}{\rho}\frac{D\rho}{Dt}. \qquad (14.103)$$

Substituting (14.101), (14.102), and (14.103) into (14.100) and dividing the pressure variations into hydrostatic and hydrodynamic parts as in (14.82) gives

$$C_p \frac{DT}{Dt} - \frac{\alpha T}{\rho}\frac{Dp_h}{Dt} - \frac{\alpha T}{\rho}\frac{Dp'}{Dt} = \frac{1}{\rho}(k\nabla^2 T + J + \Phi). \qquad (14.104)$$

This becomes the Boussinesq equation (14.32) if the second, third, and sixth terms can be neglected. We compare each in turn with the first term, remembering that we are considering steady flow.

Firstly

$$\frac{\alpha T \, Dp_h/Dt}{\rho C_p \, DT/Dt} = \frac{\alpha T \mathbf{u} \cdot \nabla p_h}{\rho C_p \mathbf{u} \cdot \nabla T} \sim \frac{\alpha T_0 p_h}{\rho_0 C_p T'} \sim \frac{g \alpha L T_0}{C_p T'}. \qquad (14.105)$$

Supposing that (14.93) applies (again assuming the result and looking for consistency) the criterion for neglecting the second term of (14.104) is (14.81), i.e. that D is small. We consider what may be done when D is not small after completing discussion of (14.104).

Secondly,

$$\frac{\alpha T \, Dp'/Dt}{\rho C_p \, DT/Dt} = \frac{\alpha T \mathbf{u} \cdot \nabla p'}{\rho C_p \mathbf{u} \cdot \nabla T} \sim \frac{\alpha T_0 p'}{\rho C_p T'}. \qquad (14.106)$$

p' can be estimated from (14.97) and so the criterion for the neglect of the third term in (14.104) is

$$\frac{g \alpha L}{C_p} \alpha T_0 \ll 1. \qquad (14.107)$$

Since αT_0 is around 1 for gases and is small for other fluids, (14.107) will be fulfilled whenever (14.80) is fulfilled.

We may note parenthetically that the above discussion provides the rigorous justification for the choice of C_p as the specific heat in Section 14.2. By using T and V as the independent thermodynamic variables one can analogously formulate an equation for $C_V \, DT/Dt$, but an extra term arises that is negligible only when the difference between C_V and C_p is itself negligible.

Finally, the reduction of (14.104) to the Boussinesq form requires that the viscous heating should be negligible. Viscous energy dissipation has been considered only in terms of the example of a laminar two-dimensional jet (Section 11.7). However, this indicates the essential information that we need for an order of magnitude estimate, i.e. that the dissipation per unit volume is proportional to the viscosity and to the square of the velocity gradient (see eqn (11.60)). In fact, generalization of the ideas in Section 11.7 gives the result that

$$\Phi = \mu \frac{\partial u_i}{\partial x_j} \left(\frac{\partial u_i}{\partial x_j} + \frac{\partial u_j}{\partial x_i} \right) \tag{14.108}$$

(with the summation convention for repeated suffixes applying). Thus

$$\Phi \sim \mu U^2 / \delta^2 \tag{14.109}$$

where δ is the length scale appropriate to viscous effects. The distinction between L and δ is made so that the discussion applies even when there is strong boundary layer formation.

In free convection, the balance between the buoyancy force and the viscous force gives

$$\mu U / \delta^2 \sim \rho g \alpha T'. \tag{14.110}$$

Hence,

$$\frac{\Phi}{\rho C_p \mathbf{u} \cdot \nabla T} \sim \frac{\rho g \alpha T' U}{\rho C_p U T' / L} \sim \frac{g \alpha L}{C_p} = C. \tag{14.111}$$

The criterion for the viscous dissipation to have a negligible effect on the thermal balance is that this should be small; i.e. it is relationship (14.80). This applies whether $\delta \sim L$ (viscous flow) or $\delta \ll L$ (viscous dissipation confined to boundary layers).

Thus neglect of the third and sixth terms of (14.104) requires small C, whilst neglect of the second requires small D. The latter is a much more severe criterion. It is thus fortunate that the need to fulfil it can be avoided by working in terms of the potential temperature

$$\theta = T - (T_a - T_0) \quad \text{where†} \quad dT_a/dz = -g\alpha T_0/C_p. \tag{14.112}$$

---

† A definition of T_a slightly different from eqn (14.54) has been chosen, so that $\nabla^2 T_a = 0$ and the heat conduction term is unaffected by the change of variable; T_a as defined in (14.112) is significantly different from the physical adiabatic temperature only in circumstances in which the approximation breaks down. There is also slight arbitrariness, depending on the exact definition of T_0, but this is similarly unimportant.

We see this as follows. When $C \ll 1$ but D is not, (14.104) becomes

$$C_p \frac{DT}{Dt} - \frac{\alpha T}{\rho} \frac{Dp_h}{Dt} = \frac{1}{\rho}(k\nabla^2 T + J). \qquad (14.113)$$

p_h and T_a vary vertically only. Hence,

$$\frac{\alpha T}{\rho} \frac{Dp_h}{Dt} = \frac{\alpha T w}{\rho} \frac{dp_h}{dz} = -g\alpha w T \qquad (14.114)$$

$$C_p \frac{DT_a}{Dt} = C_p w \frac{dT_a}{dz} = -g\alpha w T_0 \qquad (14.115)$$

where w is the vertical component of the velocity. Hence,

$$C_p \frac{D\theta}{Dt} = C_p \frac{DT}{Dt} - \frac{\alpha T}{\rho} \frac{Dp_h}{Dt} + g\alpha w T' \qquad (14.116)$$

where (14.83) has been used to give the last term. But

$$\frac{g\alpha w T'}{C_p DT/Dt} \sim \frac{g\alpha L}{C_p} \qquad (14.117)$$

and neglect of this last term requires only that C should be small. When that is the case

$$C_p D\theta/Dt = \frac{1}{\rho}(k\nabla^2\theta + J) \qquad (14.118)$$

i.e. the equation for θ has become the same as the Boussinesq equation for T.

Since T_a does not vary horizontally, this change makes no difference to the dynamic equation (14.94).

One further point does need consideration. Relation (14.93) may not be valid when θ replaces T, since the largest temperature difference may be that associated with the adiabatic gradient. However, the fractional density change due to this difference is

$$\alpha\Delta T_a \sim \frac{g\alpha L}{C_p}\alpha T_0 \qquad (14.119)$$

and, as with (14.107), this is always small when C is small.

In addition to the various effects considered above, the Boussinesq approximation, as designated in Section 14.2, requires effective constancy of the physical properties of the fluid; that is of μ, k, α, and C_p.† The

---

† For a more complete treatment of this aspect see Ref. [190]. In general, this paper presents in a more formal way the analysis of the present appendix. However, it confines attention to the case when the Boussinesq approximation applies without the potential temperature replacing the actual temperature. Hence, (14.81) emerges as a principal criterion.

temperature and pressure variations must not be so large as to produce large fractional changes in these quantities. Often fulfilment of this requirement will go along with fulfilment of the requirement of small fractional density changes. However, in principle, any parameter having a much larger temperature or pressure coefficient than density needs special consideration. In practice, the temperature variation of viscosity is often the most serious problem, both in the laboratory and in studies of natural situations.

What is the physical significance of the parameters, A, B, C, and D that principally govern the applicability of the Boussinesq approximation?

The requirement of small A just says that the fractional density change produced by expansion in the imposed temperature field must be negligible.

Parameters B and C relate to the 'scale heights' of the system, the heights over which various parameters change by a fraction of order unity. For example, in an isothermal perfect gas at rest

$$dp/dz = -\rho g, \qquad \rho = p/RT \qquad (14.120)$$

giving

$$p = p_0 \exp(-gz/RT), \qquad \rho = \rho_0 \exp(-gz/RT). \qquad (14.121)$$

In this case the scale heights of pressure and density are both RT/g. More generally, $1/g\rho\beta$ and $C_p/g\alpha$ are the scale heights of isothermal pressure and adiabatic temperature. Parameters B and C are the ratios of the length scale of the system to these scale heights. The Boussinesq approximation applies only to systems that are small compared with the scale heights.

Another physical interpretation of C, in terms of thermodynamic 'efficiency', has been discussed in Section 14.7.

The parameter D is evidently related to C (although usually $D \gg C$), but it is more readily thought of as the ratio of the adiabatic temperature gradient to a typical imposed temperature gradient.

Parameters B and C (and for that matter D) are always exceedingly small on the laboratory scale. Departures from Boussinesq conditions can occur owing to variations of fluid properties—particularly temperature variation of viscosity and, in water, temperature variation of expansion coefficient—and, in experiments with gases, owing to A not being small enough.

The importance of B and C arises in large-scale geophysical systems, where they are not always small. Motions of the atmosphere extending throughout its depth, and motions in the interiors of planets and stars, for example, may involve processes that cannot be modelled in the laboratory. The foregoing analysis shows that when B is not small, one must additionally take into account (i) the effect of the hydrostatic pressure p_h throughout the equations of motion and (ii) the effect of the hydrodynamic pressure p' on the buoyancy force. When C is not small, one must take into account (iii) heat generation by viscous dissipation and (iv) other non-Boussinesq effects in the thermal balance (relationships (14.106) and (14.117)).

B and C are in principle thermodynamically independent parameters. In practice they tend to be of the same size. For a perfect gas

$$\alpha = 1/T, \qquad \beta = 1/\rho RT \qquad (14.122)$$

and so

$$B/C = C_p/R = \gamma/(\gamma - 1). \qquad (14.123)$$

For real gases the ratio will similarly be of order unity. For condensed phases (liquids and solids) the ratio $\alpha/\rho\beta C_p$ is known as Grüneisen's ratio, G, giving

$$B/C = 1/G. \qquad (14.124)$$

The ratio G rarely differs greatly from 1. Water is something of an exception to this (its G varies strongly with temperature but is typically 10^{-1}), but there would seem to be no application of this; the density at the bottom of the sea is increased only a small amount by the weight of the water above. On the other hand, there are applications of these ideas to condensed phases in the study of planetary interiors (see, e.g., Section 26.5).

These considerations imply that all the effects (i)–(iv) above are liable to become significant simultaneously and it is generally not useful to take one into account whilst ignoring the others. Thus situations in which the Boussinesq approximation fails become very complicated theoretically. We have already seen the problems of modelling them in the laboratory, so it is altogether difficult to ascertain the flow patterns. Numerical experiments (Section 25.5) often offer the best hope.

15

STRATIFIED FLOW

15.1 Basic concepts [37, 38]

The name stratified flow is applied to a flow primarily in the horizontal direction that is affected by a vertical variation of the density. Such flows are of considerable importance in geophysical fluid mechanics. The obvious case of the effect of vertical temperature variations on the wind near the ground is only one of a number of examples in the atmosphere, and the effects of both temperature and salinity variations play an important role in many aspects of dynamical oceanography.

The density may, in general, either increase or decrease with height. The former case gives rise to an interaction between the mean flow and the convection that would occur in the absence of mean flow. One example is the alignment of Bénard cells by a mean shear [60, 281, 320], illustrated in the laboratory by Fig. 15.1. This shows an illuminated cross-section, perpendicular to the flow, of an air channel with heated bottom and cooled top; the smoke has been introduced a long way upstream and so the pattern indicates the occurrence of regular rolls with their axes along the flow. Another example—turbulent motion originating partly from a mean flow and partly from convection—will be considered in Section 21.7.

However, in this chapter we are primarily concerned with the case of stable stratification, that is to say the density decreases with height. Vertical motions then tend to carry heavier fluid upwards and lighter fluid downwards, and are thus inhibited. This inhibition may take the form of modifying the pattern of the laminar motion or of preventing or modifying its instability.

We require a quantitative criterion for this to be a strong effect. Since

FIG. 15.1 Cross section of convection cells in channel flow; $Ra = 4.16 \times 10^3$, $Re = 8.3$. (Note: patterns at top are reflection of cells in channel roof.) From Ref. [60].

most of the experiments on stratified flows have used salt rather than heat as the stratifying agent (cf. Section 14.3) we shall retain the density variations explicitly, rather than relating them to temperature variations. We consider the case of flow outside boundary layers at high Reynolds and Péclet† numbers, so that both viscous and diffusive processes are negligible. Thus we write the momentum and density equations (for steady flow)

$$\rho \boldsymbol{u} \cdot \nabla \boldsymbol{u} = -\nabla p + \rho \boldsymbol{g} \qquad (15.1)$$

$$\boldsymbol{u} \cdot \nabla \rho = 0. \qquad (15.2)$$

We take z vertically upwards and suppose that the basic stratification consists of a uniform density gradient $(-\mathrm{d}\rho_0/\mathrm{d}z)$. Because ρ_0 does not vary horizontally, the balance between $\rho_0 \boldsymbol{g}$ and the hydrostatic pressure can be subtracted out from eqn (15.1) just as it can for an entirely uniform density (Section 14.2).

We now consider, superimposed on this basic configuration, a flow with length and velocity scales L and U, produced, for example, by moving an obstacle of size L horizontally through the fluid at speed U. This will produce a modification of the density field which we denote by ρ', related to the stratification by eqn (15.2) in the form

$$\boldsymbol{u} \cdot \nabla \rho' + w \, \mathrm{d}\rho_0/\mathrm{d}z = 0. \qquad (15.3)$$

In order of magnitude

$$\rho' \sim \frac{WL}{U} \left| \frac{\mathrm{d}\rho_0}{\mathrm{d}z} \right|. \qquad (15.4)$$

W is now restricted by the fact that the flow cannot produce buoyancy forces associated with ρ' that are larger than the other forces involved. Since the buoyancy force does not contribute directly to the horizontal components of eqn (15.1) it is convenient to work in terms of the vorticity form of this equation:

$$\rho(\boldsymbol{u} \cdot \nabla \boldsymbol{\omega} - \boldsymbol{\omega} \cdot \nabla \boldsymbol{u}) = -g \left(\hat{\boldsymbol{x}} \frac{\partial \rho'}{\partial y} - \hat{\boldsymbol{y}} \frac{\partial \rho'}{\partial x} \right). \qquad (15.5)$$

Since the order of magnitude of $\boldsymbol{\omega}$ is U/L this indicates that the order of magnitude of ρ' must remain not greater than

$$\rho' \sim \rho_0 U^2/gL. \qquad (15.6)$$

---

† For brevity, we retain the names, Péclet number and (subsequently) Prandtl number, although, when salt is the stratifying agent these now refer to UL/κ_c and ν/κ_c (the Schmidt number), where κ_c is the diffusivity of the salt.

Comparison of this with (15.4) indicates that

$$W/U \sim \rho_0 U^2 \Big/ gL^2 \left|\frac{\mathrm{d}\rho_0}{\mathrm{d}z}\right| = (\mathrm{Fr})^2. \qquad (15.7)$$

When $(\mathrm{Fr})^2$ is small the horizontal motion has only much weaker vertical motion associated with it.

Fr is called the internal Froude number; when, as at present, there is no danger of confusion with the Froude number associated with free surface effects (Section 7.4), it is simply called the Froude number. $1/(\mathrm{Fr})^2$ is sometimes known as the Richardson number (see also Section 21.7).

Similar analysis can be given for flows in which viscous and/or diffusive effects are strong. This is a matter of some complexity, since different detailed treatments are appropriate for low, intermediate and high Prandtl number. Thus we omit consideration of it; when we talk below of low Froude number flows, it is assumed that any other criterion for the flow to be strongly constrained by stratification is also fulfilled.

Often low Froude number motion can be considered to be entirely two-dimensional in horizontal planes. For example, in the relative movement between a spherical obstacle and a stratified fluid, nearly all the fluid is deflected to the sides of the sphere, little above and below it. Thus the flow pattern in a horizontal plane has a closer resemblance to unstratified flow past a cylinder than to unstratified flow past a sphere.

This is illustrated by Fig. 15.2, showing patterns produced by dye originating on a sphere in this type of flow. The view from above (Fig. 15.2(a)) is very similar to patterns observed in flow past cylinders (e.g. Figs 3.4 and 3.8). In contrast the sideview (Fig. 15.2(b)) shows none of the structure of such patterns; for it to do so, there would have to be marked vertical motions.

An indirect consequence of this is that the instability in the wake of a sphere in a stratified fluid also resembles that in the wake of a cylinder, as is illustrated by Fig. 15.3 (cf. Fig. 3.5).

(The above discussion has supposed that the geometry involves only a single length scale L. Geophysical problems for which these ideas are of interest frequently involve horizontal and vertical scales, L and D, of very different sizes; e.g. flow of a stratified ocean of depth D over topography of horizontal scale L, with $L \gg D$. It is therefore worth pointing out parenthetically that, in these circumstances, the relative importance of stratification is determined by the Froude number based on the vertical scale. In the above analysis, the relevant length scale is L at all points of the argument, except that the order of magnitude of ω is U/D. Hence, (15.7) becomes

$$W/U \sim \rho_0 U^2 \Big/ gLD \left|\frac{\mathrm{d}\rho_0}{\mathrm{d}z}\right|. \qquad (15.8)$$

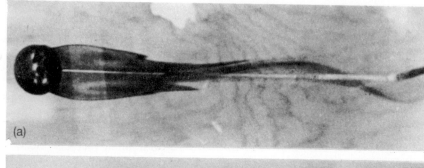

Fig. 15.2　Plan-view (a) and side-view (b) of flow past a sphere in a stratified fluid; Fr = 0.21, Re = 164. From Ref. [137].

Fig. 15.3 Wake of a sphere in a stratified fluid; Fr = 0.8, Re = 377. Photo by W. R. Debler. Reproduced by permission from 'Periodic Flow Phenomena', by E. Berger and R. Wille, *Ann. Rev. Fluid Mechanics*, Volume 4. Copyright © 1972 by Annual Reviews Inc. All rights reserved.

However, even without stratification, the geometry (via the continuity equation) constrains W/U to be of order D/L. Stratification is important when it provides a constraint at least as strong as this; i.e. when

$$\rho_0 U^2 \bigg/ gD^2 \left| \frac{d\rho_0}{dz} \right| \lesssim 1. \tag{15.9}$$

See also remarks at the end of Section 16.8.)

15.2   Blocking

Clearly a geometry for which the motion is two-dimensional in *vertical* planes will lead to radical differences between strongly stratified and unstratified flow. Consider, for example, relative motion between the fluid and a horizontal cylinder which extends right across the flow. No fluid can be deflected round the cylinder without vertical motion. If this is prevented, all the fluid in front of or behind the cylinder must be at rest relative to the cylinder no matter how far upstream or downstream one goes. Formally, if $v = 0$ because of two-dimensionality and $w = 0$ because of strong stratification, the continuity equation becomes

$$\partial u / \partial x = 0; \tag{15.10}$$

then for any z at which u is zero at one value of x, u is zero at all x.

Figures 15.4 and 15.5 show experimental realizations of low Fr flows past two-dimensional obstacles—a rather complex shape in Fig. 15.4 and a circular cylinder in Fig. 15.5. Such experiments are usually performed by traversing the obstacle slowly through a long channel containing salt-stratified water, but it is convenient to think about them in terms of flow past a fixed obstacle. In both figures the flow is visualized by optical techniques depending on the refractive index variations associated with the density variations. The pictures in Fig. 15.4 were obtained with a schlieren system (Section 25.4), modified so that dark regions correspond to regions of relatively large gradient of ρ' (not of $\rho_0 + \rho'$). Figure 15.5 was obtained by a holographic method (Section 25.4) such that the lines in the picture are lines of constant density; for this particular case, as we shall be seeing below, these are effectively equivalent to streamlines in the obstacle frame of reference.

Blocking is apparent in Fig. 15.4. The regions affected by the obstacle extend horizontally with little variation and it is evident that their actual length is much greater than the picture width. There is, however, a striking difference between parts (a) and (b) of the figure: in the former blocking occurs both upstream and downstream of the obstacle, almost symmetrically; in the latter it occurs upstream, but is absent downstream

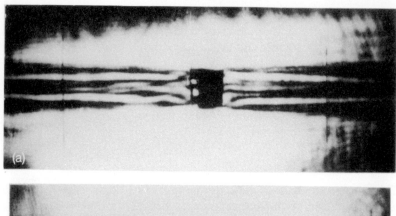

FIG. 15.4 Modified schlieren images (see text) of density field produced by motion of a two-dimensional obstacle through salt stratified water. (a) Fr = 3.9×10^{-4}, Re = 0.035; (b) Fr = 0.014, Re = 1.3. (In (a), optical system was being used near to its limit; the irregularities in the image are due to optical imperfections and not to turbulence in flow.) From Ref. [79]

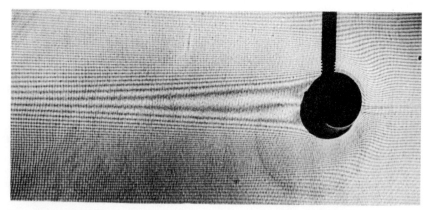

FIG. 15.5 Hologram of density field produced by motion of a cylinder through salt stratified water. Fr = 0.069, Re = 6.8. Photo by W. Debler and C. Vest, University of Michigan, similar to ones in Ref. [138].

except for a layer much thinner than the obstacle. We need to understand why this difference arises and, in particular, how there can ever be a low Fr flow without downstream blocking. The key to this understanding is in fact to be found in what happens upstream of the obstacle [102, 185].

The simple theory leading to eqn (15.10) provides a useful understanding of why blocking occurs but little information about the detailed motion. In any real situation, the disturbance produced by relative motion of fluid and an obstacle can extend only a finite, even if very large, distance ahead of the obstacle. Fluid initially far ahead of the obstacle must ultimately pass above or below it into the downstream region. This requires terms neglected in the simple prediction of blocking to play a role. In our discussion of boundary layers (Sections 8.3, 11.2) we saw how a difference in length scales in different directions can arise, with the effect that terms that appear to be negligible on a simple scaling analysis become significant. A similar thing happens here. Now, however, the spontaneously arising scale, the length of the blocked region, is large compared with the imposed scale, the height L of the obstacle (Fig. 15.6). For the same reason as with eqn (15.5) we analyse this situation in terms of the vorticity equation; we make the simplifications appropriate to two-dimensional flow, but, for a reason which will become apparent, we now include the viscous term:

$$\boldsymbol{u} \cdot \nabla \eta = \nu \nabla^2 \eta + \frac{g}{\rho} \frac{\partial \rho}{\partial x} \quad (\boldsymbol{\omega} = \hat{\boldsymbol{y}} \eta). \tag{15.11}$$

We suppose that diffusion of the stratifying agent is negligible. This rather obviously requires the Prandtl number to be high—and we note that most of our experimental information comes from work with salt in water giving a Prandtl (Schmidt) number of about 700—but we shall need to return to the question of just how high. On this supposition every fluid particle conserves its density throughout its motion (the meaning of eqn (15.2)). The horizontal density differences produced by some fluid particles rising a distance $\sim L$ will be $\sim L |d\rho_0/dz|$. Remembering also that $\eta \sim U/L$, we can write down the orders of magnitude of each of the terms in eqn (15.11), using the same method as for the analysis of

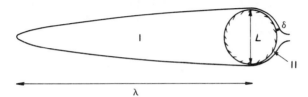

FIG. 15.6 Length scales L, λ, and δ in stratified flow past an obstacle (see text). I, blocked region ahead of obstacle; II, boundary layer region behind obstacle.

boundary layers in Section 11.2:

$$\boldsymbol{u}\cdot\nabla\eta \sim \frac{U^2}{\lambda L}; \quad v\nabla^2\eta \sim \frac{vU}{L^3}; \quad \frac{g}{\rho}\frac{\partial\rho}{\partial x} \sim \frac{gL\,|\mathrm{d}\rho_0/\mathrm{d}z|}{\rho\lambda}. \tag{15.12}$$

These are in the ratios

$$(\mathrm{Fr})^2 : \lambda(\mathrm{Fr})^2/\mathrm{Re}\,L : 1, \tag{15.13}$$

where Fr is defined by eqn (15.7) and Re is the Reynolds number UL/v. We see that, no matter how large λ/L is, the inertia force remains small compared with the buoyancy force when Fr is small. But the viscous force can become comparable with the buoyancy force if

$$\lambda/L \sim \mathrm{Re}/\mathrm{Fr}^2. \tag{15.14}$$

This result serves two purposes. Firstly it provides a quantitative estimate of the length of the blocked region. As expected it is long compared with the obstacle size when Fr is small (provided Re is not too small).† Incidentally such a long upstream effect of an obstacle is sometimes called an upstream wake.

Secondly, the deduction of (15.14) indicates the physical mechanism by which fluid particles pass the obstacle. Small viscous effects come into play far ahead of the obstacle and very gradually lift fluid particles (or lower them in the case of fluid particles starting below the centre plane) so that, by the time they reach the obstacle, they are able to pass over (or under) it.

Before considering the downstream behaviour we need to formulate the criterion for the assumption that density diffusion is negligible to be valid. This criterion is not obvious because of the different length scales involved; very weak diffusion may produce a significant effect over the long time it takes a fluid particle to traverse the distance λ. Although the flow is steady it is convenient to write the density equation in the form

$$\mathrm{D}\rho/\mathrm{D}t = \kappa\nabla^2\rho \tag{15.15}$$

showing that the change in density of a fluid particle in time t is of the order of $\kappa t(\nabla^2\rho)_T$, where $(\nabla^2\rho)_T$ denotes a typical value of $\nabla^2\rho$. For the above formulation to be applicable, a fluid particle must pass the obstacle with a density much closer to its original density than to the ambient density at its new level; i.e. the diffusively produced density change during the time of order λ/U that it is changing level, must be small compared with the background density difference between the two levels.

---

† The full theory [102, 185] shows that numerical factors accumulate so that, although the above argument correctly indicates that $\lambda/L \propto \mathrm{Re}/\mathrm{Fr}^2$, the constant of proportionality is rather small.

One thus requires

$$\frac{\kappa\lambda}{U}(\nabla^2\rho)_T \ll L\left|\frac{\mathrm{d}\rho_0}{\mathrm{d}z}\right|. \tag{15.16}$$

The background density gradient $\mathrm{d}\rho_0/\mathrm{d}z$ is significantly modified by the upstream wake in a region of smaller length scale L (as can be seen in Fig. 15.5). Hence

$$(\nabla^2\rho)_T \sim \frac{1}{L}\left|\frac{\mathrm{d}\rho_0}{\mathrm{d}z}\right| \tag{15.17}$$

and (15.16) becomes

$$\kappa\lambda/UL \ll L. \tag{15.18}$$

Substituting (15.14) and rearranging, the criterion is

$$\mathrm{Pr} \gg 1/(\mathrm{Fr})^2. \tag{15.19}$$

Since we are concerned with low Fr flows this is a much more severe criterion than just requiring the Prandtl number, Pr, to be large.

When (15.19) is fulfilled, the particles coming over the obstacle have retained their original density, which is markedly different from that appropriate to the level they now occupy. They thus tend to drop back to their original level very rapidly. Inertia and/or viscous forces come into play here through the appearance of a local length scale δ, small compared with L (Fig. 15.6).

Figure 15.5 illustrates this sequence of events; although (15.19) is not strongly fulfilled, $\mathrm{Pr}(\mathrm{Fr})^2$ is large enough for the main features to be seen. Since density diffusion is weak and the flow is steady, the constant density lines are also approximately streamlines. One can thus see clearly the gradual lifting and rapid dropping-back of the fluid respectively in front of and behind the cylinder.

The argument leading to (15.19) can be reversed to show that if

$$\mathrm{Pr} \ll 1/(\mathrm{Fr})^2 \tag{15.20}$$

diffusion can act on the density in the upstream wake sufficiently that particles passing over the obstacle differ in density only slightly from other particles at the same level. The dropping-back process does not then occur. There is no longer a mechanism inhibiting downstream blocking. One may then expect the flow to be much more nearly symmetrical upstream and downstream. An analogy with Taylor columns to be discussed in Section 16.4, implies such symmetry specifically when $\mathrm{Pr} = 1$. However, the fact that, for sufficiently small Fr, (15.20) may be fulfilled even when Pr is a lot larger than 1 allows symmetrical flow to be observed even in salt-stratified water. This is the situation in Fig. 15.4(a).

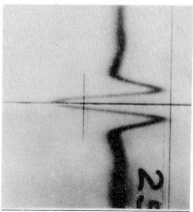

FIG. 15.7 Pattern formed by initially vertical dye line in upstream wake for conditions similar to Fig. 15.4 (b). From Ref. [79].

In contrast (15.20) is not fulfilled in Fig. 15.4(b) (although nor is (15.19)) and there is a marked contrast between the upstream and downstream regions.

The detailed structure of blocked regions is often more complicated than can be inferred from simplified discussion like that above. As an example Fig. 15.7 shows the distortion of an initially vertical dye line by an upstream wake. The velocity profile must be similar; in the frame of reference in which the obstacle moves through stationary ambient fluid, there are alternations of flow direction—a result that has also been obtained theoretically [258]. (The white bands within the dark regions of both parts of Fig. 15.4 are a consequence of the density field having a similar structure.)

15.3 Lee waves

Stratified fluids can support a variety of types of wave motion, which have no counterparts in unstratified fluids. The reason is basically the tendency for vertical motion to be suppressed: a fluid particle that does get displaced vertically tends to be restored to its original level; it may then overshoot inertially and oscillate about this level. We shall be examining the consequences of this in the simplest possible context in Section 15.4, but, before we leave the topic of flow past obstacles, mention should be made of lee waves. These have been extensively studied, partly because of the important meteorological application to flow behind hill ranges [40, 69, 350].

FIG. 15.8 Lee wave formation.

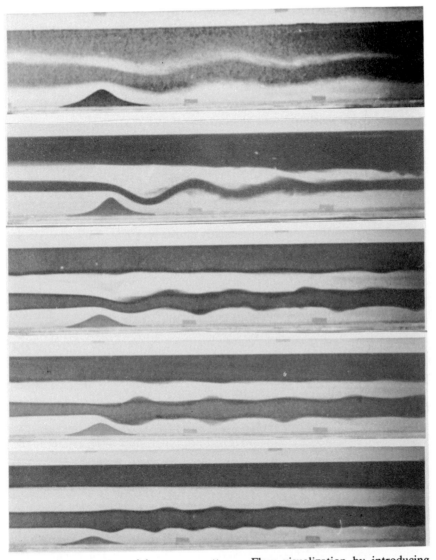

FIG. 15.9 Examples of lee wave patterns. Flow visualization by introducing acidic and basic layering into water containing thymol blue indicator (this produced negligible density change and boundaries between light and dark regions are *not* density interfaces). Different cases correspond to different oncoming velocity and density profiles as well as different obstacle shape and Froude number. Photos by L. R. Pattinson.

Lee waves occur as a result of heavy fluid being raised and/or light fluid being lowered by motion past an obstacle; the fluid drops (or rises) towards its original level, overshoots and oscillates (Fig. 15.8). Although each fluid particle oscillates up and down as it travels downstream, the overall flow pattern is steady in the frame of reference of the obstacle. This mechanism implies the importance of both buoyancy forces and inertia forces. Lee waves thus occur primarily when $\text{Fr} \sim 1$. (In Section 15.2 we looked at a particular case of low Fr flow in which inertial 'drop-back' occurred behind the obstacle. A hint of lee waves can in fact be seen in Fig. 15.5, but they are rapidly damped by viscosity.)

Both two-dimensional and three-dimensional obstacles can generate lee waves. The typical Froude number at which they occur most markedly is higher for the latter because low Fr flow then goes round rather than over and under the obstacle.

Figure 15.9 shows some examples of lee waves produced by two-dimensional obstacles. They follow the common practice (originating from the meteorological application) of having the obstacle on the channel floor, so that the flow goes only above it. Figure 15.9(a) is a case where the main features relate quite closely to the above simple explanation (Fig. 15.8). The other examples illustrate the fact that detailed patterns of lee waves are very varied. Some of the complexities in these cases are related to variations in the velocity and density profiles of the oncoming fluid. Even without such variations, changes in Froude number and geometry lead to a wide variety of patterns [38, 74]. In the laboratory, the overall depth of the channel as well as the height of the obstacle usually influences the structure of the flow [38, 72].

15.4 Internal waves

Some general remarks about waves in stratified fluid were made at the beginning of the last section. In this section we examine the properties of such waves in the simplest context of a small-amplitude wave in an expanse of fluid that is otherwise at rest.

An important parameter associated with this situation may be introduced (and its physical significance illustrated) by considering first the behaviour of a small fluid particle that is displaced vertically. Suppose its density, which it conserves, is $\rho_0(0)$. If it is displaced a distance Δz upwards, the density of the fluid surrounding it is $\rho_0(0) + \Delta z \, d\rho_0/dz$ and the net gravitational force on it is $-g\Delta z \, d\rho_0/dz$. Hence, its motion is governed by the equation

$$\rho_0 \frac{d^2 \Delta z}{dt^2} = g \frac{d\rho_0}{dz} \Delta z \qquad (15.21)$$

and it oscillates about its original position with an angular frequency

$$N = \left(-\frac{g}{\rho_0}\frac{d\rho_0}{dz}\right)^{1/2} \qquad (15.22)$$

($d\rho_0/dz$ being negative).

N is called the Brunt–Väisälä frequency. When the density variations are due to temperature variations,

$$N = (g\alpha\, dT_0/dz)^{1/2}. \qquad (15.23)$$

When, in addition, the adiabatic temperature gradient (Section 14.6) is significant,

$$N = \left[g\alpha\left(\frac{dT_0}{dz} + \frac{g\alpha T_0}{C_p}\right)\right]^{1/2}. \qquad (15.24)$$

The above analysis does not, of course, describe any actual fluid dynamical situation. For that we must turn to the full equations of motion. We consider a wave of small amplitude in an inviscid, non-diffusive fluid. Hence, we require the equations of unsteady motion, but we omit the non-linear terms on the basis that these must be negligible when the amplitude is small enough:

$$\rho_0 \partial u / \partial t = -\nabla p + g\Delta\rho \qquad (15.25)$$

$$\nabla \cdot u = 0 \qquad (15.26)$$

$$\partial \Delta\rho / \partial t + w\, d\rho_0/dz = 0, \qquad (15.27)$$

where $\Delta\rho$ is the departure of the density from its basic distribution, $\rho_0(z)$, and p is the corresponding departure from the hydrostatic pressure. We note that, although the problem is linearized, the term $u \cdot \nabla\rho$ in the density equation still enters through the interaction between the vertical velocity component and the basic stratification.

We look for wavelike solutions, periodic in both space and time:

$$u = U \exp i(\omega t - k_x x - k_y y - k_z z) \qquad (15.28)$$

$$p = P \exp i(\omega t - k_x x - k_y y - k_z z) \qquad (15.29)$$

$$\Delta\rho = Q \exp i(\omega t - k_x x - k_y y - k_z z) \qquad (15.30)$$

(where, of course, the real parts correspond to the physical quantities). Substitution of these into eqns (15.25)–(15.27) gives

$$i\omega\rho_0 U = ik_x P \qquad (15.31)$$

$$i\omega\rho_0 V = ik_y P \qquad (15.32)$$

$$i\omega\rho_0 W = ik_z P - gQ \qquad (15.33)$$

$$ik_x U + ik_y V + ik_z W = 0 \qquad (15.34)$$

$$i\omega Q + \frac{d\rho_0}{dz} W = 0 \qquad (15.35)$$

where (U, V, W) are the components of U. Elimination of U, V, W, P and Q from these homogeneous equations shows that they are consistent when and only when

$$\omega = N\left(1 - \frac{k_z^2}{k_x^2 + k_y^2 + k_z^2}\right)^{1/2} \tag{15.36}$$

which may also be written

$$\omega = N(k^2 - k_z^2)^{1/2}/k \tag{15.37}$$

or

$$\omega = N |\sin \theta| \tag{15.38}$$

where θ is the inclination to the vertical of the vectorial wave number $\mathbf{k} = \hat{\mathbf{x}}k_x + \hat{\mathbf{y}}k_y + \hat{\mathbf{z}}k_z$ and $k = |\mathbf{k}|$.

Thus waves exist for any value of the angular frequency from zero up to the Brunt–Väisälä frequency N. Above N there are no wavelike solutions. (Since, in an unstratified fluid, $N = 0$, this gives confirmation that the waves are essentially a consequence of the stratification.) When $k_z = 0$, corresponding to a wave pattern without vertical variation, $\omega = N$; an array of vertical columns, with the velocity varying in the horizontal direction, oscillates at the Brunt–Väisälä frequency, as might be expected from its simple derivation above. When $k_x = k_y = 0$, corresponding to a wave pattern without horizontal variation, $\omega = 0$; this corresponds to the blocking phenomenon in steady flow discussed in Section 15.2.

The phase velocity of the waves is in the direction of $\mathbf{k}$ and has magnitude ω/k.

Of greater physical importance is the group velocity, indicating the speed and direction with which kinetic and potential energy are transmitted through the fluid. The waves are dispersive in a rather unusual way; the frequency does not depend on the magnitude of the wave number but it does depend on its direction. The standard result that for a wave in one dimension the group velocity is $d\omega/dk$ may be extended to three dimensions giving

$$\mathbf{c}_g = \hat{\mathbf{x}}\frac{\partial \omega}{\partial k_x} + \hat{\mathbf{y}}\frac{\partial \omega}{\partial k_y} + \hat{\mathbf{z}}\frac{\partial \omega}{\partial k_z}. \tag{15.39}$$

For the present situation

$$\mathbf{c}_g = \frac{Nk_z}{k^3(k^2 - k_z^2)^{1/2}}[\hat{\mathbf{x}}k_xk_z + \hat{\mathbf{y}}k_yk_z - \hat{\mathbf{z}}(k^2 - k_z^2)]. \tag{15.40}$$

Since the properties of the waves are obviously axisymmetric about the vertical, we discuss them in terms of the case for which there is no variation in the y-direction; that is $k_y = 0$ Then

$$c_g = \frac{N \cos \theta}{k} (\hat{x} \cos \theta - \hat{z} \sin \theta). \tag{15.41}$$

This is perpendicular to the wave number

$$k = k(\hat{x} \sin \theta + \hat{z} \cos \theta) \tag{15.42}$$

and thus to the phase velocity. Energy is thus transmitted along lines in the planes of the wavefronts. It is this that gives the waves their rather unfamiliar character—although we shall see in Section 16.6 that analogous waves occur in a rotating fluid.

The other direction that is of interest in providing an understanding of the structure of the wave motion is the direction of motion of the fluid particles. Equation (15.34), which may be written

$$U \cdot k = 0 \tag{15.43}$$

shows that this motion is always perpendicular to the wave number. The waves are essentially transverse. Moreover, since when $k_y = 0$, then $V = 0$ (eqn (15.32)), the motion is along the same line as the group velocity. (More generally, eqns (15.31), (15.32), and (15.40) show that the x- and y-components of U and c_g are in the same ratio k_x/k_y.)

Figure 15.10 summarizes the above properties of the waves.

The applicability of this theory has been demonstrated experimentally. Equation (15.41) shows that the group velocity is directed at an angle θ to the horizontal, independently of k. Since θ is directly related to ω, the angular frequency (eqn (15.38)), all energy associated with a single frequency is transmitted at this angle. In the experiments a long horizontal cylinder was oscillated in a stratified fluid in the horizontal

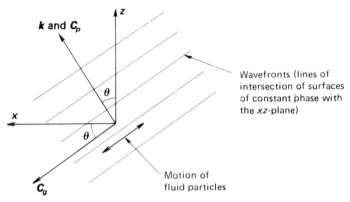

FIG. 15.10 Summary of properties of internal wave in stratified fluid.

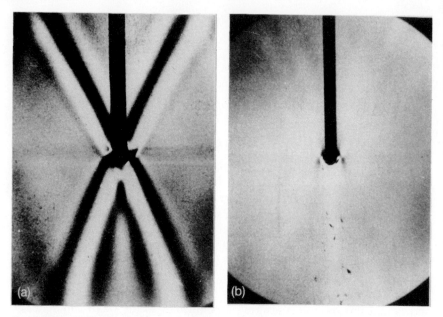

FIG. 15.11 Waves produced by vibrating cylinder in a stratified fluid (dark vertical line is cylinder support). (a) $\omega/N = 0.90$; (b) $\omega/N = 1.11$. Ref. [282].

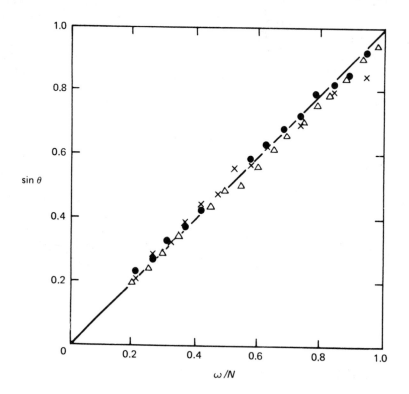

direction perpendicular to its axis. This produced a two-dimensional wave pattern which was observed by a schlieren optical method (Section 25.4) working on the refractive index changes associated with the density changes. (A special system was used to eliminate the effect associated with the basic stratification and to show only the changes produced by the wave perturbation.) Since the source is localized and the frequency fixed, energy radiates from the cylinder only in four narrow bands at angle θ (above and below) to the horizontal. This expectation was confirmed by the experiments, as is illustrated by Fig. 15.11(a). Experimental confirmation of the relationship between the oscillation frequency and the orientation of the rays, as given by eqns (15.38) and (15.41), is shown in Fig. 15.12. When ω exceeds N, no waves are produced, as is shown by the contrast between Fig. 15.11(a) and (b).

The light and dark bands in Fig. 15.11(a) are lines of constant density perturbation and thus, in an instantaneous photograph, lines of constant

FIG. 15.13 Pattern produced by motion of a horizontal cylinder through a stratified fluid (dark vertical line is cylinder support). Ref. [356].

FIG. 15.12 Experimental verification of relationship between frequency and ray angle for waves in a stratified fluid. Different symbols correspond to different amplitudes of the driving oscillation. From Ref. [282].

phase. These are seen to run along the ray as in Fig. 15.10. Observations of the changing pattern showed that they moved across the ray in agreement with the theoretical direction of the phase velocity.

No prediction about the wave numbers present is given by the theory. For a single ω, all values of k are possible and all transmit in the same direction. In the experiments, there will have been a range of wave numbers present that Fourier synthesize to give the narrow rays with undisturbed fluid elsewhere. The detailed structure depends on the motion in the immediate vicinity of the oscillating cylinder.

Superposition of waves of various frequency and wave number can give rise to a variety of patterns. Such patterns may be observed so long as the fluid velocities remain small enough for non-linear interactions to be negligible. We shall not consider this in detail, but we will look briefly at one example. Figure 15.13 shows the pattern produced in an apparatus similar to the one used for the above oscillating cylinder experiments when a horizontal cylinder was moved uniformly in a direction at 10° to the horizontal. One sees a wake and a wave pattern. The flow in the immediate vicinity of the cylinder is governed by the full non-linear equations of motion. But the motion also produces a far-field disturbance that would not exist in a system that did not support waves. This can be understood in terms of the above linear wave theory.

16

FLOW IN ROTATING FLUIDS

16.1 Introduction [39, 201]

This chapter is concerned with the dynamics of fluids in rotating systems. This branch of fluid mechanics has developed rapidly in recent years as an obvious consequence of interest in geophysical flow problems. Evaluation of the parameters shows that the motions, particularly on the large scale, of the Earth's atmosphere, oceans, and core and of stars and galaxies will all exhibit the effects discussed in this chapter. The rotation gives rise to a range of new phenomena; here we consider a small selection of these. For the most part we return to considering constant density flows, although we shall look briefly at rotating flows with buoyancy forces in Sections 16.8 and 16.9.

The whole subject could be formulated as seen by an observer external to the rotation. Since, however, all boundary conditions will be specified in terms of the rotating frame of reference it is easier to modify the equations of motion so that they apply in such a frame.

If one takes a body of fluid and rotates its boundaries at a constant angular velocity $\mathbf{\Omega}$, then at any time sufficiently long after starting the rotation, the whole body is rotating with this angular velocity, moving as if it were a rigid body. There are then no viscous stresses acting within the fluid. Any disturbance—i.e. anything that would produce a motion in a non-rotating system—will produce motion relative to this rigid body rotation. This relative motion can be considered as the flow pattern; it is the pattern that will be observed by an observer fixed to the rotating boundaries.

16.2 Centrifugal and Coriolis forces

The effect of using a rotating frame of reference is well known from the mechanics of solid systems; there are accelerations associated with the use of a non-inertial frame that can be taken into account by introducing centrifugal and Coriolis forces. That statement may be expressed in a form appropriate to fluid systems by

$$\left(\frac{\mathrm{D}\boldsymbol{u}}{\mathrm{D}t}\right)_I = \left(\frac{\mathrm{D}\boldsymbol{u}}{\mathrm{D}t}\right)_R + \mathbf{\Omega} \times (\mathbf{\Omega} \times \boldsymbol{r}) + 2\mathbf{\Omega} \times \boldsymbol{u}_R. \tag{16.1}$$

The subscripts I and R refer to inertial and rotating frames of reference. $(Du/Dt)_I$ is thus the actual acceleration that a fluid particle is experiencing and so $\rho(Du/Dt)_I$ is the quantity to be equated with the sum of the various forces acting on the fluid particle. $(Du/Dt)_R$ is the acceleration relative to the rotating frame and can thus be expanded in the usual way:

$$\left(\frac{Du}{Dt}\right)_R = \frac{\partial u_R}{\partial t} + (u \cdot \nabla u)_R. \tag{16.2}$$

Dropping the subscript R, as all velocities will be referred to the rotating frame throughout the rest of this chapter, the equation of motion is

$$\frac{\partial u}{\partial t} + u \cdot \nabla u = -\frac{1}{\rho}\nabla p - \Omega \times (\Omega \times r) - 2\Omega \times u + \nu\nabla^2 u. \tag{16.3}$$

The second and third terms on the right-hand side of (16.3) are, of course, respectively the centrifugal and Coriolis forces.

In many problems the centrifugal force is unimportant. This is because it can be expressed as the gradient of a scalar quantity,

$$\Omega \times (\Omega \times r) = -\nabla(\tfrac{1}{2}\Omega^2 r'^2) \tag{16.4}$$

where r' is the distance from the axis of rotation (Fig. 16.1). Hence replacing the pressure by $(p - \tfrac{1}{2}\rho\Omega^2 r'^2)$ reduces the problem to one that is identical except that the centrifugal force is absent. This is entirely analogous to the procedure of subtracting out the hydrostatic pressure to remove the effect of gravitational forces, as discussed in Section 14.2. The centrifugal force is balanced by a radial pressure gradient which is present whether or not there is any flow relative to the rotating frame and which does not interact with any such flow. The limitations to this statement are just the same as for the gravitational case. Firstly, the pressure must not appear explicitly in the boundary conditions. Secondly (since ρ has been taken through ∇), the density must be constant; centrifugal force variations associated with density variations do give rise to body forces that can alter or even cause a flow.

It should be emphasized that the centrifugal force under discussion here is that associated with the rotation of the frame of reference as a

FIG. 16.1   Definition sketch.

whole. In other contexts (e.g. rotating Couette flow) it is sometimes convenient to talk about the centrifugal force associated with circular motion relative to the frame of reference (either inertial or rotating). This is then a way of discussing physically effects that are contained mathematically in one or both of $(u \cdot \nabla u)$ and the Coriolis term.

16.3 Geostrophic flow and the Taylor–Proudman theorem

We thus see that the cause of differences between the dynamics of non-rotating and rotating fluids is the Coriolis force. It is instructive to consider flows that are dominated by the action of Coriolis forces and thus see these differences in their extreme form. Let us suppose therefore that the Coriolis effect is large compared with both the inertia of the relative motion and viscous action. Restricting attention to steady flow this means

$$|u \cdot \nabla u| \ll |\Omega \times u| \qquad (16.5)$$

and

$$|v\nabla^2 u| \ll |\Omega \times u|. \qquad (16.6)$$

Expressing these in terms of scales in the usual way,

$$U^2/L \ll \Omega U \quad \text{and} \quad vU/L^2 \ll \Omega U \qquad (16.7)$$

or

$$U/\Omega L \ll 1 \quad \text{and} \quad v/\Omega L^2 \ll 1. \qquad (16.8)$$

The quantities $U/\Omega L$ and $v/\Omega L^2$ are known respectively as the Rossby number and the Ekman number; they indicate the ratios of inertial forces to Coriolis forces, and viscous forces to Coriolis forces, respectively. As in a non-rotating fluid, viscous effects may become important, even when simple order-of-magnitude considerations indicate otherwise, in boundary layers close to surfaces. We are here considering flows outside such boundary layers; that is, flow in a region for which Euler's equation would apply if the system were not rotating.

When both the Rossby number and the Ekman number are small, the equation of motion becomes

$$2\Omega \times u = -\frac{1}{\rho}\nabla p \qquad (16.9)$$

where the pressure has been modified by the centrifugal pressure as discussed above. Flows in which this balance between Coriolis force and pressure force pertains are called geostrophic flows.

An important property of such flows is immediately evident. The Coriolis force is always perpendicular to the flow direction. Hence, the pressure gradient is also perpendicular to the flow direction. This means that the pressure is constant along a streamline—in marked contrast to the behaviour in non-rotating systems where one is accustomed to think of the pressure variations along a streamline (e.g. Bernoulli's equation).

This feature of geostrophic flows is familiar in the interpretation of weather maps. These are usually compiled primarily from information about the pressure distribution. Isobars are drawn and, because of the dominating influence of the Earth's rotation, these are then taken to be also the lines along which the wind is blowing (strictly speaking, the wind at some height rather than the wind near ground level).

Another interesting property of geostrophic flow is discovered by taking the curl of eqn (16.9),

$$\nabla \times (\mathbf{\Omega} \times \mathbf{u}) = 0. \tag{16.10}$$

This expands to

$$\mathbf{\Omega} \cdot \nabla \mathbf{u} - \mathbf{u} \cdot \nabla \mathbf{\Omega} + \mathbf{u}(\nabla \cdot \mathbf{\Omega}) - \mathbf{\Omega}(\nabla \cdot \mathbf{u}) = 0. \tag{16.11}$$

$\mathbf{\Omega}$ is not a function of position, so the second and third terms are zero. The continuity equation is unaltered by the rotation, so

$$\nabla \cdot \mathbf{u} = 0 \tag{16.12}$$

and eqn (16.11) becomes

$$\mathbf{\Omega} \cdot \nabla \mathbf{u} = 0. \tag{16.13}$$

If axes are chosen so that $\mathbf{\Omega}$ is in the z-direction, this is

$$\Omega \partial \mathbf{u} / \partial z = 0 \tag{16.14}$$

i.e.

$$\partial \mathbf{u} / \partial z = 0. \tag{16.15}$$

There is thus no variation of the velocity field in the direction parallel to the axis of rotation. This result is known as the Taylor–Proudman theorem. In component form

$$\frac{\partial u}{\partial z} = \frac{\partial v}{\partial z} = \frac{\partial w}{\partial z} = 0. \tag{16.16}$$

If one is dealing with a system with solid boundaries perpendicular to the rotation axis so that $w = 0$ at some specified value(s) of z then this implies

$$\frac{\partial u}{\partial z} = \frac{\partial v}{\partial z} = 0, \quad w = 0 \quad \text{everywhere} \tag{16.17}$$

and the theorem says that the flow is entirely two-dimensional in planes perpendicular to the axis of rotation.

16.4 Taylor columns

The Taylor–Proudman theorem has simple and striking consequences, illustrating the fact that rotating fluids exhibit a range of phenomena not found in non-rotating fluids. The principal of these is the formation of 'Taylor columns'. These occur when there is relative motion between an obstacle and the fluid in a strongly rotating system. We consider first the case in which this motion is perpendicular to the axis of rotation (giving what is called a transverse Taylor column). The fluid is deflected past the obstacle. Since the flow must be two-dimensional this deflection also occurs above and below the obstacle (visualizing the axis of rotation as vertical). There are thus columns of fluid, extending parallel to the axis from the obstacle, round which the fluid is deflected just as if the solid walls themselves extended there.

Although Taylor columns realizable in the laboratory involve complications not described by this simple explanation, prediction of their occurrence has received abundant experimental confirmation. Figure 16.2 illustrates this with a sequential pair of photographs showing the development of dye streaks originating inside and outside a Taylor column. Two features are to be noted. Firstly, the dye streak originating in the column is much shorter than those originating outside, showing that the fluid there is moving very slowly (all the dye streaks were initiated at the same instant). Secondly, the dye streaks meeting the Taylor column are deflected around it much as if it were a solid obstacle; it is important to realize that the dye lines are well above the top of the obstacle and would just pass straight over it in the absence of rotation.

Full appreciation of Fig. 16.2 requires some description of the method by which it was obtained. The basic apparatus consists of a rotating cylindrical tank, with an arrangement for pumping water in at the periphery and out on the axis. This produces the relative flow between the body of the fluid and an obstacle fixed in the tank. Because the tank is rotating, this flow is not radial but azimuthal, fluid particles moving round on circles concentric with the tank; the pumping can be thought of as producing a radial pressure gradient which is balanced by the Coriolis force associated with the azimuthal motion. The obstacle is a truncated cylinder on the floor of the tank and extending about one-third of its depth. Dye can be released (by the pH technique described in Section 25.4) in four places, of which the positions in plan are shown in the sketch (Fig. 16.3); in elevation, these dye sources are about midway

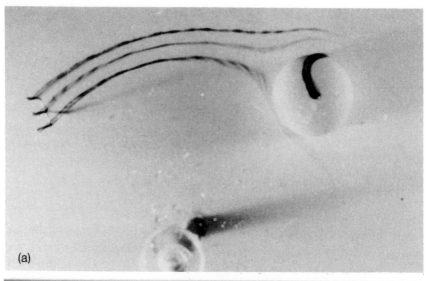

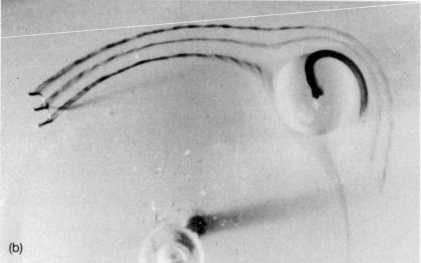

FIG. 16.2 Streakline development in and around transverse Taylor column; see text for details. Photos by C. W. Titman.

between the top of the obstacle and the top of the tank. One source is over the obstacle and thus in the Taylor column. The other three are placed so that the dye from them meets and passes round the Taylor column.

This experiment is, in principle, a repetition of the pioneering experiment by Taylor [372], first demonstrating the validity of the

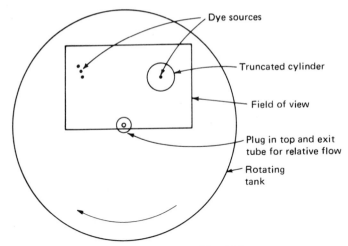

Dye sources

Truncated cylinder

Field of view

Plug in top and exit
tube for relative flow

Rotating
tank

FIG. 16.3 Key to Fig. 16.2.

surprising predictions of the Taylor–Proudman theorem. Subsequent work [196, 197, 399] has provided much information about the quite complex detailed structure of flows of this type; brief mention of further features will be made at the end of this section.

A different type of Taylor column, a longitudinal Taylor column, arises when an obstacle moves along or parallel to the axis of rotation. To understand this directly in terms of the Taylor–Proudman theorem, one needs to consider the fluid as extending to infinity in both directions parallel to the axis of rotation. Then the obstacle pushes a column of fluid moving with its own speed ahead of it and pulls a similar column behind it. In practice, as one might expect, Taylor columns are very long but not infinite when the Rossby number is very small but not zero. Moreover, for reasons connected with the local non-geostrophic regions mentioned below, part of a Taylor column can be observed even when the apparatus is relatively short. Experimental observation of this type of Taylor column is thus possible. Figure 16.4 gives an example. A sphere is rising along the axis of a rotating tank. Water near the bottom of the tank is dyed, and the sphere has risen into a region of clear water above. Dye carried into this region by the motion demonstrates the presence of Taylor columns ahead of and behind the sphere. With this technique, not all the fluid in the forward Taylor column is necessarily dyed, since some may have started above the dyed region; but the presence of dye so far ahead of the sphere clearly demonstrates the existence of the column, and is in marked contrast with what would happen in the absence of rotation.

FIG. 16.4   Longitudinal Taylor columns, shown by method described in text. The sphere is seen as a bulge in the column. $Ro = u_0/\Omega d = 0.136$; $Ek = v/\Omega d^2 = 0.00134$. From Ref. [270].

There is an evident similarity between a longitudinal Taylor column and the blocking phenomenon in stratified flow discussed in Section 15.2. In both cases, fluid is pushed ahead of a moving obstacle at distances much larger than it would be in the absence of rotation and stratification. There is a similar extended influence behind the obstacle—always in the rotating fluid, sometimes in the stratified. Blocking is produced only by cylindrical obstacles (two-dimensional flow) whereas a longitudinal Taylor column is produced by any shape (two- or three-dimensional flow). However, if the Taylor column is also produced by a cylindrical obstacle then there is a rather close analogy between the two flows, particularly when the stratified fluid has a Prandtl number around unity [400]. This analogy was invoked in Section 15.2 to indicate that blocking must occur both upstream and downstream when $\text{Pr} \sim 1$.

In developing an understanding of the behaviour of rotating fluids, it is evidently going to be valuable to formulate a physical description of the processes underlying the Taylor–Proudman theorem. This is not in fact easy, just as it is not easy to develop a ready physical intuition about the behaviour of solid gyroscopes. A reading of Taylor's original papers makes it evident that he was surprised by the theoretical predictions and perhaps half expected that his experiments would not confirm them. However, some comments may be made.

As a starting point we note that the Coriolis force is independent of the actual location of the axis of rotation. The phenomena are thus not a result of rotation about a *particular* axis but rather the result of the flow occurring in the presence of uniform background vorticity of value 2Ω. At low Rossby number any vorticity associated with the relative motion is small compared with this. The relative motion cannot interact with the background vorticity in a way that would alter it so that it no longer corresponded just to a uniform rotation. It is motions that would do this that are excluded by the Taylor–Proudman theorem. (Note that the derivation of the theorem, starting with taking the curl of the momentum equation, indicates that it is a statement about the action of vorticity.)

One can say that there are two types of such motion: flows that violate

$$\partial u / \partial z = \partial v / \partial z = 0 \qquad (16.16a)$$

and that would cause twisting of the background vorticity; and flows that violate

$$\partial w / \partial z = 0 \qquad (16.16b)$$

and that would cause stretching of the background vorticity. (These ideas depend on the result (Section 6.6) that vortex lines always consist of the same fluid.) The two processes are not in general independent but it is convenient to consider them separately.

Vortex twisting would occur, for example, if the fluid approaching an obstacle were slowed down but that approaching the region above it were not (Fig. 16.5—the sort of flow that would occur in the absence of the background vorticity). This would produce a large vorticity component not in the z-direction, associated with which would be relative motions large compared with the original relative motion. The background vorticity thus resists the twisting and, in the case of low enough Rossby number, prevents it altogether. This is closely analogous to the action of a solid gyroscope in resisting turning, except that it is each individual vortex line that has the gyroscopic action and not the system as a whole.

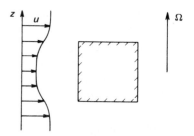

FIG. 16.5 Flow violating Taylor–Proudman theorem through vortex twisting in regions of non-zero $\partial u/\partial z$.

Vortex stretching would occur if fluid in a tank of finite depth moved from above an obstacle to some other station (Fig. 16.6). This would locally increase the vorticity (by the process described in Section 6.6) again causing a relative motion large compared with the original relative motion. Thus this is also resisted. The analogy with a gyroscope is not quite so simple here; if a gyroscope were made of a deformable material certain deformations would be resisted by gyroscopic action. It should be noted that, although the above discussion has been given in terms of vortex stretching, it applies equally to vortex compression; a local reduction in the vorticity also corresponds to a large relative motion.

The Taylor–Proudman theorem indicates that, under circumstances specified above, the flow is two-dimensional. It is natural to enquire next about the pattern of this two-dimensional flow. For example, we have seen that, according to the theorem, there is no transfer of fluid between the inside and outside of a Taylor column; but there will be separate two-dimensional flows in each of these regions. However, no further information about these flows can be obtained from the geostrophic flow equation. This may be seen as follows. The equation now has two

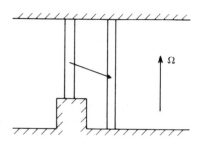

FIG. 16.6 Motion of fluid from above obstacle, violating Taylor–Proudman theorem.

components:

$$-2\Omega v = -\frac{1}{\rho}\frac{\partial p}{\partial x} \tag{16.18}$$

$$2\Omega u = -\frac{1}{\rho}\frac{\partial p}{\partial y}. \tag{16.19}$$

Also u and v are linked by the continuity equation

$$\frac{\partial u}{\partial x} + \frac{\partial v}{\partial y} = 0 \tag{16.20}$$

apparently providing three equations for the three unknowns u, v, and p. However, differentiating (16.18) with respect to y and (16.19) with respect to x and then eliminating $\partial^2 p/\partial x\,\partial y$ gives eqn (16.20) again. Continuity, therefore, does not give any independent information and there are effectively only two equations for the three variables. Any solenoidal two-dimensional motion satisfies the geostrophic equations. The Taylor–Proudman theorem is the limit of the information to be obtained in this way.

How then is the motion, for example inside a Taylor column, determined? The answer is that the flow develops local regions where the geostrophic equations do not apply. Both the Rossby number $(U/\Omega L)$ and the Ekman number $(\nu/\Omega L^2)$ are increased if the length scale is decreased. If, therefore, there are regions in which the flow parameters vary over a distance small compared with the imposed length scales, then inertia forces and/or viscous forces may be locally important. There may be local violation of the Taylor–Proudman theorem. An obvious place for such a development to occur is between the two flows which, according to the theorem, do not interact; that is at the edge of the Taylor column. Thin shear layers are observed here. The whole structure of the detailed flow in the Taylor column (such as the slow motion seen in Fig. 16.2) is determined by these shear layers together with the boundary layer on the obstacle's surface. Because the two regions are separated by a layer in which the Taylor–Proudman theorem does not apply, there is in fact some interchange of fluid between the interior and the exterior of a transverse Taylor column. It is the shear layers also that govern the length of a Taylor column (either transverse or longitudinal) so that, although it is long at low Rossby and Ekman numbers, it does not extend to infinity as predicted by the Taylor–Proudman theorem.

A discussion of the structure of the shear layers is beyond the scope of this book [280]. However, it is interesting to note that we now have two classes of flows that spontaneously develop regions of small length scale. In boundary layer formation it happens in order that all the imposed conditions should be fulfilled; there is no solution of the equations when all terms that are small on the first approximation are neglected. Here, it happens as a process that selects a particular flow pattern; there is a multiplicity of solutions when all apparently small terms are neglected.

16.5 Ekman layers

We consider now a flow in a rotating fluid in which viscous forces are important but which is much simpler than the shear layers mentioned above. This is the Ekman layer, the boundary layer between a geo-strophic flow and a solid boundary at which the no-slip condition applies. This turns out to be actually simpler than the corresponding problem in a non-rotating fluid. The results have the added interest of rather direct application to the atmosphere and the oceans.

We suppose that a body of fluid rotating about the z-axis has a boundary in the xy-plane. A geostrophic flow is occurring within the main body of the fluid and this is taken to be a uniform unidirectional flow, u_0, in the x-direction. In practice this means that the length scale of variations in the geostrophic flow must be large compared with the boundary layer thickness that emerges from the following analysis. Associated with u_0 is a uniform pressure gradient in the y-direction, the equations of motion in the geostrophic region being

$$2\Omega u_0 = -\frac{1}{\rho}\frac{\partial p_0}{\partial y} \tag{16.21}$$

$$0 = -\frac{1}{\rho}\frac{\partial p_0}{\partial x}. \tag{16.22}$$

Between this geostrophic flow and the boundary is a region in which, as usual, viscous forces are brought into play by the boundary condition. We look for a solution for this boundary layer with the velocity field uniform in both the x- and y-directions. This is best justified *a posteriori*: one finds such a solution. However, it is at first sight a rather surprising procedure, since one knows that no solution of this type exists in the absence of rotation (Section 11.4). The reason for the difference is associated with the fact that a stress acts between the boundary and the fluid. Without rotation, this stress extracts momentum from the flow; in zero pressure gradient, the total flow momentum must decrease as one goes downstream, i.e. the boundary layer must grow. In a rotating fluid the stress can come into equilibrium with the integrated Coriolis force associated with the motion; once the boundary layer is formed no further growth need occur. Since the Coriolis force acts at right angles to the motion, this explanation implies that the boundary layer necessarily involves flow in both directions parallel to the wall.

Accordingly, we take u and v both non-zero but

$$\frac{\partial u}{\partial x} = \frac{\partial u}{\partial y} = \frac{\partial v}{\partial x} = \frac{\partial v}{\partial y} = 0 \qquad (16.23)$$

and, as is then required by continuity,

$$w = 0. \qquad (16.24)$$

The three components of the momentum equation become

$$-2\Omega v = -\frac{1}{\rho}\frac{\partial p}{\partial x} + v\frac{\partial^2 u}{\partial z^2} \qquad (16.25)$$

$$2\Omega u = -\frac{1}{\rho}\frac{\partial p}{\partial y} + v\frac{\partial^2 v}{\partial z^2} \qquad (16.26)$$

$$\frac{\partial p}{\partial z} = 0. \qquad (16.27)$$

Equation (16.27) implies that

$$\frac{\partial p}{\partial x} = \frac{\partial p_0}{\partial x} = 0; \quad \frac{\partial p}{\partial y} = \frac{\partial p_0}{\partial y}. \qquad (16.28)$$

Hence, substituting (16.21), the equations to be solved are

$$-2\Omega v = v\frac{\partial^2 u}{\partial z^2} \qquad (16.29)$$

$$-2\Omega(u_0 - u) = v\frac{\partial^2 v}{\partial z^2}. \qquad (16.30)$$

The boundary conditions are

$$u = v = 0 \quad \text{at} \quad z = 0 \qquad (16.31)$$

$$u \to u_0 \quad \text{and} \quad v \to 0 \quad \text{as} \quad z \to \infty. \qquad (16.32)$$

Eliminating u or v gives a soluble fourth-order differential equation. However, the equation can be kept second-order by working in terms of the complex variable

$$Z = (u + iv)/u_0. \qquad (16.33)$$

Multiplying eqn (16.30) by i and adding it to eqn (16.29) gives

$$2i\Omega(u + iv) - v\frac{\partial^2}{\partial z^2}(u + iv) - 2i\Omega u_0 = 0 \qquad (16.34)$$

i.e.

$$v\frac{d^2Z}{dz^2}-2i\Omega(Z-1)=0 \tag{16.35}$$

with

$$Z=0 \quad \text{at} \quad z=0 \tag{16.36}$$

and

$$Z\to 1 \quad \text{as} \quad z\to\infty. \tag{16.37}$$

The solution of this is

$$Z=1-\exp\left[\left(\frac{2i\Omega}{v}\right)^{1/2}z\right] \tag{16.38}$$

or, taking real and imaginary parts,

$$u=u_0[1-e^{-z/\Delta}\cos(z/\Delta)] \tag{16.39}$$

$$v=u_0e^{-z/\Delta}\sin(z/\Delta) \tag{16.40}$$

where

$$\Delta=(v/\Omega)^{1/2} \tag{16.41}$$

This solution is often displayed in the form of a polar diagram, Fig. 16.7, known as the Ekman spiral. Different points along this correspond to different values of z/Δ (not uniformly spaced) and the velocity at each height is then given in magnitude and direction by the line from the origin to the corresponding point on the spiral. Noteworthy features are, firstly, that close to the boundary $v/u=1$ and so the flow is at 45° to the

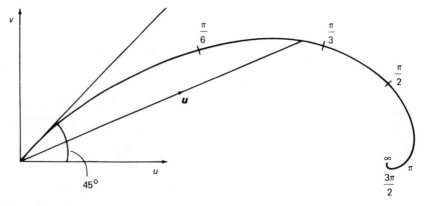

FIG. 16.7 The Ekman spiral: polar diagram of velocity vector $\boldsymbol{u}$ in Ekman layer. Numbers along spiral are values of z/Δ.

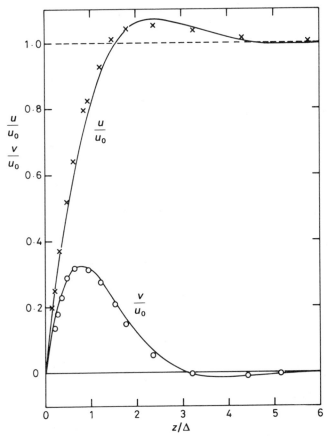

Fɪɢ. 16.8 Theoretical and experimental velocity distributions in an Ekman layer. Data from Ref. [110].

geostrophic flow; and secondly, that at $z/\Delta = \pi$ the flow is in the same direction as the geostrophic flow but approximately 1.1 times as fast.

Figure 16.8 shows eqns (16.39) and (16.40) plotted more directly. Experimental measurements, made with a hot-wire anemometer (Section 25.2) in a rotating air duct, shown in the figure agree excellently with the theory.

From eqn (16.41), $\Delta \to \infty$ as $\Omega \to 0$; in the absence of rotation the above solution collapses, as implied by the preceding physical discussion. In that case the problem of physical interest is always the growth of the boundary layer.

As well as providing the above comparison between theory and observation, laboratory work on Ekman layers has revealed two types of instability [110, 156]. Both can be seen in Fig. 16.9, a dye pattern

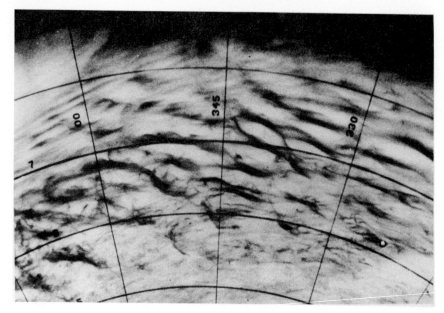

FIG. 16.9   Ekman layer instabilities. From Ref. [156].

generated from crystals on the boundary. One type produces the shorter waves at the top left of the picture; the other produces the less regular, larger-scale pattern in the middle.

Turbulent Ekman layers have also been studied. An Ekman spiral again occurs, but because Reynolds stresses (Section 19.3) replace viscous stresses, the turning angle is less than 45°, typically about half that [111, 202]. This is the case of most direct meteorological application: the wind at ground level and the wind aloft are commonly observed to be in different directions. At a general latitude, it is the vertical component of the Earth's angular velocity that is important. In the northern hemisphere the wind close to the ground blows to the left of the geostrophic wind, as in Fig. 16.7.

An Ekman layer below a geostrophic flow of variable velocity has an additional important property, which leads to a significant interaction with the geostrophic region. The process is illustrated schematically by Fig. 16.10. The solid arrows indicate the geostrophic flow and the broken arrows the transverse motion in the associated Ekman layer on a boundary in the plane of the page. This transverse motion involves divergence (or convergence) and thus gives rise to a flow out of (or into) the geostrophic region. This process is known as Ekman layer suction (or injection). The signs are such that the Ekman layer sucks when the

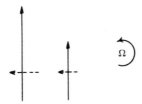

FIG. 16.10   Sketch indicating reason for Ekman layer suction: see text.

relative vorticity of the geostrophic flow has opposite sign to the basic rotation. The suction changes the boundary condition experienced by the geostrophic region. Because of the Taylor–Proudman theorem this has an effect throughout the geostrophic flow, and, in certain circumstances, Ekman layers can exert a controlling influence on the development of this. Transfer of fluid between boundary layers and the body of the fluid in this way is, for instance, the process by which a tank of fluid is brought to the angular velocity of its boundaries (as mentioned in Section 16.1). For most geometries 'spin-up' in this way occurs much more rapidly than it would through viscous diffusion [39].

16.6   Intrinsic stability and inertial waves

Rotating fluids have an intrinsic stability, in the sense that if a fluid particle is displaced there is a tendency for it to return in a way that would not occur in a non-rotating fluid. Consider an isolated particle of unit mass which, as a result of some disturbance, moves with a speed q in any direction perpendicular to the axis of rotation. (We are not, of course, considering here a genuine fluid motion or even a particular physical situation, but merely illustrating a general feature in the simplest way.) A Coriolis force of magnitude $2\Omega q$ acts on it, always at right angles to its direction of motion. It thus moves on a circular path of radius r given by

$$q^2/r = 2\Omega q \tag{16.42}$$

that is

$$r = q/2\Omega. \tag{16.43}$$

It goes once round the circle in a time

$$t = 2\pi r/q = \pi/\Omega \tag{16.44}$$

independent of q. It thus returns to its original position periodically, twice during every revolution of the fluid. It is thus often said that a rotating fluid has an intrinsic angular frequency, $2\pi/t = 2\Omega$.

FIG. 16.11 Inertial waves produced by oscillating disc in a rotating tank (lower part of pattern is produced by reflection at walls). Aluminium powder flow visualization (see Section 25.4). From Ref. [39].

In a complete system, the effect of this constraining tendency acting on every fluid particle is that rotating fluids can support wave motions, known as inertial waves, that would not arise in the absence of rotation. Inertial waves are in many ways closely similar to internal waves in a stratified fluid, as is illustrated by comparison of Fig. 16.11, showing inertial waves generated by an oscillating disc in a rotating tank, with Fig. 15.11(a). Just as internal waves occur with values of the angular frequency from zero up to the Brunt–Väisälä frequency, so inertial waves occur with the values from zero up to 2Ω; the counterpart of (15.38) is

$$\omega = 2\Omega \, |\cos \theta| \qquad (16.45)$$

where θ is now the angle between the wavenumber vector and the rotation axis. Again the frequency is related to the orientation and not the magnitude of the wavenumber. Consequently, the property of the group velocity being perpendicular to the phase velocity is also exhibited by inertial waves [39, 304]. We can thus allow this analogy to indicate the nature of inertial waves without entering into a separate detailed analysis of them.

16.7 Rossby waves [307]

Instead we consider Rossby waves by way of a fuller illustration of waves resulting from rotation of the system. These occur in particular geometrical arrangements. Their origin can be seen rather directly in terms of the Taylor–Proudman theorem. Also they are of meteorological and ocean-ographic importance as we shall discuss briefly at the end of this section.

The simplest configuration in which Rossby waves occur is a layer of fluid bounded by two planes both almost perpendicular to the axis of rotation but not quite parallel to one another so that there are slight variations in the depth of the layer. Wave motions can then arise from 'twanging' the constraint produced by the Taylor–Proudman theorem. If a flow in the layer were exactly geostrophic, then it would have to follow contours of constant depth (from a line of reasoning exactly parallel to that given for the formation of Taylor columns). Correspondingly, if a fluid particle is displaced to a position of different depth, there will be a tendency for it to return to its original station. If it has inertia it will overshoot and oscillate. A complete flow pattern of such oscillations constitutes a Rossby wave.

The inertia involved here is that associated with the unsteadiness of the wave, $\partial u / \partial t$. That associated with $u \cdot \nabla u$ is still neglected—by restricting attention to waves of small enough amplitude. Viscous effects are also assumed negligible. The dynamical equation is thus

$$\frac{\partial u}{\partial t} + 2\Omega \times u = -\frac{1}{\rho} \nabla p. \tag{16.46}$$

The nature of Rossby waves can be shown by considering the simplest case (Fig. 16.12) in which the depth, h, varies uniformly and in one direction only (chosen as the y-direction):

$$\frac{\partial h}{\partial x} = 0; \quad \frac{dh}{dy} = \gamma, \quad \text{a constant.} \tag{16.47}$$

γ is taken to be small.

We suppose that the waves retain the properties of geostrophic flow that

$$\partial u / \partial z = \partial v / \partial z = 0 \tag{16.48}$$

but that $\partial w / \partial z$ becomes non-zero. (This apparently arbitrary procedure can be shown, a posteriori, to involve no inconsistency; the pressure variation obtained by integrating the z-component of eqn (16.46) produces negligible modification to the x- and y-components when γ is small.) Then the first two terms in the continuity equation

$$\frac{\partial u}{\partial x} + \frac{\partial v}{\partial y} + \frac{\partial w}{\partial z} = 0 \tag{16.49}$$

are independent of z and so

$$\frac{\partial w}{\partial z} = \text{const. with respect to } z. \tag{16.50}$$

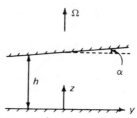

FIG. 16.12   Definition diagram for simple Rossby wave theory.

It now simplifies the analysis to choose the z-axis to be exactly normal to one of the boundaries, $z = 0$ say (Fig. 16.12). If γ is small, this makes negligible difference to the dynamical equations. At this boundary $w = 0$. Hence

$$\partial w/\partial z = w_h/h \qquad (16.51)$$

where w_h is the value of w at the top boundary, $z = h$. Here the flow must be parallel to this boundary (since it is supposed separated from the no-slip condition by a thin boundary layer). Thus

$$w_h = v_h \tan \alpha = \gamma v_h = \gamma v \qquad (16.52)$$

in view of eqn (16.48). Combining eqns (16.49), (16.51), and (16.52),

$$\frac{\partial u}{\partial x} + \frac{\partial v}{\partial y} + \frac{\gamma v}{h} = 0. \qquad (16.53)$$

The Rossby waves are given by this form of the continuity equation taken in conjunction with the x- and y-components of eqn (16.46),

$$\frac{\partial u}{\partial t} - 2\Omega v = -\frac{1}{\rho}\frac{\partial p}{\partial x} \qquad (16.54)$$

$$\frac{\partial v}{\partial t} + 2\Omega u = -\frac{1}{\rho}\frac{\partial p}{\partial y}. \qquad (16.55)$$

From the physical ideas discussed earlier we expect to find waves involving oscillations up and down the depth gradient. We look for ones in which the velocity is constant in this direction; i.e. u and v are not functions of y. In so doing, we are losing generality even for the particular geometry considered. However, this provides an adequate context in which to see the principal properties of the waves.

The pressure must still vary in the y-direction because of the Coriolis force. Cross-differentiation of eqns (16.54) and (16.55) now gives

$$\frac{\partial^2 v}{\partial x\,\partial t} + 2\Omega\frac{\partial u}{\partial x} = 0. \qquad (16.56)$$

Substituting from eqn (16.53) (in which the second term is now zero),

$$\frac{\partial^2 v}{\partial x\, \partial t} - \frac{2\Omega\gamma}{h} v = 0. \tag{16.57}$$

For scales over which the depth variation is small enough, $2\Omega\gamma/h$ may be treated as a constant, and we have a linear equation for which we may look for a solution of the form

$$v = v_0 \exp\{i(\omega t - kx)\} \tag{16.58}$$

If this can be found with real ω and k, the flow can be described as a wave motion. Substituting gives

$$\omega k v_0 - 2\Omega\gamma v_0/h = 0 \tag{16.59}$$

and thus

$$k = 2\Omega\gamma/h\omega. \tag{16.60}$$

The phase and group velocities are respectively

$$c_p = \omega/k = h\omega^2/2\Omega\gamma \tag{16.61}$$

$$c_g = d\omega/dk = -h\omega^2/2\Omega\gamma. \tag{16.62}$$

These waves have the following properties:
1. they are linear (the properties are independent of the amplitude so long as this is small enough for the equations to apply);
2. they are dispersive (c_p is a function of ω);
3. they are essentially progressive (the solution applies for only one sign of c_p, and so there can be no superposition to give standing waves);
4. $\partial u/\partial x$ and v are in phase (eqn (16.53) with $\partial v/\partial y = 0$) and so u and v are $\pi/2$ out of phase.

The most direct experimental method of checking a theory of this sort is to introduce a mechanical oscillator into an appropriate configuration. Such experiments have been performed [209], although not in a geometry to which the above simplest theory can be applied, and show good agreement between theoretical and observed properties of Rossby waves.

Spontaneously occurring Rossby waves arise when an otherwise geostrophic flow is deflected in a way that violates the Taylor–Proudman theorem. Figure 16.13 shows schematically the pattern produced when a Rossby wave of the type analysed above is superimposed on a uniform bulk flow (speed U) in the x-direction. The wave causes the flow to meander up and down the depth variations. The wavy lines represent the instantaneous streamline pattern produced by the combination of the bulk flow and the instantaneous Rossby wave motions (indicated by the short arrows). The streamlines have the same spacing at the crests and at

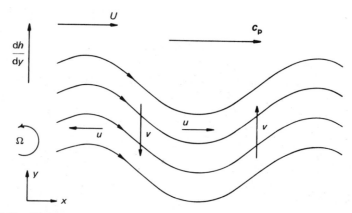

Fig. 16.13 Diagram summarizing the properties of a Rossby wave superimposed on a mean flow.

the troughs; the greater speed when u has the same sign as U is associated with each streamline then being at its minimum depth.

The whole pattern in Fig. 16.13 progresses in the x-direction with the phase velocity c_p relative to the bulk flow. If $U = -c_p$, the pattern remains stationary. This is what happens in flow past an obstacle in the appropriate direction in a rotating fluid of variable depth. The fluid is disturbed by the obstacle in a way that 'twangs' the Taylor–Proudman theorem and then oscillates as it travels down-stream. The waves select the wavelength such that $c_p = -U$, and so the pattern is a steady flow relative to the obstacle (which could be analysed as a steady flow by replacing $\partial/\partial t$ by $U\partial/\partial x$ in the above treatment). The waves propagate downstream despite the 'freezing' by the bulk flow because the group velocity is different from the phase velocity; they propagate at a speed c_g relative to the fluid and thus at a speed $c_g + U = 2U$ relative to the obstacle.

Figure 16.14 shows a wave pattern produced in this way. This again is a situation to which the above simple analysis is not really applicable, but the processes occurring are similar to those described above. The rotating annulus contains fluid of greater depth at the outer wall than at the inner. The obstacle is at the bottom left of the picture, obscured by its support, and the relative flow generating the waves was produced as a transient effect by a sudden small change in the rotation rate.

One of the reasons Rossby waves have received much attention is that they occur in the atmosphere (as illustrated in Section 26.3) and oceans [31, 40]. Obviously the laboratory geometry considered above is not directly applicable to these. One way of understanding how Rossby waves can arise is to note that the depth of the atmosphere or the oceans

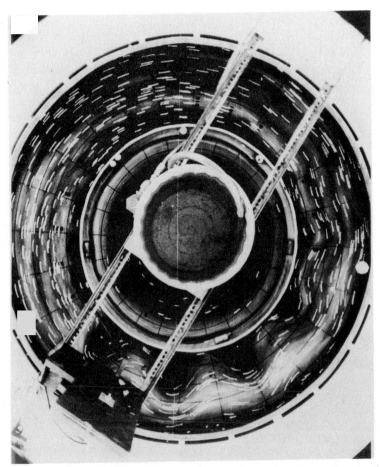

FIG. 16.14 Train of Rossby waves produced by flow past a bump (see text) and shown by trajectories over a short time interval of suspended particles. Photo by A. J. Faller from Ref. [307].

parallel to the Earth's rotation axis varies with latitude. However, the meteorologist or oceanographer usually formulates the topic in a slightly different way, and we look briefly at this. The following ideas in fact have a wider relevance to the application of basic studies of rotating fluids to natural situations than just in the case of Rossby waves. It can be shown that, because the atmosphere or the sea is shallow compared with the horizontal scale of most motions of interest, the principal effect of the Earth's rotation is produced by the vertical component of the angular velocity—of which the magnitude varies with latitude. A common way of

handling this situation theoretically is the so-called β-plane approximation. The flow is supposed to occur in a thin plane layer rotating about an
axis normal to itself; but the angular velocity of rotation is treated in the
equations as a function of position. More precisely, 2Ω in equations such
as (16.54) and (16.55) is replaced by

$$f = f_0 + \beta y \qquad (16.63)$$

with β taken constant. (If Ω is now the Earth's angular velocity,
$f_0 \doteq 2\Omega \cos \theta$ where θ is the co-latitude of the middle of the geographical
region being represented; y is the distance in the northerly direction; and
$\beta = 2\Omega \sin \theta / a$, where a is the radius of the Earth.)

The equations so formulated have properties very similar to the
equations considered earlier in this section. In particular they give rise to
Rossby waves. The simplest example is given by an analysis like that
above of eqns (16.53)–(16.55) with $\gamma = 0$ but $\beta \neq 0$. It is readily seen that
this leads to

$$k = -\beta / \omega. \qquad (16.64)$$

The similarity of this to (16.60) implies that the waves have similar
properties to those discussed. (The difference in sign merely means that
northern hemisphere Rossby waves are most closely analogous to
laboratory Rossby waves in an apparatus with a top of opposite slope
from that of Fig. 16.12.)

16.8   Rotation and stratification

We do not have space for an extended discussion of rotating fluids with
density variations, despite their evident importance in geophysical fluid
dynamics. However, we conclude this chapter by drawing attention to the
subject through some brief comments and a couple of examples. In this
section we consider the combination of rotation and stratification, and in
the next section we look at the important example of free convection in a
rotating fluid.

It is clear that the frequencies 2Ω and N (eqn (15.22)), in a rotating
fluid and a stratified fluid respectively, play analogous roles. Both can be
shown by a crude argument to be the angular frequency with which a
displaced particle oscillates; and both are the upper limit of the angular
frequency range for which waves can occur. The relationship between the
role of the Rossby number in a rotating fluid and that of the Froude
number in a stratified fluid is readily seen if we write the latter in the
form

$$\text{Fr} = U/LN. \qquad (16.65)$$

When both stratification and rotation are present, their relative importance is indicated by the parameter

$$S = N/2\Omega. \tag{16.66}$$

A value of S around 1 indicates that stratification and rotation have comparable influences; when S is small, one is dealing with a rotating flow modified by stratification, and when S is large it is the other way round.

The modifications can, however, be substantial even when S is not close to 1—a point shown by the particular flow with which we illustrate this topic. Taylor column formation in a rotating fluid is strongly reduced as S is changed from 0 to around 0.1, as may be seen in the experimental results in Fig. 16.15. This shows the length of the transverse Taylor column on a sphere as a function of S, the length being defined as the distance in which an average fluid velocity in the column relative to the ambient fluid falls to $1/e$ of the sphere's velocity.

(Since much of the interest in rotating and stratified fluids arises from geophysical applications, we extend the parenthetic comment at the end of Section 15.1 about the implications of different horizontal and vertical length scales, L and D. One should not, for example, infer from the fact

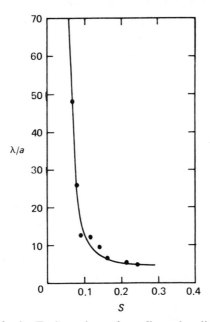

FIG. 16.15 Length λ of a Taylor column (non-dimensionalized by sphere radius a) as a function of stratification parameter S. Ro $= 1.76 \times 10^{-2}$, Ek $= 8.0 \times 10^{-4}$. From Ref. [136].

that S is commonly large in the oceans, that ocean dynamics are little effected by the Earth's rotation. Since the relevant Froude number is based on D, whereas the derivation of the Rossby number in Section 16.3 still leads to L being the appropriate length scale, the relative importance of stratification and rotation is now indicated by

$$S' = ND/2\Omega L. \tag{16.67}$$

S'^2 is sometimes called the Burger number. The minimum horizontal length scale on which rotational effects are very significant, $ND/2\Omega$, is called, for historical rather than descriptive reasons, the Rossby radius of deformation.)

16.9 Convection in a rotating annulus [198]

The configuration of our example of rotating fluid convection is shown in Fig. 16.16. An annulus of fluid rotates about its axis of symmetry, this being vertical; the outer wall is heated and the inner wall is cooled.

Very extensive experiments have been performed with this arrangement, partly because they have been a fruitful source of understanding of the dynamics of flows involving the interaction of gravitational and Coriolis forces and partly because of meteorological application (Section 26.3). However, here we take only a glimpse at this topic.

In the annulus, convection sets up a flow in which hot fluid moves upwards and inwards and cold moves downwards and outwards. Coriolis forces act on this to give an azimuthal flow in the same sense as the rotation near the top and in the opposite sense near the bottom. This becomes dominant when the rotation is fast enough.

This flow can then develop instabilities, such as that in Fig. 16.17,

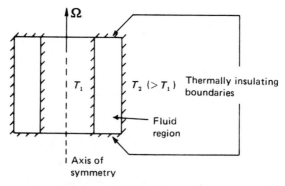

FIG. 16.16 Arrangement for convection in a rotating annulus.

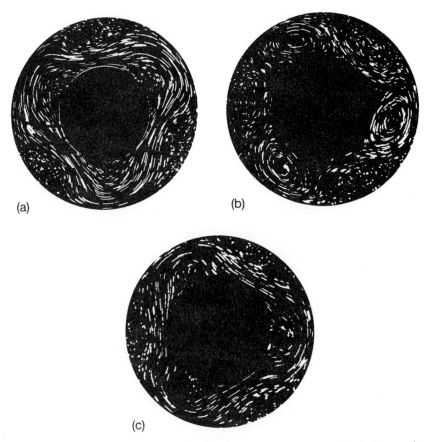

(a)

(b)

(c)

FIG. 16.17   Patterns produced by suspended particles at three levels in rotating convecting annulus: (a) near top, (b) in middle, (c) near bottom. From Ref. [148].

which shows sections through the same flow at three different depths. The development of a number of regular cells (three in Fig. 16.17) causes the azimuthal flow near the top and bottom to meander in a wavelike fashion. As a non-linear development when the amplitude of the instability becomes large, this circulation is concentrated into a narrow fast meandering stream with rather slow closed eddies in the regions between this and the walls (as can be seen in the first and third pictures of Fig. 16.17). The number of waves round the annulus varies with the parameters concerned—the geometry and the non-dimensional forms of the temperature difference and rotation rate. The waves can also develop less regular and/or less steady forms, either because the flow parameters are close to the transition from one wave number to another or because the whole wave pattern itself becomes unstable.

The instability generating the waves is often known as baroclinic instability because the slope of the constant density surfaces to the horizontal is a crucial aspect of its dynamics. Correspondingly the waves themselves are called baroclinic waves.

17

INSTABILITIES

17.1 Introduction

Figures 2.7, 3.5, 4.4, 14.6, 15.3, 16.9, and 16.17 have all shown flows undergoing transition from one type of motion to another as a result of instability of the former, and there have been a number of other references to such transitions. In Section 4.4, we considered an important case in which instability caused a transition from a state of rest to a state of motion. Instabilities are pervasive in fluid dynamics. We need to consider how they are to be understood and predicted and what are their consequences.

The central notion (expounded more fully in Section 17.4) is that a configuration is unstable when small perturbations to it tend to be amplified. Often one can identify the essential features of the destabilizing mechanism (as we did for example in the case of Bénard convection (Section 4.4) where it is just the fact that cold heavy fluid is overlying hot light fluid). When one can do this, the physical insight so given is very valuable. However, in some cases the dynamics of the destabilizing process are quite subtle; one certainly cannot identify all configurations that may become unstable just by simple physical arguments. In any case, even when one can do this, one knows only that the configuration is liable to become unstable, not quantitatively when it will do so. Theories of fluid dynamical stability thus constitute a major branch of the subject. Part of the aim of the present chapter is to give an introduction to linear stability theory.

The mathematical complexity of even this simplest type of theory is too great for us to treat any particular case in full. However, some (*only some*) of the principles of the method and of the features of the results can be illustrated by considering highly simplified systems. We look at one such system in Section 17.3. Then, after a general discussion in Section 17.4 of how the principles so illustrated are extended for a full fluid dynamical situation, we discuss rotating Couette flow instability and shear flow instability; these are important both in their own right and as examples of how one gains an understanding of instabilities. The discussion includes (in Sections 17.5 and 17.7) examples of results of linear stability analyses and their interpretation; further examples will be mentioned when we consider Bénard convection and double diffusive convection in Chapters 22 and 23.

An initial instability is normally the first stage of a sequence of changes in the flow (sometimes as one goes downstream in a given flow and sometimes as one increases a governing non-dimensional parameter). The usual final result of this sequence is that the flow becomes turbulent. The present chapter includes some description of various consequences of the instabilities, but transition to turbulence will be discussed in other chapters—Chapter 18 for shear flows and Chapter 22 for Bénard convection.

17.2 Surface tension instability of a liquid column

We preface the topics outlined above by looking at one other case of instability, as a reminder of the sort of real problem with which we are concerned before moving to the highly simplified model in Section 17.3. It comes from a branch of fluid mechanics—flows with free surfaces—generally outside the scope of this book, but it relates to familiar observations and the instability has a simple physical explanation.

The basic state is a cylindrical column of liquid held together by the action of surface tension. As time proceeds, such a column develops corrugations in its shape and ultimately breaks into discrete drops [147, 180, 188].

Theoretically, one may consider the column to be initially stationary; we are dealing with the instability of a rest configuration. In practice, of course, such a column cannot be supported. The instability is investigated experimentally in jets of liquid emitted from a circular hole. If the stress exerted on the jet by its surroundings is negligible its velocity is uniform both across and along the jet. The instability develops as the fluid travels away from the hole, resulting in a break up some distance downstream. (Such jets differ from those considered in Section 11.6, in that we were there concerned with jets emerging into the same ambient fluid—for example air into air or water into water—whereas we are now concerned with jets emerging into surroundings with which there is negligible interaction—for example water into air.)

The destabilizing mechanism may be understood in terms of Fig. 17.1, showing the type of disturbance that grows. The column retains its circular cross-section but with diameter varying along its length. The

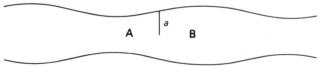

Fig. 17.1   Instability of water column—schematic.

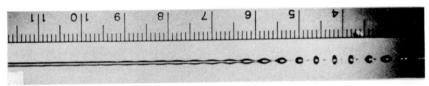

FIG. 17.2 Initial development of capillary instability of water jet. From Ref. [180].

changed surface curvature will produce a change in the pressure within the jet needed to balance the action of surface tension. Provided that the wavelength is long enough the azimuthal curvature dominates over the longitudinal curvature; i.e. the radius of curvature is effectively the local radius a. Hence, the pressure is a maximum where the radius is a minimum (at, e.g., A in Fig. 17.1) and a minimum where the radius is a maximum (at, e.g., B). The pressure gradient thus pushes fluid in directions that amplify the original disturbance. (More precisely, in an inviscid situation, it accelerates fluid in a direction that increases the velocity associated with the disturbance. In a viscous situation it works against the viscous force in a way that increases the displacement associated with the disturbance.)

An alternative, equivalent, formulation of the physical cause of the instability may be made in terms of the fact that a disturbance of sufficiently long wavelength that conserves the volume of the jet decreases its surface area, and thus its surface energy.

A full theory of the instability, on the principles to be explained in Section 17.4, has been worked out [43] but we shall not be considering this.

Figure 17.2 shows an instability of this type developing in a water jet. The highly regular wavelength has been achieved in this laboratory investigation of the instability by deliberately introducing a disturbance, periodic in time, at the orifice.

The break-up of the jet into discrete drops, when the instability amplitude has become large, can be seen at the right-hand side of Fig. 17.2. The drops can form various quite complicated, but repeatable, patterns, of which an example is shown in Fig. 17.3.

FIG. 17.3 Example of subsequent development of instability in Fig. 17.2. From Ref. [180].

Although, in these pictures, the instability was promoted by a periodic disturbance, this is not an essential feature; break-up of any such jet will occur spontaneously. This relates to the break-up of a column of water from a tap (even when the flow is fast enough for the fractional changes in velocity produced by gravitational acceleration to be negligible) and to the familiar observation that the surface of water poured from a kettle is wavy. (In many 'domestic' examples, the situation is complicated by the jet not having an initially circular cross-section, in which case one does not start with an equilibrium configuration. However, the flow from a laboratory tap with a nozzle for fitting rubber tubing, for example, may correspond fairly closely to the ideal situation.)

17.3 Convection in a loop and stability of the Lorenz equations

We come now to the illustrative model mentioned in Section 17.1. Consider a vertical torus—something like the shape of an inflated bicycle inner tube—containing fluid (Fig. 17.4). This is heated at the bottom and cooled at the top in a way that does not introduce any asymmetry between the left and right sides. The fluid is thus light at the bottom and heavy at the top and might be expected to convect round the torus. However, there is no asymmetry to determine whether the circulation so arising will be clockwise or anticlockwise. That will be determined by whatever small uncontrolled disturbances happen to be present when the heating and cooling are initiated. This implies that there is an equilibrium configuration with the fluid remaining at rest. Convection arises through that equilibrium being unstable.

If dissipative processes, such as the action of viscosity, are included in the model, then the equilibrium configuration may remain stable until the heating and cooling are sufficiently strong to overcome them. One now

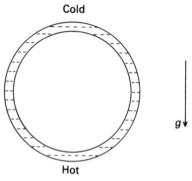

FIG. 17.4   Convection in a torus.

has a non-trivial problem, requiring some formal analysis, in deciding whether convection will occur under a given set of conditions.

There is also the possibility that a steady circulation is set up as a result of the above instability but that, with further increase in the heating and cooling, this too becomes unstable. Some more complicated behaviour will then result. Again, whether and when this happens cannot be predicted without theoretical analysis.

These questions are examined below by simplifying the problem to a set of three ordinary differential equations (eqns (17.17)–(17.19)). This obviously involves substantial assumptions and approximations. The stability analysis is much simpler and shorter than those arising typically in fluid dynamics. It does, however, provide an effective illustration of the principles and some aspects of the procedures of linear stability theory.

The principal difference between this theoretical model and any fully treated problem is the reduction from the start to ordinary differential equations in place of the partial differential equations usual in fluid dynamics. This means that the part of the analysis concerned with discovering which of many possible different patterns of motion actually arises is not exemplified by the model. We will discuss that matter in other contexts from Sections 17.4 onwards. In the present case, the simplifying assumptions essentially prescribe the spatial structure of any motion and the analysis is concerned entirely with its evolution in time. (For example, in the physical system, there is obviously the possibility that, once a circulation is established in the loop, then the velocity profile might be subject to shear instability, Section 17.6. In the model, the simplifications exclude this possibility.)

The set of equations considered is a particular case of a set known as the Lorenz equations. These have played a central role in the development of some modern ideas relating to transition to turbulence and so will be discussed further in Chapter 24. Historically, the Lorenz equations did not originate in the system of Fig. 17.4 (see Section 24.2). After it was realized that solutions of the equations had surprising properties of great significance, people looked for simple systems that might be approximately described by the equations. The system of Fig. 17.4 is a convenient context in which to visualize mentally the various types of behaviour discovered.

Experiments have been carried out with arrangements similar in essentials to Fig. 17.4 both before the theory was developed [133] and specifically to try to observe some of the theoretical results, particularly those in Section 24.2 [184]. Some of the observations can be understood through the theory, but there is not complete correspondence between theoretical and experimental results. This is not a cause for concern. The purpose of the theory is not to predict the behaviour of a particular system but to show various properties of the solutions of differential equations—properties that have much more general significance. The theory is not the sort that requires experimental verification. The discrepancies are explained by the physical situation being more complicated than the mathematical model. Experimental work can, however, provide most informative demonstrations.

We now see how a more precise specification of the system of Fig. 17.4, together with certain approximations, leads to the equations concerned. Various quantities involved are defined in Fig. 17.5. In particular, position round the loop is indicated by the angle ϕ. The external temperature is taken to vary linearly with height:

$$T_E = T_0 - T_1 z/a = T_0 + T_1 \cos \phi. \tag{17.1}$$

Within the loop, it is supposed that the variations in velocity and temperature over a cross-section may be handled by working in terms of an average speed q and average temperature T, which are functions of ϕ and time only:

$$q = q(\phi, t), \quad T = T(\phi, t) \tag{17.2}$$

(q is taken positive when flow is anticlockwise.) The Boussinesq approximation is made and so the continuity equation is simply

$$\partial q / \partial \phi = 0. \tag{17.3}$$

Thus q depends on time only. The temperature variations round the loop are taken to have the form

$$T - T_0 = T_2 \cos \phi + T_3 \sin \phi. \tag{17.4}$$

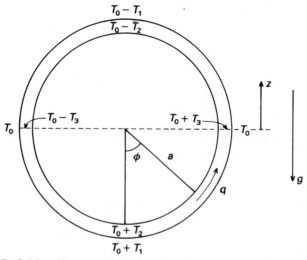

FIG. 17.5 Definition diagram. Note that T_0 and T_1 are imposed quantities, but T_2 and T_3 are variables of the convection.

The temperature difference between top and bottom is thus $2T_2$ and that between the two sides at mid-height is $2T_3$. T_2 and T_3 are, in general, functions of time.

The Navier–Stokes equation (14.9) is taken as

$$\frac{\partial q}{\partial t} = -\frac{1}{\rho a}\frac{\partial p}{\partial \phi} + g\alpha(T - T_0)\sin\phi - \Gamma q \qquad (17.5)$$

where only the last term requires comment: the action of viscosity is supposed to produce a resistance to the motion proportional to its speed; this would be the case, for example, if q is the average speed of a Poiseuille-type profile. ($\boldsymbol{u}\cdot\nabla\boldsymbol{u}$ is, of course, zero because of eqn (17.3).) There can be no pressure change when one goes right round the loop and so, after substitution of (17.4), integration of (17.5) with respect to ϕ gives

$$2\pi\frac{dq}{dt} = g\alpha\int_0^{2\pi}(T_2\cos\phi\sin\phi + T_3\sin^2\phi)\,d\phi - 2\pi\Gamma q. \qquad (17.6)$$

(Here $\partial q/\partial t$ is replaced by dq/dt because of eqn (17.3).) This gives

$$dq/dt = -\Gamma q + g\alpha T_3/2. \qquad (17.7)$$

As is to be expected, it is the horizontal temperature difference that drives the motion.

The temperature T is taken to be determined by the equation

$$\frac{\partial T}{\partial t} + \frac{q}{a}\frac{\partial T}{\partial \phi} = K(T_E - T) \qquad (17.8)$$

(as an approximation to eqn (14.12) with $J = 0$). The second term is a point at which the model is likely to differ from a real system; the average over a cross-section of $u_\phi\,\partial T/\partial\phi$ will not normally be the same as the product of the average of u_ϕ ($=q$) and the average of $\partial T/\partial\phi$. The right-hand side of eqn (17.8) supposes that heat is transferred through the walls at a rate proportional to the local difference between the external temperature and the average internal temperature. In both these terms, the assumption is really that there is good heat transfer over a cross-section so that the temperature does not vary too much, although there is negligible conductive heat transfer round the loop.

Substitution of (17.1) and (17.4) into (17.8) gives

$$\frac{dT_2}{dt}\cos\phi + \frac{dT_3}{dt}\sin\phi - \frac{qT_2}{a}\sin\phi + \frac{qT_3}{a}\cos\phi$$
$$= K(T_1 - T_2)\cos\phi - KT_3\sin\phi. \qquad (17.9)$$

Since this has to be true at all ϕ,

$$dT_3/dt - qT_2/a = -KT_3 \qquad (17.10)$$
$$dT_2/dt + qT_3/a = K(T_1 - T_2). \qquad (17.11)$$

Equations (17.7), (17.10), and (17.11) are basically the three ordinary differential equations in terms of which the dynamics are going to be studied. It is, however, convenient to introduce

$$T_4 = T_1 - T_2 \tag{17.12}$$

as a dependent variable in place of T_2. T_4 is the amount by which the fluid temperature differs from the external temperature at the top or bottom of the loop. Since T_1 is independent of time, eqns (17.10) and (17.11) become

$$dT_3/dt = -KT_3 + qT_1/a - qT_4/a \tag{17.13}$$
$$dT_4/dt = -KT_4 + qT_3/a. \tag{17.14}$$

The three dependent variables q, T_3, and T_4 are now expressed as respectively the non-dimensional variables X, Y, and Z where

$$X = q/aK; \quad Y = g\alpha T_3/2a\Gamma K; \quad Z = g\alpha T_4/2a\Gamma K \tag{17.15}$$

and time is non-dimensionalized as

$$t' = tK. \tag{17.16}$$

Equations (17.7), (17.13), and (17.14) become

$$dX/dt = -PX + PY \tag{17.17}$$
$$dY/dt = -Y + rX - XZ \tag{17.18}$$
$$dZ/dt = -Z + XY \tag{17.19}$$

where the quantity temporarily denoted by t' has now been written as t and

$$P = \Gamma/K; \quad r = g\alpha T_1/2a\Gamma K. \tag{17.20}$$

The parameters P and r are, of course, non-dimensional. The latter has in its numerator quantities associated with the causes of convection (the temperature difference between top and bottom, expansion, and gravity) and in its denominator quantities related to the action of viscosity and heat conduction. It is thus the counterpart for this model of the Rayleigh number for a fully treated convection problem. P is similarly the counterpart of the Prandtl number.

Equations (17.17), (17.18), and (17.19) are three simultaneous equations governing how the three variables X, Y, and Z evolve in time in terms of their present values. It is these equations that we are going to use to illustrate the nature of a stability analysis.

We first look for steady-state solutions,

$$dX/dt = dY/dt = dZ/dt = 0, \tag{17.21}$$

of the equations, and these are easily found to be

$$X = Y = Z = 0 \tag{17.22}$$

and

$$X = Y = \pm(r-1)^{1/2}; \quad Z = r - 1 \qquad (17.23)$$

For $r < 1$, only the first of these exists (we are interested only in real solutions); for $r > 1$, there are the two additional solutions (17.23).

Solution (17.22) corresponds to the fluid remaining at rest in the torus with its temperature the same as the external temperature at the same level. The two solutions (17.23) correspond to the fluid circulating in the loop at a constant speed and with a constant temperature distribution. The different signs in the two solutions relate to the two senses of circulation; X and Y have the same sign, so that in each case hot fluid is rising and cold fluid falling.

The stability of these solutions is investigated by superimposing a perturbation and examining its development in time. Thus we write

$$X = X_S + x, \quad Y = Y_S + y, \quad Z = Z_S + z \qquad (17.24)$$

where (X_S, Y_S, Z_S) is the steady solution of which the stability is being investigated and (x, y, z) is the perturbation. Substituting into eqns (17.17)–(17.19) and cancelling some terms because (X_S, Y_S, Z_S) is itself a solution of the equations gives

$$dx/dt = -Px + Py \qquad (17.25)$$

$$dy/dt = -y + rx - xZ_S - X_S z(-xz) \qquad (17.26)$$

$$dz/dt = -z + xY_S + X_S y(+xy). \qquad (17.27)$$

The final terms of (17.26) and (17.27) are put in brackets because the next stage is to drop them. This is the step that characterizes the stability analysis as a linear one. The bracketed terms are proportional to the square of the perturbation amplitude, whereas all other terms on the right-hand sides are proportional to the amplitude itself. Dropping the former means that we are considering only a small perturbation; we shall discuss this step more fully in the more general context of Section 17.4.

The equations now provide expressions for dx/dt, dy/dt, and dz/dt, each consisting of a sum of terms proportional to one of x, y, or z. One may thus anticipate that there are solutions of the form

$$x = x_0 \exp(\sigma t), \quad y = y_0 \exp(\sigma t), \quad z = z_0 \exp(\sigma t). \qquad (17.28)$$

Substitution of (17.28) into eqns (17.25)–(17.27) (without the bracketed terms) and division throughout by $\exp(\sigma t)$ leads to

$$-(\sigma + P)x_0 + Py_0 = 0; \qquad (17.29)$$

$$(r - Z_S)x_0 - (\sigma + 1)y_0 - X_S z_0 = 0; \qquad (17.30)$$

$$Y_S x_0 + X_S y_0 - (\sigma + 1)z_0 = 0. \qquad (17.31)$$

These three equations form a set of linear, homogeneous equations in (x_0, y_0, z_0) and are mutually consistent if and only if the determinant of their coefficients is zero:

$$\begin{vmatrix} -\sigma - P & P & 0 \\ r - Z_S & -\sigma - 1 & -X_S \\ Y_S & X_S & -\sigma - 1 \end{vmatrix} = 0. \tag{17.32}$$

To investigate the stability of the steady-state solution without any circulation, we put

$$X_S = Y_S = Z_S = 0 \tag{17.33}$$

(cf. (17.22)). Then (17.32) becomes

$$(\sigma + 1)[\sigma^2 + \sigma(P + 1) - P(r - 1)] = 0 \tag{17.34}$$

with the three roots

$$\sigma = -1 \quad \text{or} \quad \sigma = \tfrac{1}{2}\{-(P + 1) \pm [(P + 1)^2 + 4P(r - 1)]^{1/2}\}. \tag{17.35}$$

Associated with each root is a particular pattern of perturbation; i.e. one can determine the ratios y_0/x_0 and z_0/x_0.

However, it is σ that is of most interest as indicating how the perturbation develops in time. In general, σ may be complex:

$$\sigma = \sigma_r + i\sigma_i. \tag{17.36}$$

Referring to eqn (17.28), positive σ_r indicates that the perturbation grows in time, and the original configuration is thus unstable with respect to it. Negative σ_r indicates that the perturbation dies away and the original configuration is stable with respect to it. When, as in the present case, there are several roots, the requirement for stability is that σ_r should be negative for all of them; if $\sigma_r > 0$ for any one root, then one has instability: if an unstable mode exists then it will tend to be amplified, regardless of the behaviour of other modes. When $\sigma_r < 0$, then σ_i is of little interest. When $\sigma_r > 0$, then σ_i is significant as an indication of the nature of the developing instability, as discussed below.

With eqns (17.35), there are no cases in which σ is complex with a positive real part. The three real values of σ are all negative when $r < 1$, but one of them becomes positive when $r > 1$. Hence, the equilibrium rest solution (17.22) is stable when $r < 1$, unstable when $r > 1$. In the latter case, it cannot actually occur; a transition must occur to some alternative behaviour.

We have found that increasing r through $r = 1$ produces two changes—the appearance of the additional steady state solutions (17.23) and the onset of instability of the rest configuration. These are two separate results; the second does not automatically follow from the first.

It is also important to note that, although the above analysis predicts the instability when $r > 1$, it gives only limited information about the consequences of the instability. The immediate consequence is an exponentially growing perturbation. (Because $\sigma_i = 0$, it just grows monotonically; cf. the behaviour for $\sigma_i \neq 0$ discussed below.) So far as the analysis is concerned the growth continues indefinitely. However, as the perturbation gets larger the neglect of the non-linear terms (the bracketed terms of eqns (17.26) and (17.27)) becomes less valid. In due course these terms significantly alter the way in which the perturbation evolves. The above theory gives no information about the form of this alteration (a point which will be discussed in a more general context around Fig. 17.6). The existence of the alternative steady-state solutions (17.23) makes it plausible that the perturbation may evolve towards one of these, and in Section 24.2 we shall mention numerical work that shows that this is indeed the case provided $(r - 1)$ is not too large. However, since we are concerned here with illustrating the significance of a linear stability analysis, it is to be emphasized that the above statement does not follow from this alone.

Also, of course, the system can settle down to one of the solutions (17.23) only if this is itself stable. We next need to investigate this. (The symmetry of the two solutions (17.23) makes it evident that these will be either both stable or both unstable, and only a single treatment is needed.) Hence, we take

$$X_S = Y_S = \pm(r - 1)^{1/2}, \quad Z_S = r - 1 \qquad (17.37)$$

(cf. (17.23)). Substitution into (17.32) gives

$$\sigma^3 + \sigma^2(P + 2) + \sigma(P + r) + 2P(r - 1) = 0. \qquad (17.38)$$

This is of the form

$$\sigma^3 + A\sigma^2 + B\sigma + C = 0, \qquad (17.39)$$

with the coefficients A, B, and C all real and positive (since the solution (17.37) applies only for $r > 1$). Such an equation has either three real roots or one real root and two conjugate complex roots. It can immediately be seen that any real root is necessarily negative, since every term in (17.38) is positive for positive σ. It is thus the complex roots that require consideration. Hence, we consider the case where the three roots are $\sigma = \sigma_1$ (real and negative) and $\sigma = \alpha \pm i\beta$. It can be shown† that the sign of α is the same as the sign of $(C - AB)$. Hence, instability occurs when

$$C - AB > 0 \tag{17.40}$$

i.e. when

$$2P(r - 1) - (P + 2)(P + r) > 0 \tag{17.41}$$

or

$$r(P - 2) - P(P + 4) > 0. \tag{17.42}$$

If $P < 2$, inequality (17.42) is never satisfied, and the steady-state solution is always stable. If $P > 2$, then it becomes unstable when $r > r_c$ where

$$r_c = P(P + 4)/(P - 2). \tag{17.43}$$

The appearance of a critical value of a governing non-dimensional parameter at which a configuration becomes unstable is highly characteristic of stability analyses. Note that $r_c > 1$ for all P. Hence, there is a range of r for which a steady flow with non-zero X, Y, and Z can occur. Beyond this range $(r > r_c)$, the steady solution still exists but will not occur because it is unstable. In the absence of any stable steady solutions, some sort of unsteady motion must occur.

The fact that $\sigma_i \neq 0$ for this instability implies that it has a different character from the instability of the rest configuration. Instead of growing monotonically, the instability consists of oscillations of increasing amplitude. Growing pulsations of the flow are superimposed on the mean circulation. As with the previous instability, analysis of the linearized equations predicts the growth of the perturbation to continue indefinitely, but this will be modified when the non-linear terms become significant; i.e. when the pulsations are no longer small compared with the mean circulation. One has to analyse the full equations to find out what actually happens when $r > r_c$. We shall in fact be looking at this in Section 24.2, but for the moment we leave the matter; we have now covered the points that help us to understand ideas in the present chapter.

† Comparison of

$$(\sigma - \sigma_1)(\sigma - \alpha - i\beta)(\sigma - \alpha + i\beta) = 0$$

with (17.39) leads to

$$A = -(\sigma_1 + 2\alpha); \quad B = 2\alpha\sigma_1 + \alpha^2 + \beta^2; \quad C = -\sigma_1(\alpha^2 + \beta^2).$$

After a little manipulation this gives

$$C - AB = 2\alpha[(\sigma_1 + \alpha)^2 + \beta^2].$$

The quantity in square brackets is necessarily positive and so $(C - AB)$ and α have the same sign. The argument can be extended to show that, when $(C - AB) > 0$, the three roots cannot all be real.

17.4 Principles of the linear theory of hydrodynamic stability [42, 43]

The simplified model considered in the last section has provided some valuable insights into the concept of stability and about how one investigates whether a system is stable or not. However, the representation of the problem by ordinary, not partial, differential equations means that it does not contain some very important aspects of a full fluid dynamical stability analysis. We now need to generalize, although we still confine attention to linear stability theory.

The ideas of the present section may seem somewhat vague. Their significance is most readily seen through examples and these will be provided by Sections 17.5 and 17.7.

The underlying notion is that transition from one type of flow to another results from spontaneous amplification of disturbances present in the original flow. One would then expect the occurrence of the transition to depend on the intensity and structure of disturbances present, and this is frequently found to be the case. Clearly, a theory covering all possibilities would be unmanageably complex;† moreover, comparison with experiment would be difficult since the nature of the disturbances is often not in complete experimental control.

Linear stability theory adopts a less ambitious objective: to ascertain when a flow is unstable to infinitesimal disturbances. It thus gives no prediction about transition promoted by sufficiently large disturbances; this may occur when the theory indicates stability. On the other hand, it should indicate conditions in which the flow cannot remain in its postulated form and must undergo transition to another type of motion. Infinitesimal disturbances are always present, even in the most carefully controlled experiment, and if these tend to be amplified, the flow will necessarily break down.

The approach is analogous to that used to understand instability of solid mechanical systems. A pin stood on its point with its centre of gravity directly above the point is in equilibrium. This equilibrium is, however, unstable because even the smallest displacement of the pin will be amplified. Since infinitesimal disturbances can never be eliminated, the pin always falls.‡

The linear stability theory for a particular flow starts with a solution (or approximate solution) of the equations of motion representing the flow. One then considers this solution with a small perturbation super-

---

† There is also the rather vexed question of how large a 'disturbance' can be and still be regarded as a disturbance rather than a change in the basic flow.
‡ Buridan's ass, which died of starvation because it was exactly midway between a bunch of hay and an equally attractive pail of water, should have found a similar resolution to its problems.

imposed, and enquires whether this perturbation grows or decays as time passes. In the way already illustrated in Section 17.3, all terms involving the square of the perturbation amplitude are neglected; it is this that limits the theory to infinitesimal disturbances.

The linearization provides a means of allowing for the many different forms that the disturbances can take. (This aspect is not illustrated at all by Section 17.3.) Any pattern of disturbance may be Fourier analysed spatially. Because of the linearity, there are no interactions between different Fourier components. The equations may thus be broken down into separate sets of equations for each Fourier component, indicating the stability or instability with respect to that component.

The exact formulation of the Fourier analysis varies from problem to problem (we shall be seeing examples in Sections 17.5 and 17.7); but, of course, it always means that one considers spatially sinusoidal perturbations. Also extension of the ideas in Section 17.3 suggests that the evolution in time will be of the form $\exp(\sigma t)$. Hence, one considers a perturbation of the form

$$\Delta Q \propto \exp[i\boldsymbol{k} \cdot \boldsymbol{r} + \sigma t] \qquad (17.44)$$

(Q being any quantity associated with the flow). As before,

$$\sigma = \sigma_r + i\sigma_i \qquad (17.45)$$

with the sign of σ_r indicating whether the disturbance grows or decays in time.

All Fourier components can, of course, be treated in a single mathematical operation, by handling the wavenumber $\boldsymbol{k}$ (indicating the component) as a parameter. Any particular Fourier component will be present to some extent in a general disturbance. If σ_r is negative for all values of $\boldsymbol{k}$, the original flow is stable to all infinitesimal disturbances. This is a necessary condition for stability. If σ_r is positive for some values of $\boldsymbol{k}$, the corresponding perturbation will be spontaneously amplified. This is a sufficient condition for instability.

Two points arising in Section 17.3 are of significance throughout stability theory. Firstly, disturbance growth as predicted by linear theory is exponential. In time the original assumption that the perturbation was small will break down. Non-linear terms in the equations will become important and alter the exponential dependence on time. The ultimate consequence of the instability is never completely determined by linear theory. The departure from exponential growth may take the form either of a levelling off (as in curve A of Fig. 17.6) or of a still faster growth (curve B). In the former case, as we shall see in Section 17.5, linear theory may in fact give a good indication of the final pattern of motion, although one cannot be sure that it will. When the more catastrophic

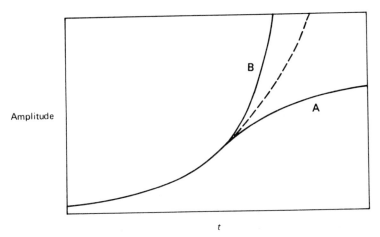

FIG. 17.6 Schematic diagram of types of departure from exponential growth.

curve B applies, there must evidently be further developments before a new type of motion is established and there is likely to be little resemblance between the form of the first instability and its ultimate consequence.

Secondly, cases with both $\sigma_i = 0$ and $\sigma_i \neq 0$ arise commonly in hydrodynamic stability theory. The distinction is an important one because it implies that the perturbation evolves in different ways. When $\sigma_i = 0$ (and $\sigma_r > 0$) it just grows continuously. When $\sigma_i \neq 0$, the development of an instability takes the form of an amplifying sinusoid as sketched in Fig. 17.7. Correspondingly, the resulting motion if non-linear effects limit the growth is an oscillatory motion. This type of behaviour is frequently known as overstability (although the name is not a good summary of the physical processes involved). Overstability tends to occur when there are present simultaneously a destabilizing influence and a feature that gives rise to wave motions, such as stratification (Section 15.4) or rotation (Section 16.6). Thus it is found for example in a horizontal fluid layer heated from below and rotating about a vertical axis

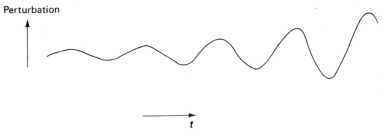

FIG. 17.7 Overstability.

[168]. It is also found in some of the double diffusive systems to be introduced in Chapter 23, and Fig. 23.5 will show experimentally observed oscillations arising from overstability.

(We shall also be meeting a situation with $\sigma_i \neq 0$ when we consider boundary layer instability in Section 17.7. However, this is due to a different type of process, a spatially periodic disturbance being advected past the point of observation, and is not normally classified as overstability.)

The above remarks about the limited relevance of linear theory to the later evolution of a growing disturbance do not imply that theory gives up at this point. Non-linear stability theory is a very extensive branch of fluid dynamics [43, 144, 218], but it is outside the scope of this book. It is nevertheless true that the relative contributions of theory and experiment to our knowledge of what happens change in favour of the latter as one follows a flow further and further beyond the first onset of instability.

In any case the extent to which even linear theory can be said to account for observed transitions from one type of motion to another varies greatly from flow to flow. We shall be meeting: cases where the agreement between linear theory and observation is strikingly good (rotating Couette flow, Section 17.5, and Bénard convection, Section 22.2); cases where agreement is not immediately apparent but where good verification of the theory is obtainable by special methods (exemplified particularly by Section 17.7 on boundary layers); and a case where the theory fails to predict the observed transition to turbulence (pipe flow, Section 18.3).

17.5 Rotating Couette flow [145]

We return to the flow between two concentric cylinders rotating at different rates (Fig. 9.1), considered in Section 9.3. There we looked at the solution applicable when the flow remains entirely azimuthal. However, this simple flow does not always occur; instability may lead to a more complex pattern developing. Four examples of such patterns are shown in Fig. 17.8, the first three being various regimes observed when the outer cylinder is at rest ($\Omega_2 = 0$) and the fourth a more complicated case to which we shall return at the end of the section.

The flow in Fig. 17.8(a) is of interest as the first development from the purely azimuthal flow as the speed of the inner cylinder is increased—not only when $\Omega_2 = 0$, but also under a wide range of conditions when the two cylinders rotate in the same sense and a more limited range when they rotate in opposite senses. The nature of this flow is shown schematically in Fig. 17.9. Vortices of alternate senses circulate between

FIG. 17.8 Flow patterns in rotating Couette flow—see text for details. Aluminium powder flow visualization (see Section 25.4). From Ref. [128].

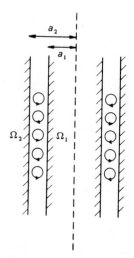

FIG. 17.9   Arrangement of Taylor cells.

the two cylinders. Each vortex extends toroidally right round the annulus, so the overall flow is axisymmetric. These vortices are known as Taylor cells.

(Only if $|\Omega_1 - \Omega_2| \ll \Omega_1$ is this a problem in rotating fluids in the sense of Chapter 16. The following discussion is thus phrased in terms of a non-rotating frame of reference. The centrifugal force considered is that acting on a particle moving on a circular path relative to this frame. The case of small relative rotation could be discussed using a rotating frame, in which case the action attributed here to the centrifugal force would appear as the action of the Coriolis force.)

The nature of the instability producing Taylor cells may be understood by considering a toroidal element of fluid (i.e. in cylindrical polar coordinates, the fluid between r and $r + dr$ and between z and $z + dz$ but at all values of ϕ). Suppose this is displaced to a slightly larger radius. If it is now rotating faster than its new environment, the radial pressure gradient associated with the basic flow will be insufficient to balance the centrifugal force associated with the displaced element. The element will then tend to move still further outwards. Similarly, an element displaced to a slightly smaller radius will tend to move still further inwards. There is thus an instability associated with some distributions of angular velocity. (The argument cannot be applied to a fluid particle localized in the ϕ-direction, as the displacement of this would introduce azimuthal pressure variations.)

The argument may be put into a quantitative form to indicate the condition for instability in the absence of the stabilizing action of viscosity; i.e. the condition analogous to the requirement for Bénard instability that the temperature must decrease with height. An elemental toroid, initially at radius ζ, and circulating at angular velocity Ω_ζ, is supposed displaced to radius η without interacting with the remainder of the fluid. Its angular momentum is then conserved and its angular velocity after displacement is

$$\Omega'_\zeta = \Omega_\zeta \zeta^2 / \eta^2. \tag{17.46}$$

Its centrifugal force will exceed that of undisturbed fluid at η, circulating with angular velocity Ω_η, if

$$|\Omega'_\zeta| > |\Omega_\eta|. \tag{17.47}$$

Hence, there is instability if

$$|\Omega_\zeta \zeta^2| > |\Omega_\eta \eta^2| \quad \text{when} \quad \eta > \zeta \tag{17.48}$$

that is if

$$\frac{\mathrm{d}}{\mathrm{d}r} |\Omega r^2| < 0. \tag{17.49}$$

This is known as the Rayleigh criterion for the instability of Couette flow [42].

When the two cylinders are rotating in the same sense the flow is either stable everywhere or unstable everywhere. Substituting in eqn (9.6), criterion (17.49) then becomes

$$\Omega_1 a_1^2 > \Omega_2 a_2^2. \tag{17.50}$$

When the two cylinders are rotating in opposite senses, the region close to the inner cylinder is unstable and that close to the outer cylinder is stable.

The Rayleigh criterion, being derived from a simple 'displaced particle' argument, is, of course, only a first indication of when the azimuthal flow is liable to instability. In particular it will be modified by the action of viscosity. To find out just when instability occurs we require a full linear stability theory along the lines of Section 17.4.

We consider this theory for the case $\Omega_2 = 0$ (although we shall later look briefly at the results of the corresponding theory for the more general case—Fig. 17.11). One has to specify at the outset the geometry in terms of the radius ratio a_2/a_1. For theoretical purposes, at least in the simplest analysis, the system is supposed to be infinitely long in the axial direction. Dynamical similarity is then determined by a Reynolds number which is conveniently based on the surface velocity of the inner cylinder and the gap width,

$$\mathrm{Re} = \Omega_1 a_1 (a_2 - a_1) / \nu. \tag{17.51}$$

The starting point for the analysis is the flow whose stability is being investigated; i.e. the velocity and pressure fields given by eqns (9.6) and (9.4) (with $\Omega_2 = 0$). A small perturbing velocity and pressure field is superimposed on this. The perturbation must, of course, satisfy the continuity equation and the boundary conditions. In accordance with the ideas in Section 17.4 it is taken to be of the form†

$$\Delta u = U(r) \exp[i(m\phi + kz) + \sigma t] \tag{17.52}$$

$$\Delta p = P(r) \exp[i(m\phi + kz) + \sigma t]. \tag{17.53}$$

The Fourier analysis is thus performed in the azimuthal (ϕ) and axial (z) directions. It is necessary, in order to satisfy all the imposed conditions, including the boundary conditions, to allow the perturbation to have general distributions in the radial direction; $U(r)$ and $P(r)$ are determined as part of the solution. m is an integer, since $\phi = 0$ and $\phi = 2\pi$ are identical. On the other hand, k may take any value (at least in the case of the theoretical infinitely long apparatus).

The linearized equations of motion are used to investigate the development of this perturbation with time, as indicated by σ. One considers first the case

$$m = 0 \tag{17.54}$$

i.e. the case in which the perturbation is axisymmetric. The analysis is lengthy and we just quote results [42, 43]. It is found that σ is always real,

$$\sigma_i = 0, \tag{17.55}$$

with implications discussed in Section 17.4. The results on the sign of σ may be presented on a graph of the non-dimensional wave number,

$$\xi = k(a_2 - a_1) \tag{17.56}$$

against the Reynolds number (17.51). This takes the form shown in Fig. 17.10.

For low values of Re, σ is negative for all ξ. All disturbances with $m = 0$ tend to die away. Since, as we shall note below, disturbances with $m \neq 0$ also die away, the original azimuthal flow is stable. For higher values of Re there is a positive σ solution for some values of ξ. Hence, some disturbances are self-amplifying and there is instability. A critical

---

† Because we are concerned more with principles than algebraic details, some simplification has been permitted here. The physical quantities are, of course, the real parts of these complex quantities. U and/or P may be complex to allow for phase differences in the spatially sinusoidal variations of the different quantities (three components of Δu and Δp). When there are sinusoidal variations in two directions, this still does not cover all possibilities, and the expressions should really each be a sum of terms with both signs of $(\pm kz)$.

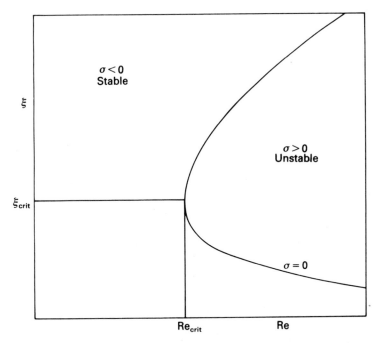

FIG. 17.10 Marginal stability curve for Couette flow. (Details depend on a_2/a_1; shape of curve is closest to that shown for small $(a_2 - a_1)/a_1$.)

Reynolds number $\mathrm{Re}_{\mathrm{crit}}$ is given by the lowest Re at which this is true (Fig. 17.10).

The details of Fig. 17.10 depend on a_2/a_1. As an example we quote the result for the 'small gap' limit, $(a_2 - a_1) \ll a_1$ [42, 145]:

$$\mathrm{Re}_{\mathrm{crit}} = 41.18[a_1/(a_2 - a_1)]^{1/2}. \tag{17.57}$$

Similar analysis for any value of $m \neq 0$ leads in a similar way to a critical Reynolds number. This is always found to be greater than $\mathrm{Re}_{\mathrm{crit}}$ for $m = 0$. Hence, the latter is the lowest Re at which there are any unstable modes and thus the value for the onset of instability.

This is one of the successes of linear stability theory. Which of the purely azimuthal motion or a flow with Taylor cells occurs, in an apparatus long compared with the gap-width, is found to depend on whether Re is below or above $\mathrm{Re}_{\mathrm{crit}}$. The transition can be seen clearly in flow visualization experiments, enabling quite a precise value of the transitional Reynolds number to be determined. A more quantitative approach is to measure the torques acting on the cylinders and observe when departures from eqns (9.8) and (9.9) occur [146] (an analogous use of change in heat transfer to detect the onset of convection cells will be shown in Fig. 22.1.) Excellent agreement is found between the observed transitional Reynolds number and the critical value given by the linear theory.

One can also ask whether the theory makes useful predictions about the flow pattern resulting from the instability. The limitation to linear stability theory noted in Section 17.4 must be remembered: the theory concerns exponential growth in time, and the neglected non-linear terms must become significant before the final effect of the instability is realized. However, this is a case to which curve A of Fig. 17.6 applies, and quite close similarities are found between the unstable flow pattern of linear theory and the ultimate steady state flow pattern occurring when Re is a little above Re_{crit}. Firstly, the fact that $m = 0$ in the relevant theory matches the observation that the Taylor cells are axisymmetric. Secondly, when $(Re - Re_{crit})$ is small, the instability occurs for only a range of ξ around ξ_{crit} (Fig. 17.10), thus indicating the axial length scale of the motion (via eqns (17.52), (17.56)). In fact, since $\xi_{crit} \simeq \pi$ (it is 3.12 for the situation to which (17.57) applies and not very different for other cases), the corresponding motion has almost equal scales in the axial and radial directions, as sketched in Fig. 17.9. The observed Taylor cells are like this.

In fact they remain like this even when Re exceeds Re_{crit} more than marginally [145], provided that Re is increased gradually. Other procedures can lead to cells of markedly different sizes [85]. A whole variety of different steady state flow patterns can be observed, all at the same Re, by different approaches to the steady state.

Increasing Re further above Re_{crit} leads to qualitative changes in the flow pattern due to further instabilities. The details vary with a_2/a_1 [351]. We shall discuss such developments for other flows in some detail in Chapters 18 and 22, so we extend the information already conveyed by Fig. 17.8 only briefly.

Figure 17.8(b) shows Taylor cells that have become wavy as a result of the inner cylinder speed being increased from its value in Fig. 17.8(a). In terms of the above analysis this is a flow with $m \neq 0$. However, its occurrence cannot be related to the fact that the azimuthal flow becomes unstable to such flows at higher Re, because the $m = 0$ instability has previously disrupted that flow. Rather one has to consider whether the fully developed Taylor cell flow is itself stable or unstable [217]. The number of waves around the annulus varies with rotation rate; again the exact pattern occurring for particular boundary conditions depends on how those conditions were approached [128].

Figure 17.8(c) shows a flow at still higher Re and one that occurs over a wide range of Re. The flow is now turbulent, in that the velocity at any point fluctuates irregularly. However, the turbulence occurs within a mean flow of highly ordered structure, similar to the Taylor cell structure of Fig. 17.8(a). The waves of Fig. 17.8(b) are suppressed by the onset of turbulence. The contrast between the spatial regularity and temporal irregularity is a very striking feature of this flow.

At the highest Reynolds numbers the regular spatial structure does disappear, and the flow can then be described as fully turbulent. However, this happens only when Re/Re_{crit} is of order 10^3 (the ratio probably depends on a_2/a_1) [349], whereas the previous developments occur over a range of order 10 in Re ratio.

Apart from the discussion of the Rayleigh criterion all the foregoing applies primarily to the case $\Omega_2 = 0$. Rotation of the outer cylinder as well as the inner leads to an almost daunting variety of flow patterns. We can only touch on these below.

However, specific mention must be made of the results of linear stability theory for the more general case, partly because of its historical importance as the first case in which a detailed comparison was made between such theory and experiment. This was done by Taylor in 1923 [373]. Figure 17.11 shows his comparison for one value of a_2/a_1 $(=1.14)$. The vertical axis corresponds to the case considered above of only the inner cylinder rotating. Points to the right of this correspond to the two cylinders rotating in the same sense, and the dotted line represents the Rayleigh criterion, eqn (17.50). The left-hand part of the figure corresponds to the cylinders rotating in opposite senses, for which the Rayleigh criterion always indicates instability (in part of the annulus).

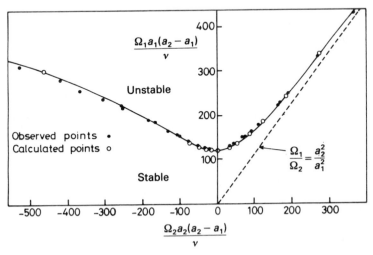

FIG. 17.11 Comparison of theoretical and experimental conditions for instability of rotating Couette flow with $a_2/a_1 = 1.14$. From Ref. [373].

The 'calculated points' are the results of theory like that leading to Fig. 17.10. (Thus the point on the vertical axis of Fig. 17.11 corresponds to Re_{crit} of Fig. 17.10 and is given approximately by eqn (17.57)—not exactly, because $(a_2 - a_1)/a_1$ is not very small.) The 'observed points' correspond to the onset of cellular motion as shown by dye visualization. As expected, there is a range in which the action of viscosity maintains stability while the Rayleigh criterion indicates instability.

Figure 17.12, plotted on the same coordinates as Fig. 17.11 but extending to larger Reynolds numbers, shows a recently proposed summary of observed flow patterns when both cylinders rotate. We cannot attempt here to explain the meaning of all the names in this diagram; it is included mainly just to illustrate the variety of phenomena to be found. Moreover, this diagram applies to a particular value of a_2/a_1 ($=1.13$) and only to flows that were achieved by a gradual increase in speed of first the outer cylinder and then the inner.

Although we are not discussing the general case in detail, one point should be made about transition from purely azimuthal motion ('Couette flow' in Fig. 17.12). In some circumstances, turbulent motion appears suddenly without the preceding stages of development. This occurs principally when the two cylinders are rotating in opposite senses, but the outer one faster, so that the unstable region according to the Rayleigh

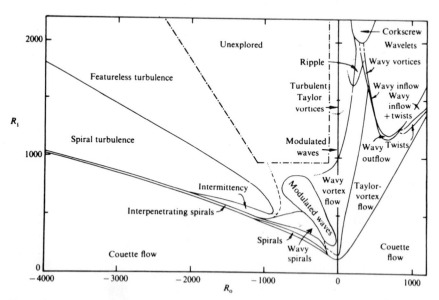

FIG. 17.12 Regime diagram for rotating Couette flow. (Note that the name 'Couette flow' on the diagram denotes the purely azimuthal motion, Section 9.3). From Ref. [61].

criterion is confined to a small fraction of the annulus close to the inner cylinder. However, the turbulent region is not so confined, and it is likely that this instability is associated more with the shear in the annulus than with the action of the centrifugal forces. Its mechanics have thus more in common with the effects to be described in Section 17.6 than with the instability giving Taylor cells. The stage preceding full turbulence is then intermittent turbulence, similar to that observed, for example, in pipe flow (Sections 2.6 and 18.3). For certain ranges of the imposed conditions, this intermittent turbulence is produced by a distinctive structure in which the laminar and turbulent regions form spiral bands as seen in Fig. 17.8(d) [61, 128, 398]. This pattern rotates at approximately the mean angular velocity of the two cylinders, so that, at a fixed position, alternately laminar and turbulent motions are observed. The motion is also alternately laminar and turbulent if one follows the trajectory of a fluid particle, so that continuous transition from laminar to turbulent motion and continuous reverse transition are involved in the mechanics of the structure.

17.6 Shear flow instability

A variety of the most important cases of the instability of fluids in motion fall into the general category of the instability of shear flows. A shear flow is one in which the velocity varies principally in a direction at right angles to the flow direction (Fig. 17.13(a)). The simplest example of this, although it does not correspond immediately to a physical situation, is a flow with a finite discontinuity in the velocity as shown in Fig. 17.13(b). The fact that this is subject to instability—the Kelvin–Helmholtz

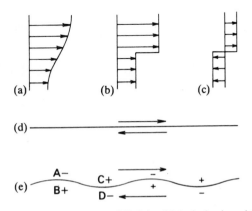

FIG. 17.13   To illustrate the cause of Kelvin–Helmholtz instability (see text).

instability—may be seen as follows. We consider the development of a disturbance in the frame of reference in which the two velocities are equal and opposite (Fig. 17.13(c) and (d)), so that, by symmetry, the disturbance does not travel with the flow. We now suppose that the boundary between the fluid moving in one direction and that in the other becomes slightly wavy (Fig. 17.13(e)). This will make the fluid on the convex side (at, e.g., A and D) move slightly faster and that on the concave side (at, e.g., B and C) slightly slower. In a steady state, Bernoulli's equation would apply. Pressure changes indicated by the + and − signs in Fig. 17.13(e) will be produced by the disturbance. Thus the steady disturbed state is not possible and the pressure gradients are in directions producing amplification of the disturbance. More precisely, the fluid cannot support a pressure discontinuity and the unsteady flow must involve motions that counteract the above action of Bernoulli's equation. These are an acceleration of fluid above A and away from it and an acceleration below B and towards it. Such accelerations amplify the disturbance.

The experimental observations most closely related to this model have been made in stratified fluids (Chapter 15), primarily because the stratification provides a convenient method of producing a shear flow with little variation in the flow direction. A long horizontal channel was filled with liquid, dense at the bottom, lighter at the top. It was then tilted slightly so that the heavier fluid flowed down the sloping bottom wall and the lighter fluid up the sloping top wall, producing a steady stratified shear flow. When the destabilizing mechanism described above is strong enough to overcome the stabilizing action of the stratification, a pattern such as that in Fig. 17.14 results. The regularity of such patterns during their development is noteworthy, but this does subsequently break down giving a turbulent motion.

Examples of shear flow instabilities in the absence of stratification appear in various places in this book. The velocity profiles (Fig. 17.15) of

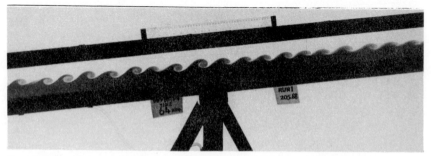

Fig. 17.14 Shear instability in stably stratified fluid, exhibited by the lower denser fluid being dyed. From Ref. [379].

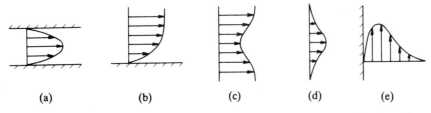

FIG. 17.15 To illustrate that the velocity profiles of (a) pipe flow, (b) a boundary layer, (c) a wake, (d) a jet, and (e) a free convection boundary layer are all shear flows.

pipe flow, a boundary layer, a wake, a jet, and a free convection boundary layer all come into this category. The profiles may be classified into two broad groups: those with a point (or points) of inflexion, such as a wake, a jet, or a free convection boundary layer; and those with no point of inflexion, such as pipe flow or a forced flow boundary layer (in zero or favourable pressure gradient).

For the first group, processes broadly similar to the Kelvin–Helmholtz instability described above come into play and the flow is always unstable at high enough Reynolds number.† The role of viscosity is primarily stabilizing, preventing the instability at low Reynolds numbers.

Flows of the second type are not subject to this instability, and there is the possibility (in theory rather than in practice) that they remain stable at all values of the Reynolds number. However, they frequently exhibit a different kind of instability that is essentially a consequence of viscous action. Viscosity now plays a dual role, not readily understood physically but indicated by linear stability theory: it provides the destabilizing mechanism at high Reynolds number, but still damps out the instability at low Reynolds number.

The role of linear stability theory (Section 17.4) in our understanding of the evolution of shear flows is best explained by considering a particular flow; the next section looks at boundary layer stability.

The consequences of shear flow instability are varied. Frequently the final result far enough downstream is fully turbulent motion, at any Reynolds number for which the laminar motion breaks down. (There is often, but not always, the complication that the Reynolds number increases with distance downstream.) However, the intermediate stages of transition to turbulence take different forms in different flows. This is the topic of Chapter 18.

A flow that does not always develop to turbulent motion far enough

† Strictly speaking, a point of inflexion is a necessary condition for instability according to linear stability theory in which viscous affects are omitted. It is not in general a sufficient condition, but it is found to be so for the sort of inflexional profiles that commonly arise.

downstream is the Kármán vortex street in a wake. Also there is room for discussion about whether this is or is not an example of shear flow instability. We shall therefore look at vortex streets in Section 17.8.

17.7 Stability theory for a boundary layer

In this section we see how linear stability theory, the principles of which have been explained in Section 17.4, is applied to a two-dimensional zero-pressure-gradient boundary layer. We also discuss the comparison of the predictions of the theory with experiment.

The breakdown of laminar flow in the boundary layer is supposed to arise essentially from the instability of the Blasius profile (Section 11.4). This profile, with given values of the free-stream velocity u_0 and boundary layer thickness δ and thus of the Reynolds number

$$\mathrm{Re} = u_0\delta/\nu \qquad (17.58)$$

is taken as the basic flow upon which the perturbation is superimposed.

A two-dimensional disturbance of the form

$$\Delta u = U(y)\exp(ikx + \sigma t) \qquad (17.59)$$

$$\Delta v = V(y)\exp(ikx + \sigma t) \qquad (17.60)$$

$$\Delta p = P(y)\exp(ikx + \sigma t) \qquad (17.61)$$

is considered (where the coordinates are the same as in Section 11.4). The fact that the basic flow is two-dimensional is not immediately a justification for taking the disturbance to be two-dimensional; a two dimensional flow can undergo a three-dimensional transition, and indeed, the later stages of development and the resulting turbulent motion are strongly three-dimensional. However, there is a result, known as Squire's theorem, that in linear stability theory the critical Reynolds number for a two-dimensional parallel flow is lowest for two-dimensional perturbations [22, 43]. We may thus restrict attention to these.

The perturbation expressed by eqns (17.59)–(17.61) is a single Fourier component in the flow direction. As for Couette flow (Section 17.5), the perturbation must have an unprescribed form in the normal direction—a feature that might be expected from the fact that the growth of the disturbance will not extend appreciably outside the boundary layer where there is no shear.

The disturbance will be carried downstream by the flow. Its periodicity in the x direction will produce a periodicity in time. We may thus anticipate that

$$\sigma_i \neq 0. \qquad (17.62)$$

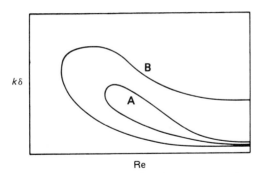

$k\delta$

Re

FIG. 17.16   General forms of stability loops for shear flows.

As before, the stability or instability is indicated by the sign of σ_r. The marginal condition is given by the locus of $\sigma_r = 0$. For shear flows in general, this locus on a Reynolds number versus non-dimensional wave number plot takes one of the forms shown in Fig. 17.16. There is instability inside either loop A or loop B and stability everywhere outside it. A loop of the form A, with the two limbs asymptoting towards one another as $\text{Re} \rightarrow \infty$ applies when the velocity profile has no point of inflexion at which the magnitude of the vorticity is a maximum. The Blasius profile is of this kind. A loop of the form B applies for a flow, such as a jet, that does have such a point of inflexion. (See Section 17.6 for a discussion of the differences in the destabilizing mechanism between these two types of flow.)

For comparison with experiment, it is more convenient to specify the particular Fourier component by its frequency rather than its wave number. The theory determines the speed at which each component travels downstream, so the two specifications are equivalent. Figure 17.17 shows results for the Blasius profile in this way, $\sigma_i \nu / u_0^2$ being the most appropriate non-dimensional form of the frequency σ_i. δ is taken as the 99 per cent thickness (eqn (8.14)). The critical Reynolds number below which no disturbances are amplified, $\text{Re}_{\text{crit}} = 1510$. (The calculations giving Fig. 17.17 ignore the variation of δ with x. Inclusion of this makes the calculations much more difficult. The effect is to reduce Re_{crit}, but it is uncertain just how much [43].)

The instability of a single Fourier component, as considered by the theory, gives rise to a wave-like motion, periodic in both space and time. Such motions are called Tollmien–Schlichting waves. Figure 17.18 shows such waves (and subsequent developments downstream, discussed in Section 18.2) in the boundary layer on an axisymmetric body. The occurrence of Tollmien–Schlichting waves in such spontaneous breakdown of laminar flow depends on the selective amplification of a particular frequency, or narrow band of frequencies. A wide range of

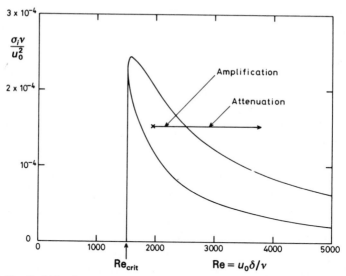

Fig. 17.17 Stability loop for Blasius profile, based on data in Ref. [81]. (Note that the two limbs asymptote to one another at much larger Re.) See text for significance of the arrow.

frequencies will be present in any general disturbance. At any given Reynolds number, only a band of these will be unstable according to Fig. 17.17, and within this band the amplification will vary. If a sufficiently narrow frequency range is amplified significantly more than other frequencies, a nearly sinusoidal disturbance develops out of the initially random disturbances.

In practice, it is difficult to make the disturbance level low enough for this selection process to give very regular waves. Sometimes it is difficult to detect Tollmien–Schlichting waves at all. Consequently, verification of the linear stability theory requires the deliberate introduction of a

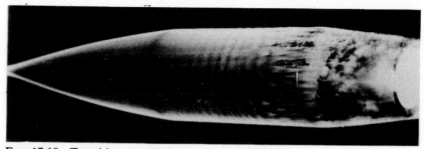

Fig. 17.18 Transition to turbulence on an axisymmetric body shown by smoke introduced as filament impinging on nose of body. Photo by T. J. Mueller, R. C. Nelson, and J. T. Kegelman, University of Notre Dame, cf Ref. [284].

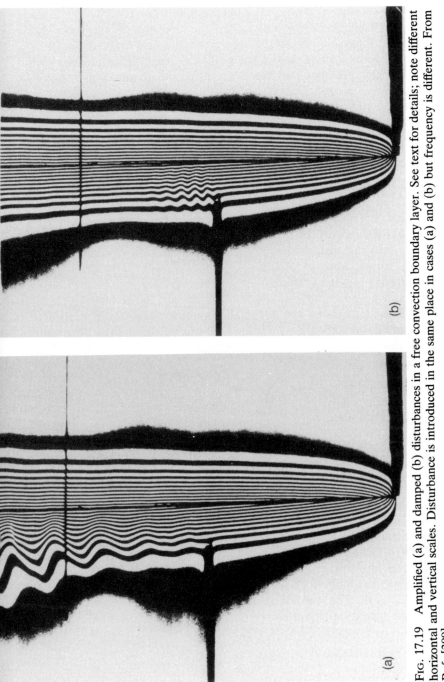

Fig. 17.19   Amplified (a) and damped (b) disturbances in a free convection boundary layer. See text for details; note different horizontal and vertical scales. Disturbance is introduced in the same place in cases (a) and (b) but frequency is different. From Ref. [309].

periodic disturbance [338]. The usual method in boundary layer studies is to stretch a fine metallic ribbon across the flow close to the wall. This is vibrated by passing an alternating current through it in the field of a magnet, just the other side of the wall. The resulting disturbance of the flow velocity can then be traced downstream, usually with a hot-wire anemometer (Section 25.2).

We illustrate this with the different example of the free convection boundary layer on a heated vertical plate (Section 14.8). The temperature variations make it easier to exhibit the waves visually. Figure 17.19 was obtained using an interferometric technique such that the light and dark bands are essentially isotherms of the two-dimensional flow. (It should be noted that the optical arrangement exaggerates the horizontal scale relative to the vertical by a factor of 6.4.) In this case, the periodic disturbance was introduced into the outer part of the boundary layer by a mechanical vibrator. The two parts of the figure contrast an amplifying and a decaying disturbance, occurring in the positive and negative σ_r regions for his flow.

Vibrating ribbon experiments in zero-pressure-gradient boundary layers have confirmed the applicability of the above theory. There is some discussion about whether small differences between theoretical and experimental results are significant [43], but essentially the observations show that whether a disturbance increases or decreases its amplitude depends on the sign of σ_r in Fig. 17.17. The Reynolds number of a laminar boundary layer increases with distance downstream, in consequence of the increasing boundary layer thickness. The development of a wave of given frequency is thus indicated by a horizontal line on the stability loop, such as the example on Fig. 17.17. A wave that is initially amplified may subsequently be attenuated as this line passes out of the loop. This behaviour is observed experimentally. Provided the amplitude remains small enough throughout, the whole history of the wave is well described by the linear theory.

If, however, a critical amplitude (0.3–2 per cent in the ratio of the root mean square velocity to the free-stream velocity, depending on the non-dimensional frequency of the wave [224, 330]) is reached, the Tollmien–Schlichting waves are only the first stage of an elaborate sequence of events. These developments are not accounted for at all by linear stability theory. This is a case represented schematically by curve B of Fig. 17.6. The further developments will be described when we consider transition to turbulence in boundary layers, Section 18.2.

17.8 Vortex streets

The occurrence of vortex streets in cylinder wakes was described in Section 3.3, where it was remarked that such streets arise rather

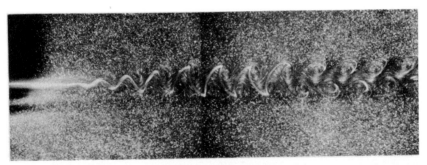

FIG. 17.20 Vortex street behind flat plate at zero incidence with Reynolds number based on plate length of 6600. Flow visualization is combination of dye released from plate and aluminium powder suspended in fluid. From Ref. [365].

commonly behind obstacles. We have seen in Section 17.6 that the velocity profile of a wake (the origin of which was considered in Sections 3.4 and 11.5) is such that it is liable to shear flow instability. However, we have also noted in Section 3.3 that vortex street formation is often described as 'eddy shedding'—a name implying a process different from shear instability. To what extent can the origin of vortex streets be understood as another example of a familiar process and to what extent do wakes have their own special processes? The same question arises over the secondary instabilities that cause the transition from vortex street flow to a turbulent wake when the Reynolds number is high enough.

Figure 17.20 shows a vortex street in the wake behind a flat plate set parallel to the oncoming flow. In this case there are no attached eddies like those immediately behind a cylinder; the wake is simply an extension of the boundary layer as illustrated by Fig. 11.3. There is unambiguous evidence that a vortex street can arise without eddies being shed (a fact that seems sometimes to be forgotten in discussions of vortex streets). The pattern in Fig. 17.20 is a consequence of the wake velocity profile being unstable as discussed in Section 17.6.†

The nature of the instability in a cylinder wake is less readily pinpointed. It varies with Reynolds number, but it is not certain whether there is a sudden transition from one mechanism to another or a gradual change in the dominant features of a complex overall process. In the range $40 < \mathrm{Re} < 100$, approximately, the attached eddies are not shed—dye injected into them remains there [176, 386]—but the instability still occurs (Section 3.3). It seems likely that it is then a shear instability [90, 418]; i.e. that Figs. 3.5 (particularly parts (d) and (e)) and 17.20

---

† Detailed analysis of the instability [267] has, however, to take into account the fact that, close behind the trailing edge of the plate, the velocity profile changes rapidly with distance downstream.

show essentially the same process. This interpretation may be an oversimplification (there is experimental evidence [176] that the attached eddies, although not shed, are still involved in the dynamics of the instability; and theoretical evidence [359] of an instability associated with the separation process) but probably not an excessive one. When Re > 100, vortex street formation is observationally an eddy shedding process, and is presumably to be understood as an instability of the attached eddy configuration. The processes involved in this have been extensively documented [176], but theoretical explanation is incomplete.

We turn now to the downstream development of vortex streets. Much of our information about this comes from experiments with circular cylinders, so we are filling out the account in Section 3.3. When the Reynolds number is less than about 200, regularity is maintained as far downstream as the street can be detected (although with some complications in narrow Reynolds number ranges [90, 165, 386]); the vortices are just gradually dissipated by viscous action. At higher Reynolds numbers the wake ultimately becomes turbulent and, as mentioned in Section 3.3, there are two types of secondary instability that can initiate this transition.

Firstly, when Re exceeds about 200, the vortex street pattern as a whole is unstable to three-dimensional disturbances [95]. Bends, such as that shown at the bottom of Fig. 3.9 for lower Re, increase in amplitude and irregularity as the vortices travel downstream. The effect can be seen in Fig. 3.10 by comparing the oscillograms at the two stations. Ultimately the irregularity becomes dominant and the wake is turbulent. Transition from a vortex street behind a flat plate parallel to the flow to turbulent motion has been observed to occur in a similar manner [332]. The later stages are very similar to the transition process in jets outlined in Section 18.4.

Secondly, in our discussion of cylinder wakes, when Re exceeds about 400, an instability occurs close to the cylinder in the two regions where the fluid is moving away from it to form the eddies [95, 230, 408]. As seen in Fig. 3.12, this produces initially small secondary vortices, and the instability can be identified [397] as a shear instability (occurring on a smaller scale than the wake width because the shear vorticity is more concentrated at higher Reynolds number; i.e. the boundary layer ahead of separation is thinner). It can thus be understood in terms of the discussion in Section 17.6. The result of the instability is that the vortices in the street are turbulent. As noted in Section 3.3, this has curiously the effect of making the Strouhal number better defined. Although the motion within the vortices involves rapid irregular fluctuations their shedding occurs with quite precise periodicity. However, as the vortices travel downstream the turbulence spreads into the regions between them and disrupts the regular periodicity, again leading finally to a fully turbulent wake.

These processes have implications for the structure of the turbulent wake and so we shall refer to them again in Section 21.4.

Other developments in flow past a cylinder at still higher Reynolds number have been described in Section 3.3, but these relate less directly to the vortex street.

18

TRANSITION TO TURBULENCE
IN SHEAR FLOWS

18.1 Introduction

All shear flows become turbulent at high enough Reynolds number. Linear stability theory can indicate when a laminar shear flow must undergo transition to some other type of motion, and the example of its application to a boundary layer was discussed in Section 17.7. However, as was pointed out there, the waves predicted by such theory are only the first stage of the evolution towards turbulent motion. What happens subsequently [48]?

Our knowledge about this comes primarily from experimental investigations. That is not to say that there are no theories, at least of some stages of the transition of some types of flow. However, the way in which the theories relate to the overall transition sequence is generally to be understood only by reference to the experimental observations. Discussion of this would lead us into great complexity and it is appropriate to give a largely phenomenological description.

Many of the experiments have used the vibrating ribbon (Section 17.7) or some similar mechanical or acoustic arrangement to introduce controlled disturbances. Although this technique was originally introduced to investigate the relevance of linear stability theory, it has proved a useful way of controlling flow development and ensuring that the same features are to be observed repeatedly at the same place. Its use has thus been extended to experiments on the essentially non-linear developments further downstream. The relationship of information so obtained to processes in naturally occurring transition is sometimes debated. It is, however, generally supposed that the same sequence of events occurs but with greater fluctuation in the position of any particular stage. If the background disturbance level is large, some of the later stages may be triggered directly without the earlier stages having appeared. We shall be seeing some comparisons of developments with controlled and uncontrolled disturbances in Figs 18.9–18.12.

The remarks so far apply to any shear flow. But the detailed dynamics of transition vary from flow to flow. There is a useful general classification into two types: flows in which turbulent motion first occurs in small patches, and flows in which the randomness characteristic of turbulence develops at a roughly equal rate throughout the transition region. Only the former type produces oscillograms like those in Fig. 2.10 with contrasted intervals of vigorously fluctuating and almost constant velocity. (We shall see oscillograms characteristic of the latter type in Figs. 18.11 and 18.12.) Most observations support the generalization that the first type of transition occurs in shear flows adjacent to a wall or walls (boundary layers, pipe flow, etc.) and the second type in shear flows away from any wall ('free flows' such as jets, wakes, etc.). But doubt has been cast on this useful generalization by observations [219] of a transition process in boundary layers, different from that to be described in Section 18.2 and much more like that usually observed in free flows.

We are evidently reaching the end to the usefulness of general remarks. Let us look at some particular cases.

18.2 Boundary layer transition

We take up the story where Section 17.7 left off. What happens after Tollmien–Schlichting waves reach the critical amplitude [368, 369]? The sequence of developments to be outlined below is becoming known as K-type transition, with the implication that there are other types [219, 229, 330]. These complications are emerging from very detailed experiments with vibrating ribbons. However, probably the K-type can be regarded as the principal type. In particular, if one wishes to infer the nature of uncontrolled transition from experiments with controlled disturbances, as considered in Section 18.1, then this type may be the most relevant. Certainly, the appearance and spreading of turbulent spots, described below as the last stages of transition, are commonly observed in uncontrolled transition (e.g. Fig. 17.18).

Most of our knowledge of K-type transition comes from the pioneering experiments by Klebanoff and his colleagues [224, 225], hence the name. Important details have been filled in by experiments in channel flow [290, 291]; transition is promoted by a vibrating ribbon close to one wall whilst flow in the other half of the channel remains undisturbed, so there is close resemblance to a boundary layer.

The first development of the Tollmien–Schlichting waves is that they become three-dimensional; their amplitude varies in the lateral direction parallel to the wall. Simultaneously, they interact with the mean flow, so that the velocity profile also has three-dimensional variations. Figure 18.1 shows one set of measurements illustrating this development.

The growth of these three-dimensionalities continues until there are transient local regions of very high shear. Figure 18.2 indicates the variation of the velocity profile inferred from measurements with

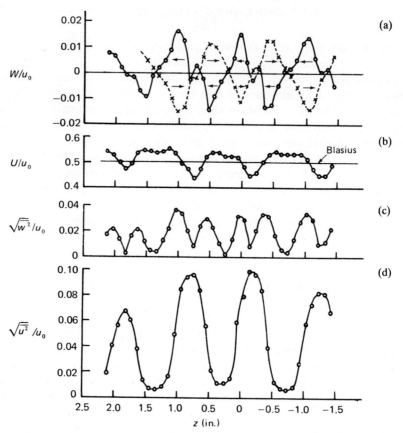

FIG. 18.1 Example of variations in z-direction (where x is flow direction, y is distance from wall) of (a) mean velocity in z-direction, (b) mean velocity in x-direction, (c) r.m.s. velocity fluctuation in z-direction, and (d) r.m.s. velocity fluctuation in x-direction, in nominally two-dimensional transitional boundary layer. Solid lines: $y/\delta = 0.31$; dotted line: $y/\delta = 0.11$. From Ref. [225].

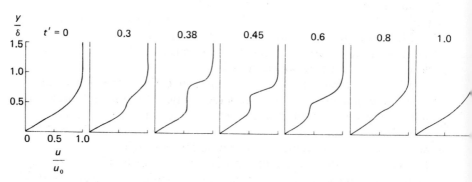

FIG. 18.2 Instantaneous velocity profiles in transitional boundary layer; t' is time in units of oscillation period. From Ref. [232].

hot-wire anemometers (Section 25.2), through one cycle of the wave at the lateral position where the wave amplitude is largest. The next development is initiated in the regions of high shear, apparent in the figure for t' between 0.38 and 0.6.

Hitherto, although the structure of the motion has become increasingly complex, the time scale of the velocity fluctuations has remained of the same order as the period of the initial wave. Now fluctuations with much shorter time scales appear, initially as spikes in the velocity vs. time oscillograms but developing rapidly to bursts of turbulent-like fluctuations. The detailed changes to the boundary layer structure during this development are complicated, although they occupy only a small fraction of the total distance over which transition occurs. They involve the formation of hairpin-shaped vortices, leading to a second region of intermittent high shear close to the wall (Fig. 18.3), where the turbulence generation occurs: we can give only this very inadequate summary here [266, 290, 291].

The upshot is that small areas of the boundary layer are turbulent. These areas are rather regularly distributed in controlled transition, more randomly in uncontrolled transition. The remainder of the transition process consists of the growth of these local regions of turbulent motion, whilst they travel downstream [142, 321]. As they grow they merge into one another. Eventually, all the non-turbulent regions have been absorbed and the boundary layer is wholly turbulent.

The localized regions of turbulence are known as turbulent spots. A spot can be generated directly without the normally preceding stages by a large localized disturbance, such as a rod or small jet momentarily

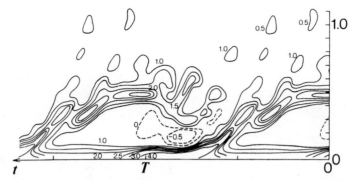

FIG. 18.3 Contours on time/distance-from-wall coordinates of instantaneous velocity gradient in boundary layer at spot formation stage. Pattern is average over many cycles of controlling oscillation and is displayed for about $1\frac{1}{2}$ cycles. Time increases to left, so that pattern resembles a spatial one in flow to right. Note existence of two regions of high velocity gradient. From Ref. [291].

FIG. 18.4 Turbulent spot seen through wall on which boundary layer has formed. Flow is from left to right. Visualization by suspended aluminium flakes (Section 25.4). (The turbulent region at the top of the picture is due to a side wall.) From Ref. [114].

injected through the wall or a small spark. Much of our extensive information on spots has come from such experiments [80, 114, 169, 265, 415].

Spots are shown in Figs. 18.4 and 18.5. The former is a view through a transparent wall on which the boundary layer is formed. The latter is a view perpendicular to the wall and parallel to the flow, along the

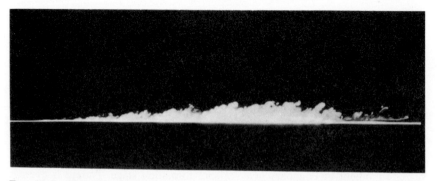

FIG. 18.5 Side-view of centre plane of turbulent spot. Flow is from left to right. Visualization by laser light illumination of fluorescent dye introduced through wall upstream. From Ref. [169].

centreline of the spot widthways. A turbulent spot is essentially a localized region of fully turbulent motion. Just after formation a spot may involve its own type of motion. But, once it has spread so that its dimensions parallel to the wall are large compared with the boundary layer thickness, the motion over most of its area closely resembles that in a fully turbulent boundary layer to be described in Sections 21.5 and 21.6 [321].

As can be seen in Fig. 18.4, spots have a characteristic shape in planes parallel to the wall, something between a triangle and a heart-shape, with the narrower part towards the downstream end. The spreading rates in different directions are such that spots retain this shape as they grow. The spreading process is one of 'sideways contamination'; transition to turbulence in laminar fluid is triggered by proximity to turbulent fluid, rather than turbulence being 'fed into' the laminar regions. The distinction is clearest if we anticipate ideas about turbulence energy in Section 19.3: the new turbulence at the edge of a spot acquires its energy not by transfer from the spot but by generation of fluctuations that can themselves extract energy from the mean motion [169, 321].

As already stated, the net effect of all these developments is a fully turbulent boundary layer. Hence, the story continues in Section 21.5.

This section has so far been concerned specifically with zero-pressure-gradient boundary layers. Similar processes often occur in the presence of a pressure gradient. The effect of an adverse pressure gradient is generally destabilizing—it decreases the Reynolds number at which corresponding developments occur—and that of a favourable gradient is generally stabilizing. These changes arise from changes in the laminar velocity profile [368].

A different transition process that can occur when the pressure gradient is adverse has already been considered in Section 12.6: the sequence of laminar separation, transition, and turbulent reattachment. Since transition occurs in the separated region, the dynamical processes are more related to those of Section 18.4. Although the initial attached laminar layer might be stable, the final reattached layer remains turbulent [368].

18.3 Pipe flow transition

We return to the topic of pipe flow; the description of transition in Section 2.6 needs filling out. Although this is in principle one of the simplest configurations and although it played such an important role in the early history of transition studies, the details of the transition process turn out to be relatively complex and are, quite possibly, not all known.

Pipe flow illustrates, in a more extreme way than other flows, the limitations to linear stability theory (Section 17.4), in providing only a sufficient condition for flow breakdown, not a necessary one.

Linear stability theory applied to Poiseuille flow indicates that the parabolic velocity profile is stable at all values of the Reynolds number; $Re_{crit} = \infty$. (This result has not been proved with full rigour, but is considered well established [43, 358].) Nevertheless pipe flow becomes turbulent at some Reynolds number between 2×10^3 and 10^5. Almost certainly, both the departures from the parabolic profile in the entry length and the response to non-infinitesimal disturbances play a role in the interpretation of this fact.

Different geometries result in different detailed flows in the entry length, but the entering fluid will often be irrotational. Then, an annular boundary layer grows on the wall, and the flow starts approximating to Poiseuille flow when the thickness of this becomes close to the radius of the pipe (Fig. 2.5). When the thickness of this boundary layer is small compared with the pipe radius, it will be prone to boundary layer instability at high enough Reynolds number [171, 331, 347]. We may anticipate that there will be a stable length right at the start of the pipe (the Reynolds number based on boundary layer thickness being small here), followed by the unstable region, this being followed in turn by another stable region as the profile approaches Poiseuille flow. If disturbances are amplified sufficiently in the unstable region, then transition to turbulence may be expected.

Evidently, this description of the start of transition is consistent with a high degree of variability in the maximum Reynolds number at which laminar flow can be maintained. In the first place, small variations in the velocity profile, due to different entry geometry, may affect the extent of the unstable region. In the second place, the maximum amplitude of velocity fluctuations reached at the end of the unstable region will be sensitive to the level of small disturbances present at its start; the appearance of local regions of turbulence probably depends on a certain amplitude being reached.†

By taking great care to minimize the disturbance level, experimenters have maintained laminar flow up to values of $Re = u_{av}d/v$ of around 1.0×10^5 [302]. It is likely that in these experiments there was a local unstable region in the entry boundary layer, but with subsequent decay of the amplified disturbances.

---

† The distinction between a change in entry geometry and a disturbance may be somewhat blurred, particularly if wall roughness enters the story. However, here the former is intended to imply a change in the detailed pattern of the laminar flow that would occur in the absence of any instability. The latter is intended to imply a small uncontrolled unsteadiness in the flow.

When turbulence develops at Reynolds numbers above about 10^4 the region of boundary layer instability is probably an important feature of the transition process. Turbulence has been observed [296, 414] to appear first of all in small spots restricted laterally as well as axially and presumably forming in the boundary layer. These spots then spread in all directions and several of them merge to form a turbulent slug.

However, the lowest values of the Reynolds number (around 2×10^3) at which transition occurs are well below the minimum value (a little above 10^4) at which theory [171] indicates the existence of an unstable region. In many experiments [92, 254, 296, 327], the entry geometry has involved sharp edges or corners and there was probably a region of separated flow. This would be unstable at a lower Reynolds number than any attached boundary layer, and rapid amplification of disturbances would occur as shown in Fig. 18.6.

However, this is probably not the only way in which transition can be initiated at the lower Reynolds numbers. Observations [254, 255] have been made, some of them with an arrangement unlikely to produce separation, of transition occurring at around 3000 in a manner similar to that described above for higher Reynolds numbers—turbulent spots originating close to the wall and then spreading to give turbulent slugs. One can only say that growing turbulent slugs can originate and so transition occur at Reynolds numbers down to about 2000, provided that there are large enough disturbances. Probably there is no unique process for the origin of slugs in this range; a large disturbance can take such a variety of forms.

The triggering disturbance can be another turbulent region. Observations have been made of one turbulent spot appearing in the boundary layer a little way downstream of an older one, whose presence has caused sufficient disturbance [254, 255].

Despite these complications in the detailed mechanics of transition, it is clear what governs the lower limit in Reynolds number for transition. Once turbulent slugs are produced, the transition process is substantially the same at all values of the Reynolds number. The turbulence

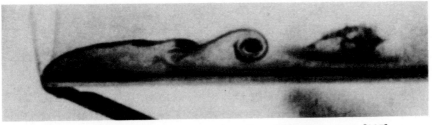

FIG. 18.6   Flow at sharp entry to pipe; Re = 5240. From Ref. [237].

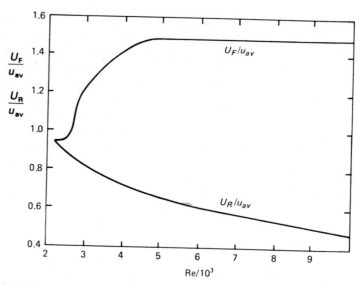

Fɪɢ. 18.7   Ratios of front and rear velocities of slugs to the mean flow speed as functions of Reynolds number, based on experimental results in Refs. [256, 296, 327] (see also [376, 414]).

propagates into the neighbouring non-turbulent regions at each end of a slug. The slugs thus grow in length at the expense of the non-turbulent region. When the front interface of one slug meets the rear interface of another, the two merge to give a single slug. Thus the intermittency factor increases with distance downstream, until all the non-turbulent regions have been absorbed and fully turbulent pipe flow results. Figure 18.7 shows the rates at which front and rear interfaces travel downstream as functions of the Reynolds number. The difference between them disappears when Re is about 2300. Below this slugs do not grow, and fully turbulent flow is not produced at any distance downstream. There is a short Reynolds number range over which localized regions of turbulence neither spread nor decay; the resulting features are called turbulent puffs and have a characteristic structure [76, 414].

There is some variability in the figure quoted for the minimum Re at which transition can be produced by large disturbances, values going down to 1800. Nevertheless, it is clear that the limiting factor is slug growth. Observations have been made of the processes leading to slug production at much lower Reynolds numbers, but transition does not then result [92].

The region of intermittently turbulent flow is often very long. This depends on the rate of slug production, which is sensitive to the disturbance level. Quantitative estimates thus relate only to particular experiments, but this region can be hundreds of diameters long.

We saw in Section 2.6 that the production of turbulent slugs can be either random or periodic in time, as illustrated by the oscillograms of Fig. 2.10. The random production might be regarded as the normal state of affairs, periodic production occurring through a rather special mechanism. Nonetheless, the latter is a phenomenon of some interest and significance and we now consider how it comes about.

The pressure gradient needed to produce a given flow rate is much larger when the flow is turbulent than when it is laminar (Fig. 2.11). Conversely, the flow rate produced by a given pressure difference between the ends is much higher for laminar flow than for turbulent flow. In most experimental arrangements, it is the pressure difference, not the flow rate, that is held constant. Hence, in transitional flow, as a turbulent slug grows the flow rate reduces. This inhibits the instability of the entry region, and the probability of another slug being produced becomes negligible. Only when the front of the slug passes out of the far end of the pipe does the fraction of the pipe length in laminar motion increase. The flow rate then starts to increase too. When it has increased sufficiently, another slug is produced near the entry, and the cycle recommences. Since the front of a slug travels downstream at about three times the speed of the rear (Fig. 18.7), there can be substantial variations in the fraction of the flow that is turbulent (Fig. 18.8) and thus in the flow rate. If these variations pass through the criterion for instability (such as the critical Reynolds number of the entry boundary layer), the instability can be renewed at just the same phase of each cycle and the flow will pulsate with a well-defined period.

This happens only over a limited Reynolds number range. At smaller Re, the difference between laminar and turbulent flow rates is too small. At higher Re the flow becomes turbulent in only a small fraction of the total length and so the variations in flow rate are again too small. As one

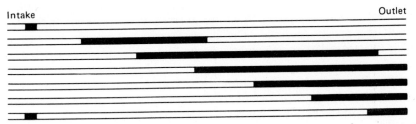

FIG. 18.8 Sequence of positions of laminar and turbulent motion (shaded regions turbulent) during periodic slug formation. Length and diameter are not to the same scale (and interfaces are therefore shown plane).

would expect, the Reynolds number range depends on the length/ diameter ratio of the pipe (and periodic transition does not occur at all when this ratio is less than about 60). One can, therefore, quote only examples of this range, and two are given below [296]. The Reynolds number ranges appear short, but this is deceptive; there is a large change in the fraction of the fluid in the pipe that is turbulent as one goes through the range, and so a small change in Reynolds number is associated with a large change in the non-dimensional pressure drop. For $l/d = 540$, the range has been observed to start† at a Reynolds number (based on a velocity averaged not only over a cross-section but also over a cycle of the pulsation) of about 4700; the intermittency factor at the outlet became unity‡ at $Re = 5900$. Corresponding values of the non-dimensional pressure gradient $(p_1 - p_2)d^3/\rho v^2 l$ were 2.1×10^5 and 5.0×10^5. For $l/d = 78$, the Reynolds number range (with the same significance) was 5500 to 5700 and the range of $(p_1 - p_2)d^3/\rho v^2 l$ was 4.5×10^5 to 5.9×10^5.

In conclusion, it is now abundantly clear why the traditional view that there is a single critical Reynolds number, below which the flow is laminar, above which the flow is turbulent, cannot describe the facts. Instead we have to introduce a set of Reynolds numbers relating to different events in the transition process. We start at the high end:

1. The theoretical linear stability limit of Poiseuille flow is $Re = \infty$. We may guess that a pipe flow with an artificially induced parabolic profile at the entry could remain laminar to even higher Reynolds numbers than flows with a normal entry length.

2. There may be a Reynolds number above which a pipe flow with a normal entry length will always become turbulent through the amplification of small disturbances. If so, this Reynolds number is at least 10^5.

3. Below this there is still a region of instability in the entry length but not a sufficiently large one to promote transition when the disturbances are small. The lowest Reynolds number at which this exists is somewhat uncertain but around 10^4.

4. When the flow is spontaneously pulsating, the Reynolds number periodically becomes larger than its average value making the entry length temporarily unstable. The lowest average Reynolds number at which this provides an instability mechanism is again uncertain but is around 5×10^3.

5. The lower limit to transition even for very large disturbances is provided by the growth or otherwise of slugs. The critical Reynolds number for this is better defined at $(2.1 \pm 0.3) \times 10^3$.

† In these experiments, the flow was laminar for Re below this.
‡ It is not certain whether this or a higher Re is the upper end of the pulsating regime. It is in principle possible for the pulsations to occur with fully turbulent flow at the outlet, so long as there is still a substantial fraction of the pipe length with intermittent turbulence.

18.4 Transition in jets

In the cases we have considered so far—boundary layer transition and pipe flow transition—the main feature of the later stages was the spreading of localized regions of turbulence. It was noted in Section 18.1 that transition can occur without this process. It was also noted that the alternative type of transition may sometimes occur in a boundary layer. However, it is in so-called free flows away from any wall that this type is the normal type. We therefore choose such a flow—an axisymmetric jet—to illustrate the difference.

The breakdown of a laminar jet usually starts quite close to the orifice, where the velocity profile still depends on the details of orifice geometry and the flow upstream of it. Hence, different experiments may not be directly comparable, and there may similarly be difficulties of comparison between experiment and theory. However, the broad features of the transition process are always much the same, and, if we omit quantitative detail, can be described without reference to any particular jet.

Disturbances first appear as approximately sinusoidal fluctuations, indicating a selective amplification process like that in a boundary layer. This stage can be satisfactorily related to linear stability theory—a stability loop of type B in Fig. 17.16 being relevant. Free shear flows become unstable at lower Reynolds numbers than shear flows next to walls; the order of magnitude of the critical Reynolds number for a jet is typically 10 (compared with 10^3 for a boundary layer) [43].

The ability of an applied periodic disturbance in the appropriate frequency range to promote the instability can have rather spectacular results in jets, as shown in Fig. 18.9. In such experiments, the disturbance is usually provided as a sound wave from a nearby loudspeaker. The acoustic wavelength is long compared with the wavelength of the instability, and the fluid thus experiences a periodic disturbance of negligible phase variation.†

---

† The observation that jets respond to sound is an old one, first made with flames. The following introduction to an early paper on the subject, written in 1858 by J. Leconte [242], makes pleasant reading: 'A short time after reading Prof. John Tyndall's excellent article, I happened to be one of a party of eight persons assembled after tea for the purpose of enjoying a private musical entertainment. Three instruments were employed in the performance of several of the grand trios of Beethoven, namely, the piano, violin and violoncello. Two "fish-tail" gas-burners projected from the brick wall near the piano. Both of them burnt with remarkable steadiness, the windows being closed and the air in the room being very calm. Nevertheless it was evident that one of them was under a pressure nearly sufficient to make it flare. Soon after the music commenced, I observed that the flame of the last mentioned burner exhibited pulsations in height which were exactly synchronous with the audible beats. This phenomenon was very striking to everyone in the room, and especially so when the strong notes of the violoncello came in. It was exceedingly interesting to observe how perfectly even the trills of this instrument were reflected on the sheet of flame. A deaf man might have seen the harmony.'

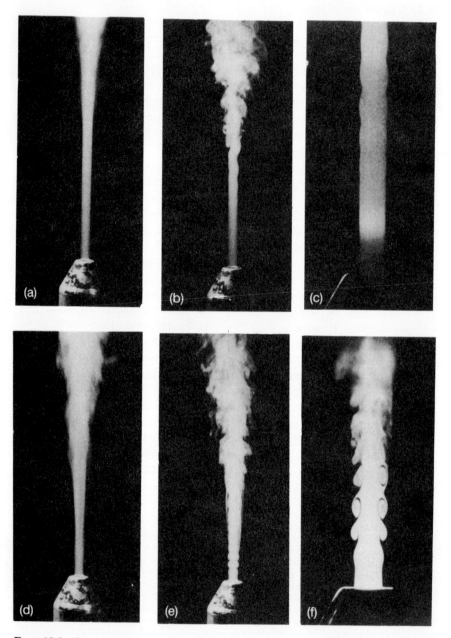

FIG. 18.9 Jet instability at Re = 1690. See text for procedures for different pictures. From Ref. [84].

FIG. 18.10 Shadowgraph flow visualization (making use of slight temperature variations) of air jets: (a) naturally occurring transition; (b) acoustically controlled transition. Re $= 6.1 \times 10^4$. Note pairing of vortices at left of each picture to give larger vortices towards right. Photos by S. I. Isataev, Kazakh State University, Alma-Ata, similar to ones in Ref. [210].

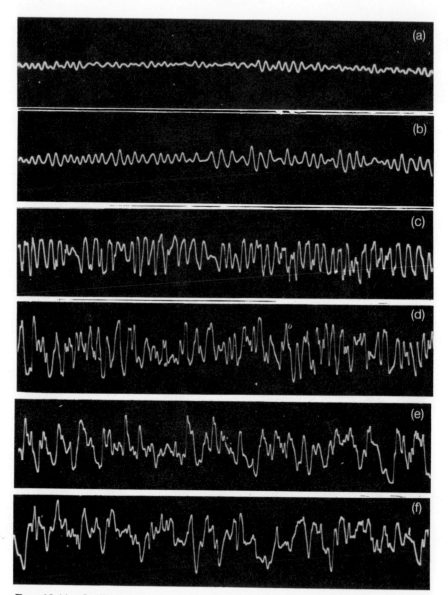

FIG. 18.11 Oscillograms of velocity fluctuations in an axisymmetric jet; Re = 750. $r/a = 0.67$ throughout; $x/a =$ (a) 0.67, (b) 1.7, (c) 3.7, (d) 5.6, (e) 8.0, (f) 10.4 (a is orifice radius).

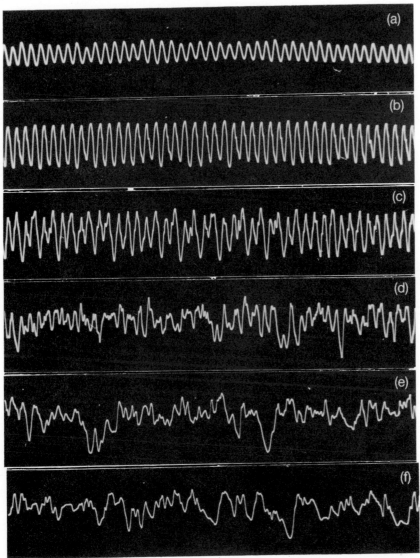

FIG. 18.12 Counterpart of Fig. 18.11 with acoustic excitation of the instability; probe positions are the same.

Pictures (a)–(c) of Fig. 18.9 show a smoke-carrying jet when acoustic disturbances were minimized. Figure 18.9(a) and (b) have general illumination, (a) with a relatively long exposure so that details of the instability are blurred and (b) as a flash so that an instantaneous pattern is seen. The close-up (c) is again an instantaneous picture, but with illumination through a slit showing one part of the jet. A spontaneously arising periodic instability can be seen. Pictures (d)–(f) show the same jet with the frequency from a loudspeaker selected to have maximum effect; (d) and (e) are the counterparts of (a) and (b). Like (c), (f) has slit illumination, but this is stroboscopic at the same frequency as the sound. Since the exposure extends over about 1800 flashes of the stroboscope, the clarity of the picture demonstrates the extreme regularity of the pattern in its initial stages.

It is clear from Fig. 18.9 and from other similar work [62, 164] that the instability can give rise to well-developed ring-shaped vortices while it is still in the periodic stage. However, the motion does not remain periodic for any great distance downstream. Breakdown to turbulent motion occurs.

An interesting feature of the transition process is vortex pairing [417], seen in Fig. 18.10. The vortices merge in pairs to form more diffuse vortices with larger overall circulation. Sequential vortex pairings may occur—giving, for example, a vortex originating as four separate vortices—but simultaneously irregular fluctuations make the regular features more difficult to detect.

Figures 18.11 and 18.12 show oscillograms of velocity fluctuations at various distances downstream, for, respectively, spontaneously arising fluctuations and acoustically stimulated ones. The difference from Fig. 2.10, which is characteristic of oscillograms produced during transition involving spot growth, is apparent. The flow becomes turbulent through the continuous development of irregularities throughout the flow, instead of the sudden local appearance of irregularities which subsequently spread.

19

TURBULENCE

19.1 The nature of turbulent motion [98, 244]

Frequent references to turbulent motion have been made in previous chapters, but detailed discussion of the nature of that motion has so far been postponed. We now take this matter up.

No short but complete definition of turbulence seems to be possible. One has rather to describe the features that are implied by the use of the name. One can formulate a brief summary, rather than a formal definition, that attempts to encapsulate the description. Perhaps the best is that turbulence is 'a state of continuous instability'.

Each time a flow changes as a result of an instability, one's ability to predict the details of the motion is reduced. (An example below will explain this statement more fully.) When successive instabilities have reduced the level of predictability so much that it is appropriate to describe a flow statistically, rather than in every detail, then one says that the flow is turbulent.

This implies that random features of the flow are dominant. One cannot, however, say that a turbulent flow is 'completely random'; to do so would define turbulence out of existence! As we shall see (particularly in Chapters 21 and 22) all flows involve organized structures. The point is just whether the randomness is sufficient for the statistical description to be the most appropriate.

This approach seems likely to leave a 'grey area' of flows that one might or might not choose to call turbulent. It is in fact a moot point whether one would expect to be able to classify all flows as either turbulent or non-turbulent—or, equivalently, whether during transition to turbulence one should be able to designate the point at which turbulent motion begins. We will see in Section 24.7 that modern ideas about chaotic dynamics suggest that such classification might be possible. However, in practice at present, there are certainly intermediate flows of which one might give a detailed description for one purpose and a statistical one for another.

The notion of loss of predictability is best understood through an example. Consider the motion in a Kármán vortex street (Sections 3.3, 17.8) in the wake of an obstacle. The velocity at a point fixed relative to the obstacle varies periodically and roughly sinusoidally. The phase of

(a)

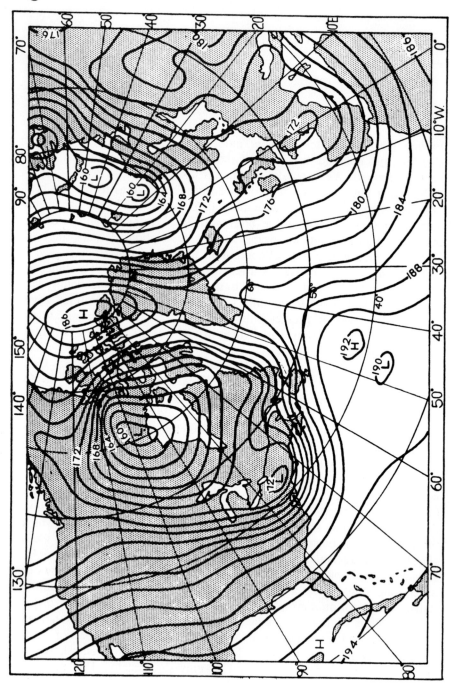

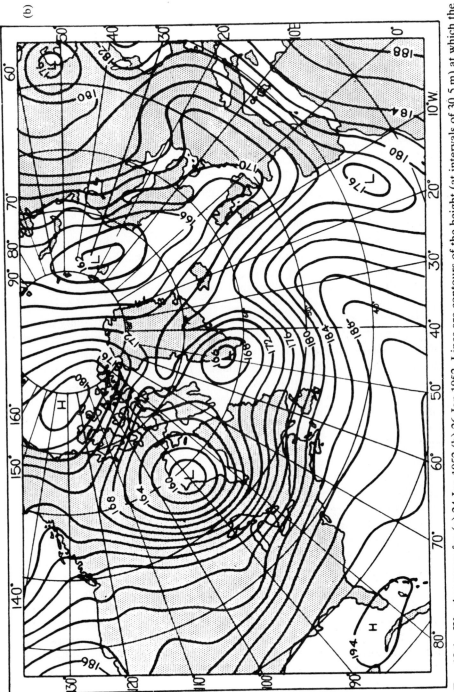

FIG. 19.1 Weather maps for (a) 24 Jan 1952 (b) 26 Jan 1952. Lines are contours of the height (at intervals of 30.5 m) at which the pressure is 500 mbar (approximately half ground level pressure); contours of pressure at a fixed height of about 5.5 km would look very similar. From Ref. [348].

this variation is arbitrary, depending on the small disturbances at the time the flow commenced. If, therefore, one asks for a prediction of the instantaneous velocity, without making any observations, this cannot be given within certain limits. This lack of predictability arises from the instability producing the vortex street; in the steady flow, from which the vortex street develops, such a prediction could be made. The degree of unpredictability is, however, small. One requires only a single observation indicating the phase of the fluctuations for all the details of the flow to be determined. When, with increasing Reynolds number, a further instability causes loss of regularity in the array of vortices, the unpredictability is increased. One can, for example, no longer say that, if one has made an observation of the velocity, then the velocity will be the same one period later. However, markedly systematic features may still exist—regions of high vorticity passing a point in a sequence, although not a completely periodic one. The flow can still be described with more emphasis on these systematic features rather than random ones. Ultimately, however, the randomness becomes so marked that such a description is no longer useful; only a statistical description gives information that applies to another experiment on the same wake (cf. Section 19.2).

There is every reason to suppose that this loss of predictability occurs as a property of the Navier–Stokes and continuity equations, although these equations contain the determinism of classical mechanics. It is not thought that the onset of turbulence represents in some sense a breakdown of these equations. The relationship of predictability to determinism will be discussed in Chapter 24; in particular reasons for the above assertions will be given in Section 24.6. In the meantime, it is useful to illustrate the essential reason why a deterministic treatment of turbulent flows is not possible. Important features of the motion (for example the large eddies to be described in Sections 21.4 and 21.6) develop from entirely insignificant perturbations. An example of this may be taken from a context in which it is rather familiar: atmospheric motions on the scale that governs the principal features of the weather. (This type of motion is in the borderland that one may or may not choose to call turbulent, but it shows sufficient of the characteristics of turbulence for the present purpose.) One of the difficulties of long-term weather forecasting is that very fine details at one time can govern major patterns at a later time. Figure 19.1 shows two weather maps, separated by two days. In the first, there are two similar-looking distortions of the pattern associated with the large cyclone over Canada, around 55°N, 62°W and around 43°N, 75°W. In the period between the two maps, the former distortion has just faded out whereas the latter has amplified and migrated to become the significant cyclonic region at the south of

Greenland in the second map. Subsequently, it merged with the other original cyclone to the north of Norway with the elimination of the high pressure ridge over the Atlantic.

19.2 Statistical description of turbulent motion

A statistical description is formulated in terms of average quantities, since these are repeatable from experiment to experiment. The detailed nature of the averaging process will be considered below, but another example may indicate the general meaning of a statistical description. Consider turbulent pipe flow. A measurement of the instantaneous velocity at some point in the pipe is, on its own, of little use. It does not indicate what the velocity will be at the same point at another instant or what it will be at any other geometrically similar point (i.e. a point at the same distance from the axis). Nevertheless, one may expect a greater similarity between two geometrically similar points than between two geometrically dissimilar points. This similarity is provided by the average quantities. The mean velocity, for example, is the same at all points at the same distance from the axis, but varies with this distance.

The ideal aim of a theory of turbulent motion would be the development of a statistical mechanics analogous to those developed in the kinetic theory of gases. Such a theory would have to be based on the equations of fluid motion instead of those of the dynamics of molecules interacting only through elastic collisions. It is thus a very difficult task. Work on these lines (beyond the scope of this book) has been attempted, but various, not necessarily valid, assumptions have to be made before progress is possible.

Consequently, much of our knowledge of turbulent flows comes from experiment. Some of the quantities that enter into any statistical theory can be measured experimentally. A physical description of the principal processes occurring within a turbulent flow can be developed from these. Generally, a rather complex interaction between theory and experiment results, theory indicating which quantities might usefully be measured and the experimental results pointing the way for further theoretical developments.

The statistical description of a turbulent flow starts by dividing the velocity and pressure field into mean and fluctuating parts. We may consider the procedure for one component of the velocity; the other components and the pressure are treated in just the same way. At each point, the velocity component is written as $U + u$, where, by definition $\bar{u} = 0$ (an overbar denoting averaging). For theoretical purposes it is sometimes convenient to think of the average as an ensemble average;

i.e. one considers a large number of identical systems and takes the average of the velocity at corresponding instants over all these systems. In practice, the average is usually a time average; one observes and averages the velocity at a point over a period long enough for separate measurements to give effectively the same result. Procedural difficulties can arise when the imposed conditions are unsteady, but we need not consider such situations here.

Thus throughout the following the average of any quantity Q signifies

$$\bar{Q} = \frac{1}{2s} \int_{-s}^{s} Q \, dt \qquad (19.1)$$

where s is large compared with any of the time scales involved in the variations of Q.

U indicates the mean motion of the fluid. Information about the structure of the velocity fluctuations is given by other average quantities, the first being the mean square fluctuation, $\overline{u^2}$. $(\overline{u^2})^{1/2}$ is known as the intensity of the turbulence component, and

$$(\overline{q^2})^{1/2} = (\overline{u^2} + \overline{v^2} + \overline{w^2})^{1/2} \qquad (19.2)$$

as the intensity of the turbulence. It is directly related to the kinetic energy per unit volume associated with the velocity fluctuations,

$$\Sigma = \tfrac{1}{2}\rho \overline{q^2}. \qquad (19.3)$$

The same intensity field can in principle be produced by many different patterns of velocity fluctuation. Before we look at the average quantities most often used to examine the more detailed structure of the velocity field, we consider briefly an alternative representation. This is in a sense the most fundamental statistical representation, although it is not the most convenient for the development of models of turbulent structure based on experimental observation. The probability distribution function $P(u)$ of a velocity component at one point is defined so that the probability that the fluctuation velocity is between u and $u + du$ is $P(u) \, du$. One thus has

$$\int_{-\infty}^{\infty} P(u) \, du = 1. \qquad (19.4)$$

The intensity is related to this,

$$\overline{u^2} = \int_{-\infty}^{\infty} u^2 P(u) \, du, \qquad (19.5)$$

but the probability distribution function contains more information than the intensity. Relationships between velocity fluctuations at different points (or times) are indicated by joint probability distribution functions.

For example a second-order function, $P(u_1, u_2)$, may be defined so that
the probability that the velocity at one point lies between u_1 and $u_1 + du_1$
and that at the other point simultaneously lies between u_2 and $u_2 + du_2$ is
$P(u_1, u_2)\, du_1\, du_2$. In principle, for a complete representation of the
turbulence, this process has to be continued to all orders.

Probability distribution functions are sometimes determined experimentally, but much more frequently further average quantities are
measured. Fuller information than is given by $\overline{u^2}$ about the fluctuations at
a single point can be obtained by measurements of $\overline{u^3}$, $\overline{u^4}$, etc.

Information about velocity fluctuations at different points (or times) is
given by correlation measurements. The correlation between two velocity
fluctuations u_1 and u_2 is defined as $\overline{u_1 u_2}$ and the correlation coefficient as

$$R = \overline{u_1 u_2} / \left(\overline{u_1^2}\, \overline{u_2^2} \right)^{1/2}. \tag{19.6}$$

In this definition u_1 and u_2 are quite general quantities; but as examples,
they could be simultaneous values of the same component of the velocity
at two different points, or two different components of the velocity at a
single point. If the fluctuations u_1 and u_2 are quite independent of one
another, then their correlation is zero. However, any turbulent flow is
governed by the usual equations and these do not allow such complete
independence, particularly for fluctuations at points close to one another.
Hence, non-zero correlations are observed.

The concept of correlations, like that of probability distributions, can
be extended to higher orders, by defining quantities such as $\overline{u_1 u_2 u_3}$. A
complete specification of the turbulence again requires one to consider all
orders up to infinity. In practice, detailed attention is usually confined to
double correlations $(\overline{u_1 u_2})$ with briefer investigation of triple correlations.

Experimental studies of turbulent flows often involve the interpretation of correlation measurements. One of the reasons for working
particularly with correlations is that those of low order lend themselves
satisfactorily to physical interpretation, in ways to be discussed in Section
19.4. We shall also be introducing later (Section 19.5) the spectrum
functions which are the Fourier transforms of correlation functions.
However, we now have enough material to examine the way in which the
equations of motion are developed for turbulent flows.

19.3 Turbulence equations

In the interests of conciseness and convention, it is necessary to use here
the suffix notation for vector equations which has for the most part been
avoided in this book (the other main exception being the appendix to
Chapter 5). For a full explanation of this notation see Refs. [26, 44, 212].

Its basic features are as follows. Each suffix can take values 1, 2 or 3, corresponding to the three coordinate directions; a vector equation can be read as any one of its component equations by substituting the appropriate value for the suffix common to every term; and the repetition of a suffix within a single term indicates that that term is summed over the three values of that suffix.

With the velocity divided into its mean and fluctuating parts, the continuity equation (5.10) is

$$\text{div}(\boldsymbol{U} + \boldsymbol{u}) = 0 \tag{19.7}$$

that is

$$\partial(U_i + u_i)/\partial x_i = 0. \tag{19.8}$$

Averaging this equation (the processes of averaging and differentiation are interchangeable in order),

$$\partial U_i/\partial x_i = 0. \tag{19.9}$$

Subtracting this result from the original equation, we have

$$\partial u_i/\partial x_i = 0. \tag{19.10}$$

The mean and fluctuating parts of the velocity field thus individually satisfy the usual form of the continuity equation.

The same division applied to the Navier–Stokes equation (eqn (5.22) with $\boldsymbol{F} = 0$) gives

$$\frac{\partial(U_i + u_i)}{\partial t} + (U_j + u_j)\frac{\partial(U_i + u_i)}{\partial x_j} = -\frac{1}{\rho}\frac{\partial(P + p)}{\partial x_i} + v\frac{\partial^2(U_i + u_i)}{\partial x_j^2}. \tag{19.11}$$

Carrying out the averaging process throughout this equation gives

$$\frac{\partial U_i}{\partial t} + U_j\frac{\partial U_i}{\partial x_j} + \overline{u_j\frac{\partial u_i}{\partial x_j}} = -\frac{1}{\rho}\frac{\partial P}{\partial x_i} + v\frac{\partial^2 U_i}{\partial x_j^2} \tag{19.12}$$

which, with the aid of the continuity equation (19.10), may be rewritten

$$U_j\frac{\partial U_i}{\partial x_j} = -\frac{1}{\rho}\frac{\partial P}{\partial x_i} + v\frac{\partial^2 U_i}{\partial x_j^2} - \frac{\partial}{\partial x_j}(\overline{u_i u_j}) \tag{19.13}$$

where, additionally, attention has been restricted to steady mean conditions by making the first term of (19.12) zero.

Equation (19.13) for the mean velocity U_i differs from the laminar flow equation by the addition of the last term. This term represents the action of the velocity fluctuations on the mean flow arising from the non-linearity of the Navier–Stokes equation. It is frequently large compared with the viscous term, with the result that the mean velocity distribution is very different from the corresponding laminar flow.

The character of this interaction between the mean flow and the fluctuations can be seen most simply in the context of a flow for which the two-dimensional boundary layer approximation applies. The turbulent fluctuations are always three-dimensional, but if the imposed conditions are two-dimensional, there is no variation of mean quantities in the third direction and terms such as $\partial(\overline{uw})/\partial z$ (that would otherwise appear in the next equation) are zero. Omitting such terms and terms that are small on the boundary layer approximation† in eqn (19.13) gives the turbulent flow counterpart of eqn (11.8); that is

$$U\frac{\partial U}{\partial x} + V\frac{\partial U}{\partial y} = -\frac{1}{\rho}\frac{\partial P}{\partial x} + \nu\frac{\partial^2 U}{\partial y^2} - \frac{\partial}{\partial y}(\overline{uv}). \qquad (19.14)$$

This equation is applied to turbulent boundary layers, jets, wakes, etc.
   Writing the last two terms of (19.14) as

$$\frac{1}{\rho}\frac{\partial}{\partial y}\left(\mu\frac{\partial U}{\partial y} - \rho\overline{uv}\right) \qquad (19.15)$$

shows that the velocity fluctuations produce a stress on the mean flow. A gradient of this produces a net acceleration of the fluid in the same way as a gradient of the viscous stress. The quantity $(-\rho\overline{uv})$, and more generally the quantity $(-\rho\overline{u_iu_j})$, is called a Reynolds stress.
   The Reynolds stress arises from the correlation of two components of the velocity fluctuation at the same point. A non-zero value of this correlation implies that the two components are not independent of one another. For example, if $\overline{uv}$ is negative, then at moments at which u is positive, v is more likely to be negative than positive; conversely when u is negative. Transferring attention to coordinates at 45° to the x- and y-directions shows that this corresponds to anisotropy of the turbulence—different intensities in different directions. Putting

$$u' = (u + v)/\sqrt{2} \qquad v' = (v - u)/\sqrt{2} \qquad (19.16)$$

gives

$$\overline{uv} = \tfrac{1}{2}(\overline{u'^2} - \overline{v'^2}). \qquad (19.17)$$

Figure 19.2 shows the geometrical significance of this.
   One can readily see how a correlation of this kind can arise in a mean shear flow. We may consider the case of positive $\partial U/\partial y$ as shown in Fig. 19.3. A fluid particle with positive v is being carried by the turbulence in

† The boundary layer approximation is used here to its fullest extent. In some studies of turbulent flows, some further terms (e.g. $\partial(\overline{u^2})/\partial x$) are retained because measurements indicate that they are not so very small.

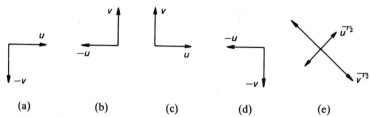

$$(a) \qquad (b) \qquad (c) \qquad (d) \qquad (e)$$

FIG. 19.2 Geometrical interpretation of Reynolds stress: if patterns of velocity fluctuations shown in (a) and (b) occur more frequently than those in (c) and (d), giving negative $\overline{uv}$, then $\overline{v'^2}$ is larger than $\overline{u'^2}$ as indicated by (e).

the positive y-direction. It is coming from a region where the mean velocity is smaller and it is thus likely to be moving downstream more slowly than its new environment; i.e. it is more likely to have negative u than positive. Similarly negative v is more likely to be associated with positive u. The process is in general (but not in detail) analogous to the Brownian motion of molecules giving rise to fluid viscosity.

The analogy has led to the definition of a quantity v_T such that

$$- \overline{uv} = v_T \partial U / \partial y \qquad (19.18)$$

v_T is called the eddy viscosity. It is important to realize that v_T is a representation of the action of the turbulence on the mean flow and not a property of the fluid. It is moreover a representation that simplifies the dynamics of that action; because of the large-scale coherent motions (Sections 21.4, 21.6), the Reynolds stress at any point depends on the whole velocity profile, not just the local gradient. Although it is sometimes useful for approximate calculations to suppose that v_T is an (empirical) constant, in general (19.18) should be regarded as the defining equation of v_T rather than an equation for $\overline{uv}$.

Further understanding of the interaction between the mean flow and the fluctuations is obtained from the equation for the kinetic energy of

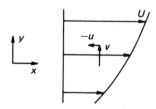

FIG. 19.3 To illustrate the generation of a Reynolds stress in a mean velocity gradient.

the turbulence. Subtracting eqn (19.12) from eqn (19.11) gives

$$\frac{\partial u_i}{\partial t} + U_j \frac{\partial u_i}{\partial x_j} + u_j \frac{\partial U_i}{\partial x_j} + u_j \frac{\partial u_i}{\partial x_j} - \overline{u_j \frac{\partial u_i}{\partial x_j}} = -\frac{1}{\rho} \frac{\partial p}{\partial x_i} + v \frac{\partial^2 u_i}{\partial x_j^2}. \quad (19.19)$$

Multiplying this by u_i and averaging

$$\frac{1}{2} \frac{\partial (\overline{u_i^2})}{\partial t} + \frac{1}{2} U_j \frac{\partial (\overline{u_i^2})}{\partial x_j} = -\overline{u_i u_j} \frac{\partial U_i}{\partial x_j} - \frac{1}{2} \frac{\partial}{\partial x_j} (\overline{u_i^2 u_j}) - \frac{1}{\rho} \frac{\partial}{\partial x_i} (\overline{p u_i}) + v \overline{u_i \frac{\partial^2 u_i}{\partial x_j^2}}$$
$$(19.20)$$

(where the rearrangement of terms has made use of the continuity equation (19.10)). Since the summation convention is being applied, the mathematics involves multiplying each component of the dynamical equation (19.19) by the corresponding velocity component and then adding the three resulting equations. For steady mean conditions the first term of eqn (19.20) is zero, but it indicates the physical significance of the equation; in view of the summation convention

$$\overline{u_i^2} = \overline{q^2} = 2\Sigma/\rho \quad (19.21)$$

and so each term in the equation represents some process tending to increase or decrease the kinetic energy of the turbulence.

With the boundary layer approximation applied to a flow which is steady and two-dimensional in the mean, eqn (19.20) becomes

$$\frac{1}{2} U \frac{\partial (\overline{q^2})}{\partial x} + \frac{1}{2} V \frac{\partial (\overline{q^2})}{\partial y} = -\overline{uv} \frac{\partial U}{\partial y} - \frac{\partial}{\partial y} \left(\frac{1}{2} \overline{q^2 v} + \frac{1}{\rho} \overline{pv} \right) + v \overline{u_i \frac{\partial^2 u_i}{\partial x_j^2}}. \quad (19.22)$$

The left-hand side and the second term on the right-hand side are terms that become zero when integrated over the whole flow. They represent the transfer of energy from place to place, respectively transfer by the mean motion and transfer by the turbulence itself. As in a laminar flow (Section 11.7), the viscous term can be divided into two parts: one is essentially negative and thus represents viscous dissipation; the other (usually small) integrates to zero and so is another energy transfer process. The input of energy to compensate for the dissipation must be provided by the only remaining term, $(-\overline{uv} \, \partial U/\partial y)$. We have seen that $\overline{uv}$ is likely to be negative where $\partial U/\partial y$ is positive, giving this term the required sign. Although local regions of positive $(\overline{uv} \, \partial U/\partial y)$ can occur, they cannot occupy the majority of the flow or the turbulence cannot be maintained.†

† This statement need not be true for systems described by a dynamical equation with additional terms to those in (19.11); for example a buoyancy force. These can give further terms in the energy equation which may represent alternative turbulence-generating mechanisms.

The equation for the energy of the mean flow contains a corresponding term of opposite sign. The term thus represents a transfer of energy from the mean flow to the turbulence. One can therefore say that the Reynolds stress works against the mean velocity gradient to remove energy from the mean flow, just as the viscous stress works against the velocity gradient. However, the energy removed by the latter process is directly dissipated, reappearing as heat, whereas the action of the Reynolds stress provides energy for the turbulence. This energy is ultimately dissipated by the action of viscosity on the turbulent fluctuations. Frequently, the loss of mean flow energy to turbulence is large compared with the direct viscous dissipation.

19.4   Interpretation of correlations

Correlation coefficients (Section 19.2) play an important role in both theoretical and experimental studies of turbulence. To illustrate how they can indicate the scale and structure of a turbulent motion, we now look at typical properties of double correlations. Some of the ideas introduced rather vaguely here will be used more specifically in Sections 20.2, 21.4 and 21.6.

When u_1 and u_2 are velocities at different positions but the same instant, $\overline{u_1 u_2}$ is known as a space correlation. Its particulars may be specified by a diagram such as Fig. 19.4(a). Most attention is usually given to correlations of the same component of velocity at points separated in a direction either parallel to that velocity component (Fig. 19.4(b)) or perpendicular to it (Fig. 19.4(c)). We may call these respectively longitudinal and lateral correlations.

The correlation will depend on both the magnitude and direction of the separation, r. Different behaviours in different directions may provide information about the structure of the turbulence, a point that will be taken up again in Section 21.4. Here we pay more attention to the variation with distance, $r = |r|$. When $r = 0$, $u_1 = u_2$ (provided they are in

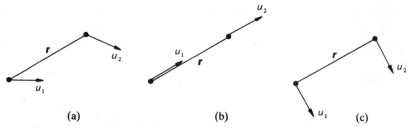

(a)   (b)   (c)

Fig. 19.4   Schematic representation of double velocity correlations.

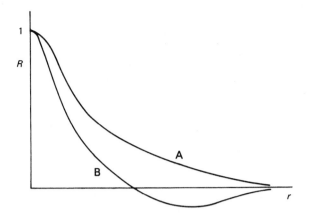

FIG. 19.5   Typical correlation curves.

the same direction) and the correlation coefficient R of eqn (19.6) is, by definition, equal to 1. At large r the velocity fluctuations become independent of one another and R asymptotically approaches 0. In consequence, the dependence of R on r takes typically one of the forms shown in Fig. 19.5. (r has a maximum value of 1 at $r = 0$ and so $(\partial R/\partial r)_{r=0} = 0$. However, the curvature at $r = 0$ is usually large and experimentally measured correlations often appear to have finite slope at $r = 0$.)

A negative region in the correlation curve implies that u_1 and u_2 tend to be in opposite directions more than in the same direction. For a longitudinal correlation this would imply dominant converging and/or diverging flow patterns. There is often no reason to expect such patterns and one may expect that longitudinal correlations will usually give a curve such as A (Fig. 19.5). On the other hand, lateral correlations may be expected to have a negative region, like curve B, since continuity requires the instantaneous transport of fluid across any plane (by the fluctuations) to be zero. Such expectations are not always fulfilled; but, when they fail, this may itself be informative about the structure of the turbulence.

A correlation curve indicates the distance over which the motion at one point significantly affects that at another. It may be used to assign a length scale to the turbulence; a length can be defined for example as $\int_0^\infty R \, dr$, or as the distance in which R falls to $1/e$, or, if the curve has a negative region, the value of r at which R is a minimum. We shall see in Sections 19.6, 20.3, and 21.3 that this concept is extended to associate a variety of length scales with the turbulence.

The correlation of the same velocity component at a single point at different instants is known as an autocorrelation. Such a correlation depends, of course, on the time separation s in a similar way to the dependence of a space correlation on r. It can be used to define a typical time scale of the turbulence.

When the turbulent motion is occurring in a flow with a large mean velocity, it is possible for the turbulence to be advected past the point of observation more rapidly than the pattern of fluctuations is changing. An autocorrelation will then be directly related to the corresponding space correlation with separation in the mean flow direction, the same curve applying for both when one puts $s = r/U$. This transformation is called Taylor's hypothesis. The extent to which it applies varies greatly between different flow situations.

A correlation between velocities measured at both different positions and different instants is called a space–time correlation. Such measurements are very useful in indicating the trajectories of certain features ('eddies', Section 19.6) associated with the turbulence, and have played an important role in some of the descriptions of the motion to be mentioned in Sections 21.4 and 21.6. However, we shall not discuss them further in this book.

In principle, of course, correlations involving the pressure fluctuations as well as the velocity fluctuations may be formulated. Equation (19.22) has already involved $\overline{pv}$—a pressure–velocity correlation. However, such quantities are difficult to measure and so have received less attention.

19.5 Spectra

Another method of discovering the time scales associated with turbulent motion is, obviously, Fourier analysis. The long-standing experimental method is to pass a turbulence signal (such as the output from a hot-wire anemometer, Section 25.2) through an electrical frequency filter. The modern method, which is largely though not totally superseding the above, is to sample the signal at frequent intervals and then carry out the Fourier analysis digitally. Either method involves subtleties, both about the definition of the quantity to be determined and about how well the procedure actually determines it. (The numerical analysis in particular contains 'traps for the unwary'.) The reader is referred to other sources [93, 288] for a proper treatment of these matters. Here just sufficient theory will be given to show the significance of the quantity measured. This is most simply done in the context of the older experimental method. The aim of the numerical analysis in the modern method is to determine the quantity $\phi(\omega)$ defined below.

We thus consider the effect of passing a turbulence signal through a frequency filter before the usual squaring and averaging. Suppose that the velocity signal is $u(t)$. Then the output from the filter is

$$\chi(t) = \int_0^\infty u(t - t')\Psi(t')\,dt' \tag{19.23}$$

where $\Psi(t)$ is the response function of the filter; i.e. $\Psi(t)$ is the output at time t when the input is a delta function at $t = 0$. $\chi(t)$ is a fluctuating function and one can measure its mean square, which is

$$\overline{\chi^2} = \int_0^\infty \int_0^\infty \overline{u(t - t')u(t - t'')}\,\Psi(t')\Psi(t'')\,dt'\,dt'' \tag{19.24}$$

where the average is over t, by the procedure of eqn (19.1), and so may be taken inside the integration with respect to t' and t''. But

$$\overline{u(t - t')u(t - t'')} = \overline{u^2}R(t' - t'') \tag{19.25}$$

where $R(s)$ is the autocorrelation coefficient for time delay s. We now introduce the Fourier transform $\phi(\omega)$ of the autocorrelation such that

$$\overline{u^2}R(s) = \int_0^\infty \phi(\omega)\exp[i\omega s]\,d\omega \tag{19.26}$$

($R(s)$ being an even function). Substituting in (19.24),

$$\overline{\chi^2} = \int_0^\infty \int_0^\infty \int_0^\infty \phi(\omega)\Psi(t')\Psi(t'')\exp[i\omega(t' - t'')]\,d\omega\,dt'\,dt''. \tag{19.27}$$

But

$$\Lambda(\omega) = \int_0^\infty \exp[i\omega t']\Psi(t')\,dt' \tag{19.28}$$

is the amplitude of the output when the input is sinusoidal with angular frequency ω. Thus

$$\overline{\chi^2} = \int_0^\infty \phi(\omega)\Lambda(\omega)\Lambda^*(\omega)\,d\omega. \tag{19.29}$$

If the filter is a good one, $\Lambda(\omega)\Lambda^*(\omega)$ is much larger over a narrow range of frequencies, centred on ω_0, than elsewhere, and eqn (19.29) reduces to

$$\overline{\chi^2} = C\phi(\omega_0) \tag{19.30}$$

where C is a calibration constant.

Thus this procedure measures the Fourier transform of the autocorrelation function. Putting $s = 0$ in eqn (19.26)

$$\overline{u^2} = \int_0^\infty \phi(\omega)\,d\omega \tag{19.31}$$

showing that $\phi(\omega)$ may be interpreted as the contribution from frequency ω to the energy of the turbulence. (We have here considered one velocity component, but eqn (19.2) provides immediate extension from the energy associated with components to the energy as a whole.) $\phi(\omega)$ is thus known as the energy spectrum.

As well as the frequency spectrum, one can define wave number spectra—Fourier transforms of the space correlations. This is more complicated because one is now dealing with three dimensions, and the reader is again referred to other sources for the theory [44, 45]. Suffice it to say here that one can define a quantity $E(k)$, where k is the magnitude of the wave number, such that

$$\Sigma/\rho = \frac{1}{2}\overline{q^2} = \int_0^\infty E(k)\,\mathrm{d}k. \tag{19.32}$$

$E(k)$ indicates the distribution of energy over different length scales. It is an important parameter in many theoretical treatments of turbulent motion. However, it cannot be measured experimentally; one would need simultaneous information from every point of the flow.

When applicable, Taylor's hypothesis (Section 19.4) can be used to derive a spatial spectrum from an observed time spectrum. However, this is a one-dimensional spectrum with respect to the component of the wave number in the mean flow direction, and so is not in general a complete determination of the spectral characteristics or of $E(k)$.

19.6 Eddies in turbulence

As we shall see in the following chapters, the division of a turbulent motion into (interacting) motions on various length scales is useful because the different scales play rather different roles in the dynamics of the motion. This is often expressed by talking of 'eddies of different sizes'. A turbulent 'eddy' is a rather ill-defined concept, but a very useful one for the development of descriptions of turbulence. The name does not necessarily imply a simple circulatory motion, but one can often identify characteristic features of particularly the large eddies. This has led to such eddies also being called 'coherent structures' (Sections 21.4, 21.6).

An eddy differs from a Fourier component in the following way. A single Fourier component, no matter how small its wave length (that is, how large the value of k), extends over the whole flow. An eddy is localized—its extent is indicated by its length scale. However, small eddies contribute to larger wavenumber components of the spectrum; the spectrum curve is often interpreted loosely in terms of the energy associated with eddies of various sizes.

For any separation, r, the correlation coefficient is determined by all eddies larger than $\sim r$. Only the largest eddies can thus be related directly to correlation measurements. Conversely, the observable spectrum function has a value at wavenumber k_x influenced by all eddies smaller than $\sim 1/k_x$ (for reasons arising from the fact that only one-dimensional spectra can be observed [130]). Hence, it is usually most convenient to use correlation measurements to provide information about the larger scales and spectrum measurements for the smaller scales.

Correlations and spectra are not the only experimental tools for elucidating the structure of turbulent flows. In Chapter 21 we shall see the importance of flow visualization. Also there are other ways of processing turbulence signals. Techniques under the general name of conditional sampling have become important in recent years [63]. These take a variety of forms and are thus difficult to summarize; basically statistical properties are measured whilst some condition is fulfilled, such as the velocity being above some level, the velocity varying particularly rapidly, or different velocity components being related so as to make a large contribution to the Reynolds stress. There is a danger that the choice of condition may prejudge the structure under investigation, but with care such measurements can reveal much about the structure of the eddies.

HOMOGENEOUS ISOTROPIC TURBULENCE

20.1 Introduction [44, 45]

Many theoretical investigations of turbulence have been developed around the concept of homogeneous, isotropic turbulence—turbulence of which the statistical properties do not vary with position and have no preferred direction. An approximation to such a motion can be obtained behind a grid, such as the one shown in Fig. 20.1, in a wind-tunnel. The theories have thus been supplemented by observation. However, we postpone consideration of experimentally based ideas about the structure of turbulence to the different contexts of Chapter 21. Here, without detail, we use the context of homogeneous isotropic turbulence to develop some ideas relevant to all turbulent flows.

The energy production term in eqn (19.20) is zero in isotropic turbulence, and so the motion must decay through viscous dissipation. In theoretical work, the turbulence is supposed to be generated at an initial instant and then to decay as time proceeds. Behind a grid, there is strong turbulent energy production for the first ten or so grid mesh-lengths along the tunnel; the turbulence then becomes substantially isotropic and decays with distance down the tunnel. (This, in principle, implies inhomogeneity, but the decay is slow enough for this to be neglected.) We shall consider certain features of the motion applying at any stage of this decay process.

20.2 Space correlations and the closure problem

The assumption of homogeneity and isotropy much simplifies the formulation of space correlations. These depend only on the distance between the two points and not on their location or the orientation of the line joining them. Moreover, it may be shown that the general correlation, as in Fig. 19.4(a), may be expressed in terms of the longitudinal and lateral correlations of Fig. 19.4(b) and (c). Thus only these two functions of r (denoted respectively by $f(r)$ and $g(r)$) are needed for complete specification of the double velocity correlations. When, additionally, the

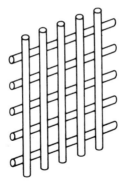

FIG. 20.1 Typical turbulence generating grid.

continuity equation is introduced a relationship between these is found,

$$f + \tfrac{1}{2}r \, df/dr = g. \tag{20.1}$$

Hence, only a single function gives the complete specification.

The mathematics behind these statements runs as follows. The correlation coefficient between velocity component u_i at one point and component u_j at a point r away is a second-order tensor $R_{ij}(r)$. When there is isotropy [45, 212]

$$R_{ij}(r) = \zeta(r)r_i r_j + \eta(r)\delta_{ij}. \tag{20.2}$$

For this to take the appropriate forms in the particular cases of longitudinal and lateral correlations,

$$\zeta(r) = (f(r) - g(r))/r^2; \quad \eta(r) = g(r). \tag{20.3}$$

Continuity gives

$$\partial u_j / \partial x_j = 0. \tag{20.4}$$

Substituting this in

$$R_{ij} = \overline{u_i(0)u_j(r)}/\overline{u^2} \tag{20.5}$$

gives

$$\partial R_{ij}/\partial r_j = 0, \tag{20.6}$$

which reduces to eqn (20.1).

Figure 20.2 shows an experimental check of eqn (20.1). Direct measurements of g are compared with ones calculated from measurements of f.

No use has been made so far of the dynamical equation. Obviously, one would like to use this to determine f as a function of r. However, one encounters what is known as the closure problem, a consequence of the

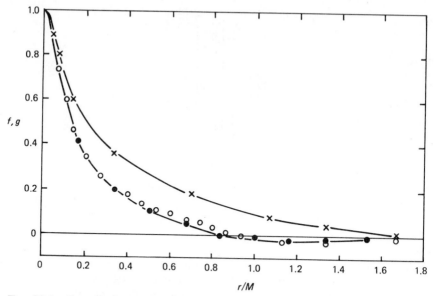

FIG. 20.2   Correlations in grid turbulence: × $f(r)$, measured; ○ $g(r)$, calculated from $f(r)$; ● $g(r)$, directly measured. From Ref. [374].

non-linearity of the equation. The mathematical analysis is somewhat complicated. The point of principle can be illustrated by much simpler equations, such as the Lorenz equations discussed in Chapter 24; the closure problem is therefore analysed in the appendix to that chapter. For turbulence the problem manifests itself as follows: when one formulates an equation for the double correlation function it involves triple correlations, an equation for these involves fourth-order ones; and so on. A variety of suggestions for an additional hypothesis to close the system have been advanced. However, our knowledge of f as a function of r remains primarily experimental (Fig. 20.2).

20.3   Spectra and the energy cascade

An important equation in the theory of homogeneous isotropic turbulence is obtained by considering the Fourier transform of the double velocity correlation. We omit the derivation [20, 44] and quote the result:

$$\partial E(k, t)/\partial t = F(k, t) - 2\nu k^2 E(k, t). \tag{20.7}$$

E is the energy spectrum of eqn (19.32); the dependence on t is shown as a reminder that we are dealing theoretically with a time-dependent situation. The closure problem has come through in the appearance of another function F in the equation for E; F is related to the Fourier transform of the triple correlation. However, it can be given its own physical interpretation through its role in eqn (20.7).

The left-hand side represents the rate of change of the energy associated with wavenumber k. The second term on the right-hand side is a negative term involving the viscosity and is thus the energy dissipation. It can be shown that

$$\int_0^\infty F \, \mathrm{d}k = 0 \tag{20.8}$$

and so the first term on the right-hand side of eqn (20.7) represents the transfer of energy between wavenumbers.

Figure 20.3 shows graphs of E and k^2E for a typical situation.† The latter has large values at much higher k than the former. Hence the viscous dissipation is associated with high wavenumbers; i.e. it is brought about by small eddies. This is a consequence of the fact that turbulent flows normally occur at high Reynolds number. The action of viscosity is slight on a length scale of the mean flow (e.g. a grid mesh-length). Yet much more dissipation occurs than in the corresponding laminar flow. This requires the development of local regions of high shear in the turbulence; i.e. the presence of small length scales.

The small dissipative eddies must be generated from larger ones. The effect of this on the energy spectrum is contained in the second term of eqn (20.7); i.e. one expects the transfer of energy to be primarily from low wavenumbers to high. This inference is confirmed experimentally by observations in grid turbulence of changes in the spectrum with distance downstream.

This interpretation of the behaviour of the terms of eqn (20.7) allows the development of a model of turbulence which has relevance not just to homogeneous isotropic turbulence but to most turbulent flows.

Energy fed into the turbulence goes primarily into the larger eddies. (In grid turbulence this happens during the initial generation; in other flows to be considered in Chapter 21 it happens throughout the flow.) From these, smaller eddies are generated, and then still smaller ones. The process continues until the length scale is small enough for viscous action to be important and dissipation to occur. This sequence is called the energy cascade. At high Reynolds numbers (based on $(\overline{u^2})^{1/2}$ and a length scale defined in a way indicated in Section 19.4) the cascade is

† Figure 20.3 is based on experimental data. For homogeneous isotropic turbulence, a relationship can be derived between $E(k)$ and the measurable one-dimensional spectrum, thus overcoming the non-measurability of $E(k)$ mentioned in Section 19.5. However, the conversion involves differentiation of experimental curves and so $E(k)$ is determined with rather poor accuracy.

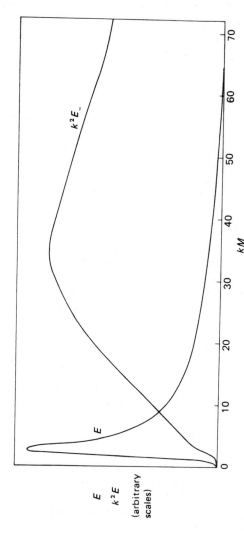

FIG. 20.3 Energy and dissipation spectra in turbulence behind a grid (distance downstream from grid = 48 M, where M = grid mesh-length). Data from Ref. [396].

long; i.e. there is a large difference in the eddy sizes at its ends. There is then little direct interaction between the large eddies governing the energy transfer and the small dissipating eddies. The dissipation is determined by the rate of supply of energy to the cascade by the large eddies and is independent of the dynamics of the small eddies in which the dissipation actually occurs. The rate of dissipation is then independent of the magnitude of the viscosity. An increase in the Reynolds number to a still higher value—conveniently visualized as a change to a fluid of lower viscosity with all else held constant—only extends the cascade at the small eddy end. Still smaller eddies must be generated before the dissipation can occur. (Since, as we shall see, the total energy associated with these small eddies is small, this extension has a negligible effect on the total energy of the turbulence.) All other aspects of the dynamics of the turbulence are unaltered.

This inference, that the structure of the motion is independent of the fluid viscosity once the Reynolds number is high enough, has important implications and will be considered again in Section 21.1. It is given the (somewhat misleading) name of Reynolds number similarity.

The dynamics of the energy cascade and dissipation may be supposed to be governed by the energy per unit time (per unit mass) supplied to it at the large eddy (low wavenumber) end. This is, of course, equal to the energy dissipation, ε. This suggests (but see below) that the spectrum function E is independent of the energy production processes for all wavenumbers large compared with those at which the production occurs. Then E depends only on the wavenumber, the dissipation, and the viscosity,[†]

$$E = E(k, \varepsilon, \nu). \tag{20.9}$$

If the cascade is long enough, there may be an intermediate range (the inertial sub-range) in which the action of viscosity has not yet come in; that is

$$E = E(k, \varepsilon). \tag{20.10}$$

Dimensional analysis then gives

$$E = A\varepsilon^{2/3}k^{-5/3} \tag{20.11}$$

where A is a numerical constant. This is a famous result, known as the Kolmogorov $-5/3$ law.

There is, however, a problem over the derivation of the result, for a rather subtle reason connected with the structure of the smaller-scale

[†] There is no contradiction between (20.9) and the inclusion of a dependence of E on t at the beginning of this section. For time-dependent situations (such as theoretical homogeneous, isotropic turbulence) (20.9) contains this through ε varying with t.

motions. Experiments show clearly that the energy dissipation does not occur roughly uniformly throughout the turbulence; there are patches of intense small eddies involving high dissipation and other patches where there is little dissipation [64, 235]. We shall see in the next section why this should be. The patchiness develops throughout the energy cascade; as the eddies get smaller, so the fraction of the volume in which they are active decreases (though the size of a patch is at each stage large compared with the corresponding eddy size). This implies that *within the patches* the rate of energy transfer per unit mass increases with wave number, thus complicating the derivation of the Kolmogorov law. The fractional volume of activity depends on how far down the cascade one is;

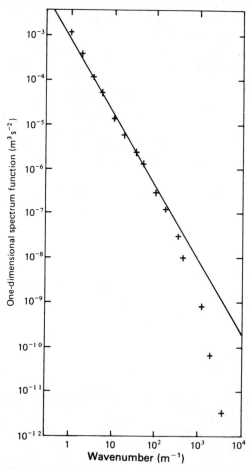

FIG. 20.4   One-dimensional spectrum measured in a tidal channel; line has slope of $-\frac{5}{3}$. Data from Ref. [187].

i.e. on k/k_0, where k_0 is a typical wave number of the turbulence energy production. This indirectly modifies (20.10) to

$$E = E(k, \varepsilon, k_0) \tag{20.12}$$

and so upsets the dimensional argument. There has been extensive theoretical discussion about how this affects the Kolmogorov result [35, 64, 166], suggesting that $E \propto k^{-n}$, with n a little larger than 5/3 but not exactly known.

Any theory which changed n substantially would be inconsistent with the observations. The Kolmogorov law is well supported experimentally. Such experiments have to be performed at very high Reynolds number in order that the inertial sub-range should be of significant length. Some of the best verifications thus come from natural flows rather than laboratory experiments. Figure 20.4 shows a spectrum obtained in an oceanographic channel flow, produced by tides, with a Reynolds number of 4×10^7; the axes are logarithmic and the line has a slope of $-5/3$. (This is actually the measurable one-dimensional spectrum, but it can be shown that if this is proportional to k^{-n}, then so is the energy spectrum.)

20.4 Dynamical processes of the energy cascade

The discussion in Section 20.3 of the energy cascade did not consider the mechanism by which the transfer of energy from large scales to small scales occurs. It is instructive to consider first the related but more readily understood process of the mixing of a scalar contaminant [130]. Suppose a blob of fluid, as shown in Fig. 20.5(1), is marked in some way, for

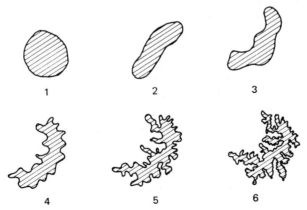

FIG. 20.5 Schematic representation of successive configurations of marked blob of fluid within turbulent motion.

example by being heated or dyed. If this blob is in a turbulent flow, its distribution will change with time in the way indicated schematically by successive configurations in Fig. 20.5 (see also Fig. 21.6). Its distribution in space becomes more and more contorted by the velocity fluctuations, but so long as molecular diffusion plays no role, the marked fluid is always just the same fluid. If the Péclet number is high, diffusion is negligible at first. However, the highly contorted patterns involve very steep gradients of the marker and so ultimately diffusion becomes significant, causing previously unmarked fluid to become marked. The higher the Péclet number, the more contorted the pattern must become before this happens.

This process is analogous to the cascade because it involves the appearance of smaller and smaller length scales until this is limited by molecular effects (diffusion or viscosity). However, the energy cascade is a more complicated matter because it involves the interaction of the velocity field with itself instead of with a quantity not involved in the dynamics. Any brief account of it must involve oversimplification, but three (not wholly independent) processes may be identified.

The first is the process of repeated instability considered in Section 19.1. Each stage may give rise to motions not only of greater complexity but also involving smaller scales than the previous stages. For example, one stage may produce local regions of high shear that can themselves be unstable.

Secondly, turbulence of a smaller scale may extract energy from larger-scale motions in a way analogous to the extraction of energy from a mean flow by the turbulence as a whole (Section 19.3).

Thirdly, there is vortex stretching [375]. The random nature of turbulent motion gives a diffusive action; two fluid particles that happen to be close together at some instant are likely to be much further apart at any later time. The turbulence will have carried them over different paths. This applies to two particles on the same vortex line. The process of vortex stretching, considered in Section 6.6, will thus be strongly present—although occurring in a random fashion. This increases the magnitude of the vorticity, but because of continuity also reduces the cross-section of the vortex tube. There is thus an intensification of the motion on a smaller scale; that is a transfer of energy to smaller eddies.

This process may also be seen as a cause of the patchy distribution of dissipation mentioned in Section 20.3. The vorticity intensification process will be strongest where the vorticity already happens to be large. At any instant the production of small eddies is thus occurring vigorously in some places and only weakly in others.

21

TURBULENT SHEAR FLOWS

21.1 Reynolds number similarity and self-preservation

Shear flows constitute by far the most important class of turbulent flows. We have traced the processes by which they become turbulent in Sections 17.6–17.8 and Chapter 18. With the background of general ideas about turbulence in Chapter 19, we now look at the structure of fully turbulent shear flows. In this and the next section we consider general ideas and illustrate them with various flows (depending largely on the best illustrations available). Subsequently we shall focus particular attention on wakes and boundary layers.

Studies of turbulent flows centre on laboratory experiments. Probably more than in any other branch of fluid dynamics, our knowledge and understanding would be slight without such experiments. As always, it is necessary to know the range of applicability of any measurement—whether it is relevant to other similar flows or whether it is peculiar to the particular situation investigated. Since the number of measurements needed for a reasonably full understanding of any turbulent flow is large, it is highly desirable to make observations of general applicability. Two ideas help here.

The first is the concept of Reynolds number similarity already introduced in Section 20.3. The larger eddies and the mean flow development are independent of the viscosity (so long as this is small enough to make the Reynolds number large). This is true of the mean flow because the last term of eqn (19.14) normally dominates the last-but-one term and because the Reynolds stress is produced by the larger eddies. (The contribution of different length scales to the Reynolds stress may be investigated by applying spectral analysis not only to the energy as in Section 19.5, but also to the Reynolds stress [129, 240]. Figure 21.1 makes a comparison between the energy and Reynolds stress (one-dimensional) spectra in turbulent channel flow; the much more rapid fall-off of the latter at high wavenumbers is apparent.) Hence, experiments at any (sufficiently high) Reynolds number provide information applicable to all values of the Reynolds number. We shall see in Section 21.5 that some qualification of this concept is needed for the motion in the vicinity of a solid boundary.

The second idea is that of 'self-preservation' [46]. This is the counterpart for turbulent flow of the occurrence of similar velocity

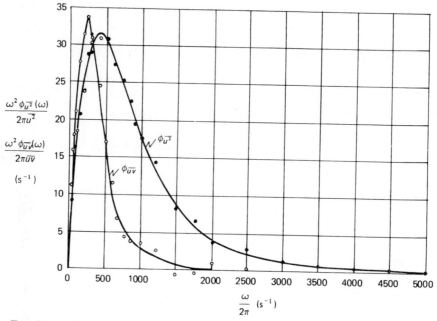

FIG. 21.1 Comparison of intensity and Reynolds stress spectra in channel flow, showing former extending to much higher frequencies. (Detailed form of coordinates is unimportant except that ordinate is proportional to (frequency)^2 × spectrum function.) From Ref. [240].

profiles at different distances downstream in laminar flow (as discussed in Sections 11.4 and 11.6). However, self-preservation requires not only that the mean velocity distribution should be similar, but also that all the parameters associated with the turbulence should have similar distributions at different stations.

The equations of motion permit self-preservation only in particular cases. For example, it can occur in a wake only sufficiently far downstream for the mean velocity deficit at the wake centre to be small compared with the total mean velocity. Such flows approach asymptotically to the self-preserving form, although often very slowly; for reasons to be seen in Section 21.4, turbulent flows have a long 'memory' of upstream conditions. When a flow is self-preserving, a description of the turbulent motion developed from measurements at one station can be applied to the whole flow.

The ideas of Reynolds number similarity and self-preservation taken in conjunction allow certain features of turbulent flows to be derived rather simply. We may illustrate this most readily by considering the development of the mean flow of a two-dimensional jet. As for a laminar flow, the jet is characterized, far enough downstream from the orifice, by M, the momentum transported per unit time over each cross-section. Ignoring a small contribution from the non-linear effects of the velocity fluctuations,

$$M = \int_{-\infty}^{\infty} \rho U^2 \, dy. \tag{21.1}$$

Putting, analogously with (11.37),

$$U_{\max} \propto x^m; \quad \Delta \propto x^n \tag{21.2}$$

(the origin of x being adjustable) conservation of momentum again gives

$$2m + n = 0. \tag{21.3}$$

Secondly, the concept of Reynolds number similarity implies that ν does not enter into the counterpart of relationship (11.45) and so

$$\Delta = \Delta(x, M, \rho). \tag{21.4}$$

Dimensional requirements now imply that

$$\Delta \propto x, \tag{21.5}$$

that is

$$n = 1. \tag{21.6}$$

Fig. 21.2 Two-dimensional turbulent jet; schlieren (see Section 25.4) picture in water, obtained by having jet at slightly different temperature from ambient. Photo by C. W. Titman.

Equation (21.3) now gives

$$m = -\tfrac{1}{2}. \tag{21.7}$$

These results have been obtained more simply than the corresponding results for laminar flow. In contrast, however, the treatment of laminar flow can be continued to indicate the detailed form of the velocity profile (eqn (11.54)), whereas nothing more can be derived about the turbulent flow without additional assumptions or experimental observations.

Figure 21.2 shows an experiment verifying the result (eqn (21.5)) that a turbulent jet spreads linearly.

This result is true also of axisymmetric jets, just the same derivation applying (although now $m = -1$). Thus such jets spread out in a conical shape.

21.2 Intermittency and entrainment

In many situations turbulent motion occurs in only a limited region—that region in which high shear has been generated. Turbulent jets and wakes are usually surrounded by non-turbulent fluid; a turbulent boundary layer usually occurs beneath an inviscid irrotational flow. In such cases the interface between the turbulent and non-turbulent regions is sharp. It has, however, a highly irregular shape with bulges and indentations of various sizes, as shown in Fig. 21.3. The bulges and indentations are carried downstream by the flow. At the same time the detailed shape of the boundary is changing; each bulge and indentation can be identified over only a limited time.

At a fixed point (such as A in Fig. 21.3), random alternations between turbulent and non-turbulent motion occur. Figure 21.4 shows oscillo-grams of the velocity fluctuations at different distances from the centre line of a wake. It is possible to define quite accurately the fraction of the time that the motion is turbulent. This quantity is called the intermittency

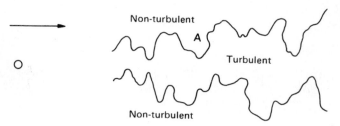

FIG. 21.3 Sketch of a wake, illustrating sharp irregular interfaces between turbulent and non-turbulent fluid.

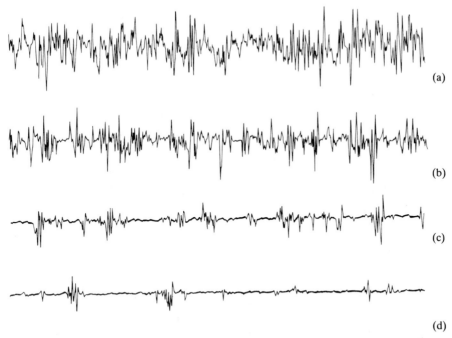

(a)

(b)

(c)

(d)

Fig. 21.4   Oscillograms in turbulent wake (Re = 6500) at $x/d = 28$, showing increasing intermittency at edge. $y/d =$ (a) 0.87, (b) 2.25, (c) 3.4, (d) 4.2. (Note: these traces were obtained with electronic boosting of high frequencies; this improves the contrast between laminar and turbulent periods by showing a quantity related to the vorticity.)

factor γ. Figure 21.5 shows its distribution in a two-dimensional wake. At the centre of the wake γ is 1; the motion is always turbulent. Outside the wake γ is 0; turbulent motion never penetrates there. But over a substantial fraction of the wake width, turbulent and non-turbulent motion alternate. Similar intermittency distributions can be determined for other flows.

When dye (or smoke) has been introduced far upstream within the turbulent region, the boundary of the dyed region closely approximates to the boundary of the turbulent region. The effect of intermittency can thus be seen in, for example, Figs 3.11, 18.9(b) and (e), and 21.6. (It is less apparent in Fig. 21.2, because the optical technique averages in the direction normal to the picture and the bulges and indentations are three-dimensional.)

At the interface between turbulent and non-turbulent fluid, the turbulence spreads. Fluid just in the non-turbulent region will a short time later be in turbulent motion. This spreading is part of the process of

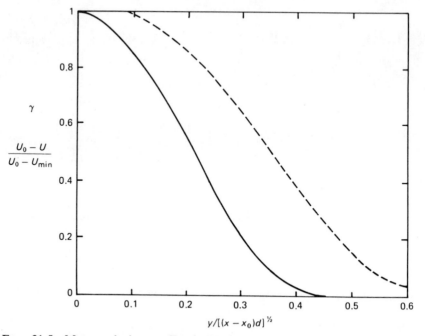

Fɪɢ. 21.5 Mean velocity profile (solid line) and intermittency distribution (broken line) in a two-dimensional wake. Abscissa is distance from centre of wake non-dimensionalized in the way appropriate to self-preservation; velocity ordinate corresponds to self-preservation because $U_0 - U_{min} \approx 0.9 U_0 [(x - x_0)/d]^{1/2}$. Both curves are experimental, although the data points are not shown. Measurements at Re = 1360, $x/d > 500$ [381]. Results for Re = 2700, $x/d = 400$ [155] are closely similar.

entrainment for a turbulent flow. Entrainment by laminar flows—the fact that a jet, for example, draws fluid into itself from the sides—has been discussed in Sections 11.6 and 11.7. In particular, through relationships (11.64)–(11.66), the entrainment was related to energy dissipation. Turbulent flows have much higher entrainment rates than the corresponding laminar flows, a fact that may be related to the additional dissipation by the velocity fluctuations. Associated with the entrainment of new fluid into a flow such as a jet there must be a process by which new fluid becomes turbulent. Otherwise a decreasing fraction of the flow would be turbulent, and, for example, self-preservation could not occur. This process is the spreading at the interface.

We can now see the reason why the interface is sharp. Associated with the velocity fluctuations of any turbulent motion are intense vorticity fluctuations. (The importance of vorticity in the dynamics of turbulence is apparent from the discussion in Section 20.4.) We will suppose that the flow in the non-turbulent region is irrotational, as is usually the case in the examples that have been mentioned. The acquisition of vorticity by initially irrotational fluid can only be brought about through the action of viscosity (Section 10.3). Thus the spreading of the turbulence at the interface involves the action of viscosity and must be effected by those eddies for which viscosity is significant; i.e. the small eddies. The length scale over which the change from turbulent to non-turbulent (rotational to irrotational) motion occurs is the size of these eddies, and so the interface appears very sharp on the scale of the flow as a whole.

The shape of the interface is, on the other hand, influenced by eddies of all sizes. The irregular and changing corrugations occur with a wide range of length scales. In particular, the largest bulges and indentations are produced by the large eddies to be described in Sections 21.4 and 21.6; these eddies are responsible for the fact that the region in which γ is non-zero but less than 1 extends over a sizeable fraction of the flow (see, e.g., Fig. 21.5).

The flow outside the interface, although called non-turbulent, does involve velocity fluctuations [303]. These are produced by the neighbouring turbulent region. Evidently the motion of a bulge in the interface changes the pattern of irrotational motion outside it. However, these velocity fluctuations are entirely irrotational and are dynamically quite different from turbulent fluctuations. They attenuate rapidly with distance from the interface. We now see that the intermittency factor is most appropriately defined as the fraction of the time that vorticity fluctuations are occurring.

The processes at the interface constitute a subject of study in their own right [231]. In particular, one wants to understand the way in which the turbulence spreads. We shall not consider this topic here except to make the following point. Although the action of viscosity is essential to the spreading process, the concept of Reynolds number similarity implies that the spreading rate does not depend on the magnitude of the viscosity. Experimentally this appears to be the case. However, the earlier discussion of Reynolds number similarity (Section 20.3) does not extend to this process. The detailed dynamics are uncertain, but, presumably, the situation is analogous with the energy cascade; although the spreading is brought about by the small eddies its rate is governed by the larger eddies. The total area of the interface, over which the spreading is occurring at any instant, is determined by these larger eddies [328].

Figure 21.6 illustrates and summarizes the entrainment process described in this section. Two sources, side by side, emit water into the uniform flow in a water channel. The water from the lower source is

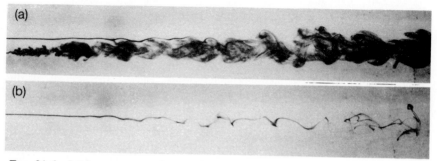

FIG. 21.6   Jet in concurrent flow and neighbouring dye streak: (a) jet also dyed;
(b) jet not dyed.

emitted sufficiently faster than the main flow that it forms a turbulent jet.
The water from the upper source is emitted at the same speed as the
main flow; there is thus no shear and no turbulence generation. This
source merely provides a way of marking fluid with minimum dynamical
effect. In Fig. 21.6(a), both sources emit dyed water. The jet is visible (it
has a rather different structure from jets considered elsewhere in this
book because of the motion of the surrounding fluid). The way in which
its spreading brings the dye from the other source into turbulent motion
can be seen. In Fig. 21.6(b), only water from the passive source is dyed,
although the jet is still present; this shows more clearly the effect of
entrainment on the initially laminar fluid. The figure also illustrates the
processes shown schematically in Fig. 20.5; the longer the dye streak has
been inside the turbulent region, the wider is the range of length scales
on which it has been distorted.

21.3   The turbulent wake

The understanding of the dynamics of turbulence in shear flows depends
primarily on the measurement and interpretation of the parameters
introduced in Chapter 19. Further discussion thus has to focus attention
on particular flows, and we now consider the two-dimensional wake to
exemplify the procedures and the ideas that come from them. The wake
is typical of shear flows away from any solid boundary, known as free
shear flows. We shall see in Section 21.5 that additional ideas are needed
for flows adjacent to walls.

   As usual the Cartesian coordinate system used to specify directions has
x in the mean flow direction, y across the wake, and z parallel to the axis
of the body producing the wake.

   Obvious first measurements, after the mean velocity profile of Fig.

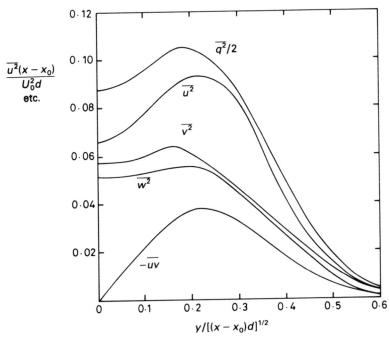

$$\frac{\overline{u^2}(x - x_0)}{U_0^2 d}$$
etc.

$$\frac{y}{[(x - x_0)d]^{1/2}}$$

FIG. 21.7 Distributions of intensity and Reynolds stress in a two-dimensional wake. Coordinates are normalized in the way appropriate to self-preservation; $y = 0$ corresponds to wake centre. Curves represent experimental data for Re = 1360, $x/d > 500$ [381] with modifications (particularly to $\overline{v^2}$ curve) for consistency with more recent data for Re = 2700, $x/d = 400$ [155].

21.5, are $\overline{q^2}$, the separate contributions to this ($\overline{u^2}$, $\overline{v^2}$, and $\overline{w^2}$), and the Reynolds stress $-\overline{uv}$. Figure 21.7 shows the distributions of these across a wake.

The Reynolds stress is zero right at the middle of the wake, by symmetry, and, of course, it falls to zero outside the wake. Its maximum is in the vicinity of the maximum of $\partial U/\partial y$ (cf. Fig. 21.5), in line with the general analogy between the Reynolds stress and a viscous stress mentioned in Section 19.3. $-\overline{uv}$ is much larger than $v\, \partial U/\partial y$. It causes the wake to increase in width and decrease in velocity deficit with distance downstream.

The interpretation of these observations is assisted by information on the energy balance; which processes are supplying energy to the turbulence at each point and which are removing it? Figure 21.8 shows distributions of each of the terms in eqn (19.22) measured in a wake.

The maximum energy production is close to the position of maximum Reynolds stress. The turbulence is being kept going by the working of this against the mean velocity gradient. At the centre of the wake, the production is zero (since both $\overline{uv}$ and $\partial U/\partial y$ are zero there). The dissipation, on the other hand, has a maximum at the centre. This is balanced primarily by the advection term $-\frac{1}{2}U\,\partial\overline{q^2}/\partial x$. As the mean velocity deficit decreases with x, so, in accordance with self-preservation, the scale of $\overline{q^2}$ decreases; thus the fluid at the centre of the wake is losing turbulent energy as it travels downstream.

At the outer edge of the wake, on the other hand, the advection term has the opposite sign. This corresponds to the supply of turbulent energy to previously non-turbulent fluid as the wake spreads. Figure 21.8 shows that this energy is supplied by the outward transport of energy by the fluctuations themselves from the region where the production is large.

When measurements such as those described above are supplemented by measurements of correlations and spectra and by flow visualization

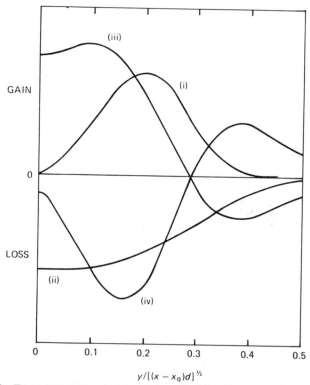

FIG. 21.8 Experimentally determined turbulent energy balance in a two-dimensional wake. Abscissa scale as in previous graphs, ordinate scale arbitrary (Re = 1360, $x/d > 500$). (i) Production, $-\overline{uv}\,\partial U/\partial y$ (ii) dissipation, $\nu\overline{u_i\partial^2 u_i/\partial x_j^2}$ (iii) advection, $-\frac{1}{2}U\partial\overline{q^2}/\partial x$ (note $\frac{1}{2}V\partial\overline{q^2}/\partial y$ is negligible in a wake); (iv) turbulent transport, $-\partial(\frac{1}{2}\overline{q^2 v}+\overline{pv}/\rho)/\partial y$. From Ref. [381].

experiments, some ideas may be developed about the role of eddies of different sizes in the dynamics of the turbulence. We have already seen that the smaller eddies contribute relatively less to the Reynolds stress than to the energy (illustrated for a different flow by Fig. 21.1). More generally, the smaller eddies have a structure that is less characteristic of the particular flow than the larger eddies and the description of their behaviour developed from observations in grid turbulence (Chapter 20) may be applied. The later stages of the energy cascade are similar in all flows. It is because of this that observations in shear flows may be used for experimental investigations of the Kolmogorov law, eqn (20.11).

The larger eddies, which do play a role in the generation of the Reynolds stress, must be more characteristic of the particular flow, since the Reynolds stress distribution necessarily varies from flow to flow. So long as the eddies are still relatively small compared with the length scale of the mean flow (e.g. wake width), their features are adequately described by the considerations of Section 19.3. The anisotropy associated with Reynolds stress production (Fig. 19.2) arises essentially in the way summarized by Fig. 19.3. Usually, the largest contribution to the energy of the turbulence is made by eddies large enough to be oriented in this way but small enough to be within one part of the mean velocity profile.

Still larger eddies extend across much of the flow and there will be appreciable variation of the mean shear within their length scale. The ideas of Fig. 19.3 are then too great an oversimplification. The large eddies play a role out of proportion to their contribution to the turbulent energy, both in the interaction between the mean flow and the turbulence and in the turbulent energy transfer process involved in Fig. 21.8. We consider these large-scale motions in the next section.

21.4 Coherent structures: in general and in a wake

We are continuing to use the two-dimensional wake as a context for ideas that apply to other flows as well. Some general comments about the large eddies are needed before we consider a wake specifically. In recent years the larger eddies of turbulent shear flows have become known as coherent structures [113, 208]. (It is a matter of some controversy whether the introduction of this name into the subject was essentially giving a new name to an older concept or whether it involved a significantly new point of view. In my own opinion it was largely the former.) The name emphasizes that it is never adequate to describe the fluctuations in a turbulent flow as totally random. There is much evidence that the fluctuation field involves distinctive patterns of motion. In any given flow, similar, although not identical, patterns occur repeatedly. Each such coherent structure is identifiable for only a limited time. There may sometimes be a tendency for a few of the structures to occur in a group or with a temporary periodicity, but, in general, the random aspect of turbulence is manifested in when and where coherent structures are to be found.

Figure 21.9 shows a simple example of the evidence for the existence of a high degree of organization in the larger scale motions. It shows measurements in a two-dimensional wake of the correlation of the y-components of the velocity at two points separated firstly by a distance r_x in the x-direction and secondly by a distance r_z in the z-direction (both measured at a distance from the centre plane corresponding to $(U_0 - U)/(U_0 - U_{min}) = 0.65$, cf. Fig. 21.5). In isotropic turbulence these two curves would be identical. It is seen that this is far from the case; in particular $R_{yy}(r_x)$ has a marked negative region, whereas $R_{yy}(r_z)$ is

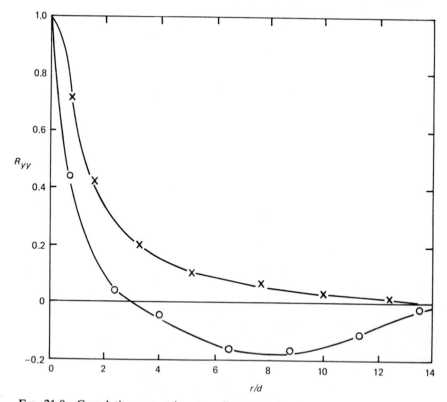

FIG. 21.9   Correlation curves in a two-dimensional wake for the y-component of the velocity fluctuation and points separated in the x($\bigcirc$) and z($\times$) directions ($Re = 1300$, $x/d = 533$, $y/d = 4$). Data from Ref. [186].

always positive. This implies that the return flow for velocity fluctuations towards or away from the centre of the wake occurs dominantly in the upstream and downstream directions and little in the transverse (z and $-z$) directions. Such a marked contrast between the curves seems unlikely to arise unless there are coherent features in the large eddy motion.

The accumulated evidence of measurements of statistical quantities (including conditional sampling, Section 19.6), flow visualization, and numerical experiments (Section 25.5) supports this inference for all flows that have been investigated in depth. It also, of course, indicates what the coherent structures are like. Different flows have similarities and differences. Some of the structures observed in two-dimensional jets [65, 293, 377] are, for example, similar to those to be described for wakes below, and a unified account for all flows can be attempted [46]. It is, however, still found necessary to investigate each flow of interest separately.

The structures are commonly described in terms of models, such as those illustrated by Figs. 21.10 and 21.11 below or those described in Section 21.6 for a boundary layer. The word 'model' implies a physical description that does not attempt to be complete or accurate in detail but that encapsulates what are thought to be the central features of the phenomenon. Although the existence of coherent structures is well established, there is often controversy as to the most appropriate models. Also, new data still often lead to significant revisions. Thus current models may be superseded in due course.

The large eddies are the longest-lived features of a turbulent flow. Their existence is thus the reason for the long memory and slow approach to self-preservation mentioned in Section 21.1. The main discussion below of wake structure is intended to apply primarily to a self-preserving wake. First, however, we consider briefly what may happen further upstream, partly to illustrate how the long memory comes about and partly to make a link with earlier discussions of wakes; i.e. we look at the 'intermediate wake', (intermediate to the 'near wake' immediately behind the obstacle and the 'far wake' which has become self-preserving). We have seen in Sections 3.3 and 17.8 that, over a wide Reynolds number range, a turbulent vortex street is generated as a stage in the transition process of a cylinder wake. There is no sharp distinction between the later stages of transition and the earlier ones of turbulent motion. Thus the residual vortices of the street form coherent structures in the turbulent intermediate wake [222, 297]. This implies that changes in the transition process, due to a change in the Reynolds number or to a different obstacle geometry, will lead to changes in the form of the coherent structures.

In a self-preserving flow the structure should be independent of such changes. It is uncertain just how far downstream this independence is achieved [143, 297]. It thus needs to be said that, although one expects that the models described below will apply to the wakes of other obstacles, they are based primarily on observations with circular cylinders.

There appear to be two distinct types of coherent structure in a two-dimensional far wake. These are shown in highly schematic ways in Figs. 21.10 and 21.11. These models were introduced quite early in the history of the topic [186] (incidentally pre-dating the name 'coherent structure'). Ideas about other flows, particularly boundary layers to be considered in Section 21.6 have evolved greatly in the same period. This may suggest that our ideas about wakes should undergo similar evolution, but recent experiments have tended to confirm the earlier picture [104, 286, 385].

The type of coherent eddy structure illustrated by Fig. 21.10 consists of a jet-like flow outwards from near the centre of the wake, producing a bulge on the interface between the turbulent and non-turbulent fluid. As it travels outwards, the fluid tends to curve round in the flow direction. The return flow towards the centre of the wake is more diffuse. The figure shows schematically the pattern of motion produced by several such eddies. There is in fact some disagreement about how much longitudinal periodicity is associated with this type of motion and whether

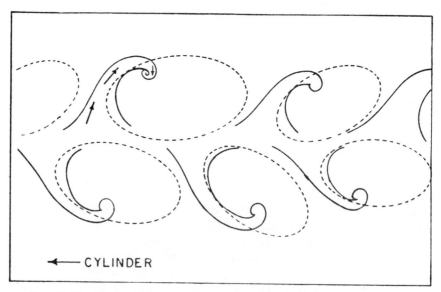

CYLINDER

FIG. 21.10   Sketch of 'jet-type' large eddies in a wake. From Ref. [186].

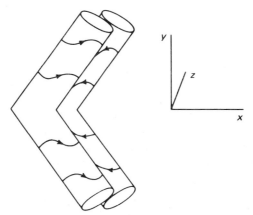

FIG. 21.11   Schematic representation of double roller eddy in a wake. [186].

the eddies on opposite sides are correlated in position [82, 104, 385]. Probably several eddies commonly occur simultaneously, as shown in the figure, with an approximately but not exactly even spacing.

It is these eddies that relate particularly to the brief discussion above of the correlation measurements in Fig. 21.9.

Also, these eddies are particularly apparent in experiments with dyed wakes. One needs cine-film to learn much from such flow visualization, but the eddies can be seen in Fig. 3.11.

Figure 21.11 is a model of the other type of coherent structure. In each half of the wake the eddy consists of two parallel vortex tubes of opposite sense. There is some flow along these tubes as well as around them, so that a spiralling motion results. The tubes may not be straight, but one does not have enough information to assign any other shape to them. Each tube has its axis in an xy-plane, with the part near the edge of the wake further downstream than the part near the centre, and the two tubes are separated in the z-direction. Such motions occur simultaneously on both sides of the wake so that the structure as a whole has the shape indicated in the figure.

Suggested mechanisms for the origin of coherent structures include two important ideas. Firstly, the mean velocity profile may be liable to local instability, somewhat analogous to instability of laminar flow. Because it may be triggered by the vigorous smaller-scale fluctuations, such an instability will normally occur intermittently. The resemblance of Fig. 21.10 to a portion of a vortex street (Section 17.8) strongly indicates that this type of process occurs in a wake (though there is the difference that the turbulent eddies are of limited extent in the third direction). Secondly, there are processes which select certain patterns of motion in preference to others. The idea of vortex stretching (Section 6.6) is central to these processes, and one notes that the vortex tubes in Fig. 21.11 are aligned so that they are being stretched by the mean shear—cf. Fig. 6.10 (more formally, they are approximately along the positive principal axis of the mean flow rate of strain tensor). A much fuller theory has been developed of how motions of this type can be generated through the action of the mean shear on initially isotropic fluctuations, but the details are not apparent without the mathematical analysis [383]. It is very likely that the two processes summarized above are simultaneous processes within an overall mechanism that is not yet fully understood. Correspondingly, the two eddy types in Figs. 21.10 and 21.11 may be related in some way rather than occurring independently [286]. It is perhaps significant that stretching of the vortex tubes in Fig. 21.11 could be produced by the structure in Fig. 21.10 as well as by the mean flow, but this is a matter requiring further investigation.

21.5 Turbulent motion near a wall [328]

The ideas developed so far provide a very incomplete story when the flow is adjacent to a solid boundary. At such a boundary (at rest), the boundary condition that the fluid velocity is zero applies at every instant. Thus it applies to the mean velocity and to the fluctuations separately,

$$U_i = 0; \quad u_i = 0. \tag{21.8}$$

The fact that the fluctuations drop to zero at the wall has the particular implication that the Reynolds stress is zero:

$$-\overline{uv} = 0. \tag{21.9}$$

The only stress exerted directly on the wall is the viscous one.

Away from the wall, on the other hand, the turbulence generates a Reynolds stress, large compared with the viscous stress, in the usual way. The total stress

$$\tau = \mu \, \partial U/\partial y - \rho \overline{uv} \tag{21.10}$$

(we are still considering a two-dimensional flow for which the boundary layer equations apply). This cannot vary rapidly with y without producing a very large mean acceleration and thus requiring an improbable mean flow distribution. (For example, in channel flow with no variation of mean quantities in the x-direction, τ varies linearly across the channel as it does for laminar motion—the first integral of eqn (2.6).) Consequently, the viscous stress close to the wall must match up with the Reynolds stress further out. Although τ varies only slowly, $\mu \, \partial U/\partial y$ and $-\rho \overline{uv}$

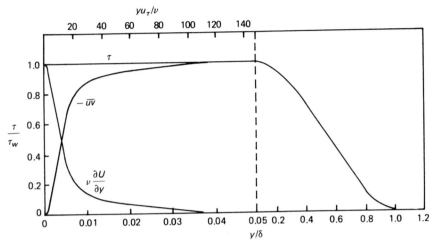

FIG. 21.12 Distributions of the total stress, Reynolds stress, and viscous stress across a boundary layer ($Re_\delta = 7 \times 10^4$). Note 30-fold change in abscissa scale at $y/\delta = 0.05$. Based on data in Refs. [223, 337].

each vary rapidly, the former being much larger at the wall than in laminar flow but becoming very small away from the wall, the latter being large away from the wall and zero at the wall.

Figure 21.12 illustrates this by showing distributions of τ and of its two contributions measured in a zero-pressure-gradient boundary layer. The region where the viscous stress makes a large contribution is a small fraction of the total boundary layer thickness (note the change in abscissa scale). To generate this distribution of viscous stress, the mean velocity profile must rise steeply at the wall and then become comparatively flat. Figure 21.13 compares a typical turbulent boundary layer profile with a corresponding laminar one.

Parallel considerations apply to turbulent pipe flow. Figure 21.14 compares laminar and turbulent mean velocity profiles, firstly for the same flow rate and secondly for the same pressure gradient.

It is now clear that the presence of the wall causes the fluid viscosity to enter in a much more important way into the dynamics of turbulent motion than it does for free flows. The concept of Reynolds number similarity is no longer so useful.

This effect extends to the fluctuations as well as to the mean flow. Figure 21.15 shows distributions of $\overline{q^2}$ and of its components $\overline{u^2}$, $\overline{v^2}$, and $\overline{w^2}$ across a boundary layer. (The reason for the alternative coordinates will be seen later.) Large variations are to be observed with strong maxima close to the wall. (Figure 21.15 shows also the distribution of the

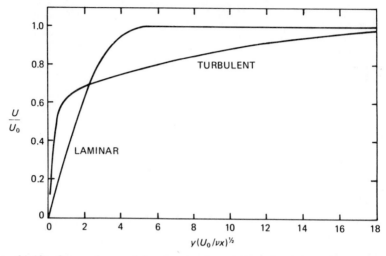

Fɪɢ. 21.13 Comparison of laminar and turbulent boundary layer velocity profiles. (Laminar: Blasius profile; turbulent: experimental profile at $Re_x = 9.2 \times 10^5$, with tripping at $x = 0$ to promote transition.)

intermittency factor and it is evident that intermittency is affecting the intensity distributions only at larger values of y/δ.)

This observation can be interpreted in connection with the turbulent energy balance, Fig. 21.16. The rate of energy production, $-\overline{uv}\,\partial U/\partial y$, also has a large peak close to the wall. The reason is related to the above discussion of the stress. Very close to the wall $-\overline{uv}$ is small and so there is little energy production; far from it, $\partial U/\partial y$ is small with the same consequence. Mathematically,

$$-\overline{uv}\frac{\partial U}{\partial y} = \frac{1}{\rho}\left(\tau - \mu\frac{\partial U}{\partial y}\right)\frac{\partial U}{\partial y} \qquad (21.11)$$

and, if τ is treated as a constant, this is a maximum when

$$\mu\,\partial U/\partial y = \tau/2. \qquad (21.12)$$

Thus the energy production is largest in the vicinity of the changeover from a predominantly viscous stress to a predominantly turbulent one.

The energy balance diagram, Fig. 21.16, shows that over the region where the production is high the other important term is the dissipation; the remaining terms in eqn (19.22) are much less significant (apart from some rather complicated transfer processes very close the wall). The inner part of the boundary layer is thus sometimes said to be in 'local equilibrium', meaning that there is a local balance between the process supplying energy to the turbulence and that removing it.

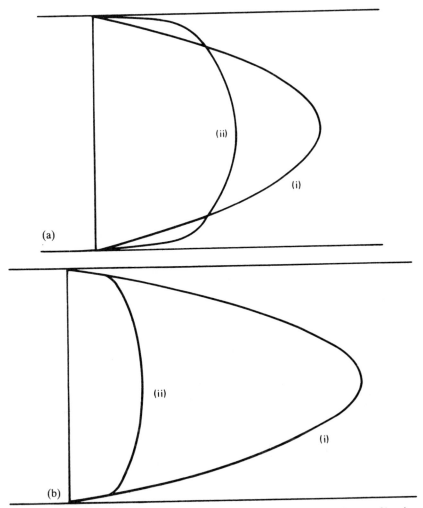

Fig. 21.14   Comparison of (i) laminar and (ii) turbulent velocity profiles in a pipe for (a) the same mean velocity, and (b) the same pressure gradient. (The diagrams correspond to a turbulent flow Reynolds number of about 4000; for higher Re the contrast is more marked, cf. Fig. 2.11.)

Towards the outer edge of the boundary layer the quantities in eqn (19.22) are much smaller, but all the terms are now important. The dissipation somewhat exceeds the production and there is also a loss by advection—corresponding, as in a wake, to the supply of energy to newly turbulent fluid. These losses are compensated by turbulent transport. The turbulence in the outer region is maintained by that in the inner region.

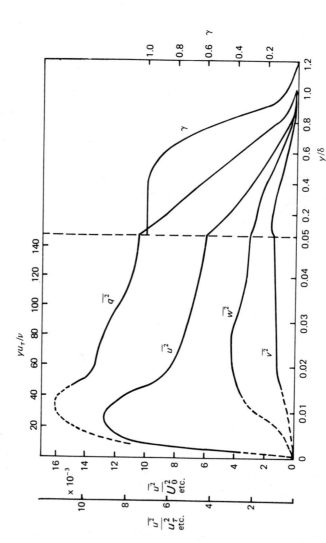

Fig. 21.15 Distributions of total and component intensities and of intermittency factor across a boundary layer ($Re_\delta = 7 \times 10^4$). Based on data in Ref. [223].

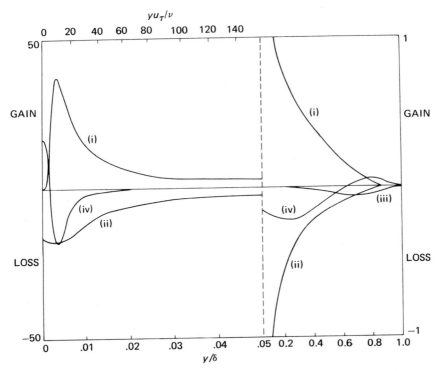

FIG. 21.16 Energy balance diagram (principal features) for a turbulent boundary layer ($Re_\delta = 7 \times 10^4$). (i) Production; (ii) dissipation; (iii) advection; (iv)
transport across boundary layer (primarily by turbulence, but including viscous
contribution at low y/δ). (Inner part based on Ref. [328]; outer part based on
Ref. [223] but modified to reduce discrepancies with Ref. [328] and with
requirement that integrated transport should be zero.)

More detailed consideration of these and related ideas has led to a
division of the turbulent boundary layer into inner and outer regions. The
dynamical processes occurring in the latter are not dissimilar from those
in a wake, but the inner region requires its own model. The ideas that
have been developed for this region can be applied to most turbulent
flows adjacent to a wall: for example, to turbulent channel and pipe flow
(although not, for example, to a boundary layer close to separation).
Because of the local equilibrium, the structure of the inner region is
substantially independent of the outer flow.

We thus want to specify the flow in the 'wall-region' by a velocity
scale characteristic of this region. This is provided by τ_W, the value of τ
at the wall, which has the dimensions of density × (velocity)^2. Hence we

define

$$u_\tau = (\tau_W/\rho)^{1/2} \qquad\qquad (21.13)$$

as our velocity scale. u_τ depends on the flow as a whole, but once it is specified, the structure of the wall region is specified. It is confirmed experimentally that the turbulent intensity distributions scale with u_τ. For example, the maximum value of $\overline{u^2}$ is always about $8u_\tau^2$.

The relationship between u_τ and the external velocity U_0 for a boundary layer depends (rather weakly) on the Reynolds number. Under typical laboratory conditions, u_τ/U_0 is in the range 0.035 to 0.05.

Conditions in the wall region are now specified by the three parameters: u_τ, the distance from the wall y, and the kinematic viscosity ν. The mean velocity U is a function of these,

$$U = f(u_\tau, y, \nu) \qquad\qquad (21.14)$$

from which dimensional considerations give

$$U/u_\tau = f(yu_\tau/\nu). \qquad\qquad (21.15)$$

It is again confirmed experimentally that various turbulent wall flows have a common profile of this form.

We have seen that the viscosity is important only very close to the wall. With increasing y it ceases to play a role long before parameters other than those in (21.14) have an influence. One can then say that the mean velocity gradient depends only on u_τ and y,

$$\partial U/\partial y = f(u_\tau, y) \qquad\qquad (21.16)$$

although one cannot say the same for U because it is separated from its origin by a region in which ν is important. Dimensional analysis applied to (21.16) gives

$$\partial U/\partial y = u_\tau/Ky \qquad\qquad (21.17)$$

where K is a universal constant (the Kármán constant) of turbulent wall flows. It is found experimentally that $K = 0.41$. Integration of (21.17) gives, in a form corresponding to (21.15),

$$\frac{U}{u_\tau} = \frac{1}{K}\left[\ln\left(\frac{yu_\tau}{\nu}\right) + A\right] \qquad\qquad (21.18)$$

where A is another constant.

This logarithmic profile is one of the most famous results in the study of turbulent flows [315]. In Fig. 21.17, the velocity profile of Fig. 21.13 is plotted with log–linear coordinates and the straight line corresponding to eqn (21.18) is evident.

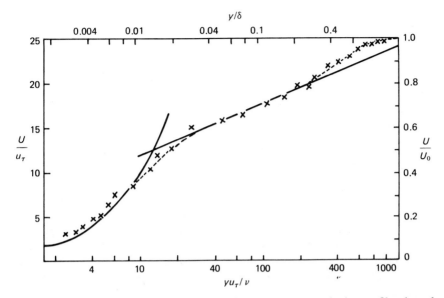

FIG. 21.17   Example of a turbulent boundary layer mean velocity profile plotted on log-linear coordinates ($Re_\delta = 2.3 \times 10^4$). Solid lines correspond to eqns (21.18) and (21.24); broken line shows trend of data.

The profile departs from the logarithmic form when $yu_\tau/\nu < 30$, because of the importance of viscosity in this region. Very close to the wall viscous action is dominant. At the wall

$$u = v = w = 0 \tag{21.19}$$

and, taken in conjunction with the continuity equation

$$\frac{\partial u}{\partial x} + \frac{\partial v}{\partial y} + \frac{\partial w}{\partial z} = 0 \tag{21.20}$$

this gives

$$\partial v/\partial y = 0 \quad \text{at} \quad y = 0. \tag{21.21}$$

Hence,

$$\frac{\partial(\overline{uv})}{\partial y} = \frac{\partial^2(\overline{uv})}{\partial y^2} = 0 \quad \text{at} \quad y = 0 \tag{21.22}$$

and treating τ as a constant in eqn (21.10)

$$\frac{\partial^2 U}{\partial y^2} = \frac{\partial^3 U}{\partial y^3} = 0 \quad \text{at} \quad y = 0. \tag{21.23}$$

We thus expect there to be a significant region right next to the wall in which the velocity profile is linear,

$$U/u_\tau = yu_\tau/\nu. \tag{21.24}$$

The curve corresponding to this is included in Fig. 21.17 (the logarithmic plotting now serving to expand this region) and it is seen that the experimental points fall on it for $yu_\tau/\nu < 8$.

This thin region is known as the viscous sub-layer. Sometimes it is called the laminar sub-layer, but that is less accurate as it contains large velocity fluctuations; $\overline{u^2}$ and $\overline{w^2}$ are free to rise more rapidly from the wall than $\overline{uv}$. Although the viscous sub-layer is very thin, a substantial fraction of the total mean velocity change across the boundary layer occurs within it.

The upper limit in y of the logarithmic profile occurs where dynamical processes relating to the boundary layer as a whole become significant. This is observed experimentally to occur around $y/\delta = 0.2$, where δ is the total boundary layer thickness (defined, say, so that $U = 0.99\,U_0$ at $y = \delta$). The corresponding value of yu_τ/ν (about 200 in Fig. 21.17) depends on the Reynolds number $U_0\delta/\nu$.

21.6 Coherent structures in a boundary layer

The fact that any turbulent flow involves large eddies with a coherent structure was introduced in Section 21.4. Wall flows have some special features which we illustrate by considering boundary layers. We have seen that models of these structures are synthesized from various types of measurement and observation; in the case of boundary layers, however, flow visualization has been particularly influential. Again, still pictures can provide only a limited impression of what is seen in motion, but Figs. 21.18 and 21.19 illustrate important aspects.

The first thing that may be noted is that the contrast between different parts of Fig. 21.18 provides immediate evidence for marked organization of the structures (the point that was illustrated in the case of a wake by Fig. 21.9). These photographs were obtained by illuminating a very thin layer of a smoke-filled boundary layer. In all four pictures the illuminated plane is at 45° to the flow direction (and across the flow in the z-direction); but in parts (a) and (c), it is tilted downstream (i.e. the top of the picture showing the outer part of the boundary layer is downstream of the bottom showing the part close to the wall), whereas in parts (b) and (d) it is tilted upstream. Parts (a) and (b) are from the same flow, as are parts (c) and (d)—the difference between the two flows being in the Reynolds number. The very different patterns in the different

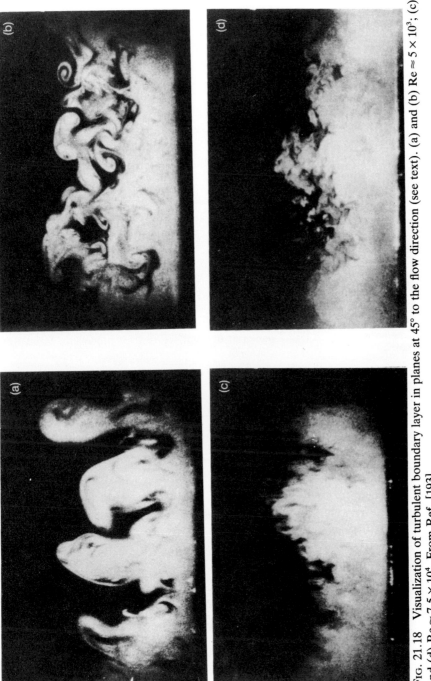

FIG. 21.18   Visualization of turbulent boundary layer in planes at 45° to the flow direction (see text). (a) and (b) Re $\approx 5 \times 10^3$; (c) and (d) Re $\approx 7.5 \times 10^4$. From Ref. [193].

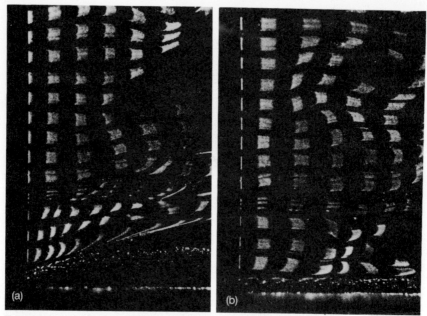

FIG. 21.19 Turbulent boundary layer (a) during eruption, (b) during inrush. Flow is from left to right, wall is at bottom; see text for procedure. From Ref. [189].

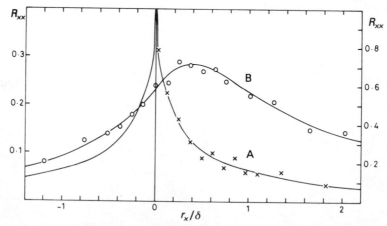

FIG. 21.20 Correlation curves in a boundary layer for the x-component of the velocity fluctuations plotted with respect to separation in the x-direction, without (curve A: right hand ordinate scale) and with (curve B: left hand ordinate scale) a separation also in the y-direction. $\mathrm{Re}_\delta = 2.3 \times 10^4$; $\times$ $y/\delta = 0.097$; $\bigcirc$ y/δ (of fixed probe) $= 0.112$, $r_y/\delta = 0.208$. Data from Ref. [387].

orientations in each flow must reflect orientation of the coherent structures.

In Figs. 21.18(a) and (c) the structures appear to be orientated approximately in the plane of the picture whereas in (b) and (d) they appear to intersect the new plane. This interpretation is confirmed by correlation measurements of which an example is given in Fig. 21.20; this shows the correlation of x-components of the velocity as a function of separation in the x-direction, and we consider curve B for which there is also a (fixed) separation in the y-direction: the asymmetry of this curve about $r_x = 0$ illustrates the tendency for the large eddies to be tilted downstream. The accumulated evidence, centred on the interpretation of the flow visualization experiments illustrated by Fig. 21.18, suggests that a predominant feature of boundary layer turbulence is the 'hairpin' vortex or 'horseshoe' vortex shown schematically in Fig. 21.21; which name more accurately summarizes the geometry depends on the Reynolds number. The smoke patterns in Figs. 21.18(b) and (d) are produced

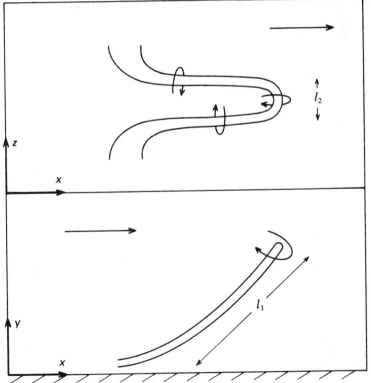

FIG. 21.21 A 'hairpin' or 'horseshoe' vortex; model of coherent structure in turbulent boundary layers.

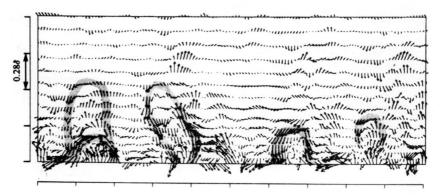

0.28δ

FIG. 21.22 Example of projection of instantaneous vorticity vectors on inclined plane (see text) in numerical simulation of turbulent channel flow. Lower boundary of picture at $y/a = 0.193$ ($yu_\tau/\nu = 123$); upper boundary at $y/a = 1$ (channel centre). From Ref. [279].

by the intersections of the vortices with the plane, and the motions associated with the two arms of the hairpin can be identified in places. Note that it is motions in this plane (and vertical in the pictures) that generate the Reynolds stress (cf. Fig. 19.2).

The hairpin or horseshoe structure is also found in numerical simulations (Section 25.5) as is illustrated by Fig. 21.22. This shows the projection on a plane tilted at 45° downstream (cf. Fig. 21.18(a) and (c)) of a computed instantaneous vorticity distribution, and patterns that correspond to the structure are picked out by shading.

The structure has similarities to one half of the double roller eddy of wakes (Fig. 21.11). The vortex stretching process outlined in Section 6.6 is undoubtedly again relevant. However, it is thought that the boundary layer vortices derive their vorticity originally from the viscous sub-layer (which has, of course, no counterpart in a wake); this is then advected and stretched by a combination of the mean shear and the turbulent fluctuations to give the hairpin shape.

The evidence for this relates to the observed length scales of the vortices; the length of the vortices (l_1 in Fig. 21.21) is of the order of the total boundary layer thickness, whereas the cross-section (l_2) is of the order of the viscous sub-layer thickness. This means that the ratio l_1/l_2 varies with the Reynolds number (is in fact approximately inversely proportional to Re). It accounts for the differences between (a) and (c) and between (b) and (d) of Fig. 21.18. There is, however, a suggestion [301] that there is a hierarchy of vortices of different sizes, of which the thinnest are the most apparent in flow visualization.

Figure 21.19 illustrates other aspects of boundary layer structure. In this figure the flow visualization used the hydrogen bubble technique (Section 25.4), enabling 'dye' to be released in initially square patches from a fine wire stretched across the boundary layer perpendicular to the wall. The two pictures show the same flow at different instants, at which respectively an 'eruption' and an 'inrush' were occurring. The model based on these two types of coherent structure—particularly the former— has been central to our understanding of turbulent boundary layers for two decades (possibly to the neglect of other aspects). The relationship to the hairpin vortices described above is not wholly clear. The most likely possibility is that a number (of uncertain size but probably Reynolds number dependent) of the hairpin vortices occur within an eruption and maybe also within an inrush—an interpretation suggested primarily by numerical simulation [221].

Another name for an eruption is a 'burst' (but this is sometimes used to mean the process generating an eruption rather than the eruption itself; conversely, it has been used to mean a sequence of an eruption plus an inrush). An eruption originates close to the wall, where there are regions, separated in the z-direction, of fluid moving faster than average downstream and fluid moving more slowly than average. Fluid in the latter regions intermittently erupts into the main body of the boundary layer; the viscous sub-layer appears to 'burst'. The motion away from the wall is apparent in Fig. 21.19(a) and the spacing of the dye patches shows also that this fluid is moving downstream more slowly than average. Eruptions have a characteristic structure [113, 220, 412]. Separate eruptions are similar to one another, but they are not identical; statistical fluctuations occur as with any feature of a turbulent flow. In particular, eruptions vary in size; some, but not all, penetrate right through the boundary layer and contribute to the intermittency at its outer edge.

In its main features an inrush is just the reverse of an eruption. Fluid moving faster than average downstream is carried in towards the wall. This gives it an even more marked excess longitudinal velocity and this feature is particularly apparent in the inner half of the boundary layer. The slight variation of the longitudinal velocity in Fig. 21.19(b), as indicated by the spacing of dye patches, implies that fluid close to the wall is moving much faster than average.

The eruptions and inrushes both have structures generating a large Reynolds stress. In fact, between them, they are probably responsible for almost the whole observed Reynolds stress. There is little interaction between the turbulence and the mean flow in the intervals between eruptions and inrushes.

It is likely that these coherent structures originate through an intermittent instability of the velocity profile. This is partly analogous with the origin of the jet-like eddies in a wake (Fig. 21.10) discussed in

Section 21.4, except that it may not be so much an instability of the mean velocity profile as an instability of a profile itself temporarily generated by the turbulent fluctuations. Because the eruptions burst from the viscous sub-layer, it is often suggested that they are produced by an instability there—particularly as some features of the motion in this region resemble features of transition to turbulence in a boundary layer (for example, the development of fast- and slow-moving regions mentioned above resembles the development of three-dimensionality illustrated by Fig. 18.1) [94]. However, this interpretation faces the difficulty that essentially identical large-scale motions occur in boundary layers on rough walls, for which the flow structure close to the wall is quite different† [189].

In Section 21.5, it was argued that the only length scale relevant to the boundary layer dynamics in the wall region but outside the viscous sub-layer is the distance from the wall y. Some of the features of the coherent structures raise doubts about this conclusion. Firstly, there is the observation noted above that the length scale l_2 in Fig. 21.21 is related to the viscous sub-layer thickness (i.e. is proportional to v/u_τ and much smaller than y in the region concerned). This makes it no longer obvious that v should be excluded from eqn (21.16). Secondly, some features of the large eddies have a length scale given by the total boundary layer thickness even when y is much less than this; an example is given by curve A in Fig. 21.20, where R_{xx} remains significant at values of r_x of the order of δ and large compared with y. This makes it no longer obvious that δ should be excluded from eqn (21.14). Yet the logarithmic profile (21.18) derived from (21.14) and (21.16) is very well established observationally [315] and there can be little doubt that the derivation of it correctly reflects the way in which the turbulence determines the mean velocity distribution. The suggestion, mentioned earlier, that there is a hierarchy of hairpin vortices was made in part to help with the first of these problems [301]. The second has led to the idea that there is a 'universal' or 'active' motion, with a length scale proportional to the distance from the wall, and an 'irrelevant' or 'inactive' motion, with a much larger length scale [97]. However, this matter cannot be considered resolved and may well be central to future development of our understanding of wall flows.

21.7 Turbulent stratified flows

There are several important ways in which turbulent shear flows may be modified—by a pressure gradient [46, 328], by wall curvature [199, 283],

---

† Evidence is lacking about whether hairpin vortices occur when the wall is rough and thus whether the interpretation above, that these derive their vorticity from the sub-layer, also faces this difficulty.

by rotation of the system [214, 389], or by stratification. As an example of such effects we consider stratified flows, which have received particular attention because of their meteorological applications. Many of the important ideas have been developed in the context of such applications, principally in relation to the lowest layers of the atmosphere as the wind blows over hot or cold ground [260, 312, 360]. This section looks briefly at some of the basic ideas and illustrates them with the results of some laboratory experiments.

We are concerned with the predominantly horizontal mean flow of a fluid whose mean density varies vertically. The mean velocity also varies vertically, and we shall confine attention to two-dimensional flow. Hence, the specification of the situation is primarily in terms of the two gradients, dU/dz and $d\rho_0/dz$.

The velocity gradient can lead to generation of turbulence in the usual way through the action of inertia forces (Section 19.3). The role of the density gradient depends on its sign. If the density increases upwards, then buoyancy forces provide an additional source of energy for the turbulence. If the density decreases upwards, then the turbulence must do work against buoyancy forces, which therefore produce a loss of turbulent energy additional to viscous dissipation; turbulence cannot persist when the density gradient is too large.

From the considerations of Chapter 15, we expect the quantitative formulation of the relative importance of inertia and buoyancy forces to be made in terms of some form of the Froude number (eqn (15.7)). The role of the velocity gradient in the dynamics of turbulent flow suggests that U/L should be replaced by dU/dz. Also it is convenient to work in terms of the reciprocal of the square of this Froude number; that is in terms of

$$\mathrm{Ri} = -\frac{g(d\rho_0/dz)}{\rho_0(dU/dz)^2}. \tag{21.25}$$

This is called the Richardson number (sometimes the gradient form of the Richardson number to distinguish it from other forms defined somewhat differently). As was to be expected, it depends on the sign of the density gradient but not on that of the velocity gradient.

Negative Richardson number corresponds to a destabilizing density gradient; both shear and buoyancy give rise to turbulence generation. When $-\mathrm{Ri}$ is small, the former is dominant and the motion is essentially of the type we have been considering hitherto. When $-\mathrm{Ri}$ is large, the latter is dominant and the turbulence may be more like the free convection turbulence to be described in Section 22.7.

Positive Richardson number corresponds to a stabilizing density gradient; turbulent motion cannot be sustained when Ri becomes large.

Various experiments have been carried out to discover the more detailed dynamics of these processes. The simplest configuration in principle (although not in practice) is a flow with uniform velocity and density gradients. Such a configuration has been achieved using a special wind-tunnel with graded heated elements and flow resistance grids at its entry [407]. It was used for investigations with positive Richardson number. The numerical details of the results may not be generally applicable, since the flow was still developing in the downstream direction at the observing station (and the Reynolds number may not have been high enough for Ri to be the only parameter); but the results should indicate well the general trends. The damping action of the stratification on the turbulence is illustrated by Fig. 21.23, which shows the variation of the temperature fluctuations relative to the temperature gradient as a function of Richardson number (since the geometry was constant, the fact that the ordinate is not non-dimensional is unimportant). The turbulence is almost completely suppressed when Ri reaches 0.45.

Changes in the structure of the turbulence also occur. Figure 21.24 shows the variations of the correlation functions $-\overline{uw}/(\overline{u^2}\,\overline{w^2})^{1/2}$ and $-\overline{w\theta}/(\overline{w^2}\,\overline{\theta^2})^{1/2}$ (where θ is the temperature fluctuation); the former is the normalized Reynolds stress, whilst the correlation $-\overline{w\theta}$ plays a role in the transport of heat by the turbulence corresponding to the role of the Reynolds stress in momentum transport. The latter falls off more rapidly, indicating that the turbulence changes in a way that makes it

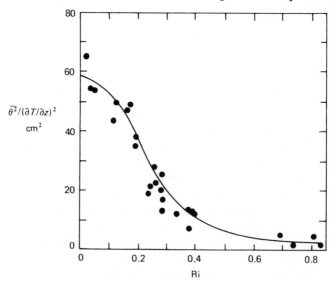

FIG. 21.23 Variation of intensity of temperature fluctuations with Richardson number in a stratified shear flow. From Ref. [407].

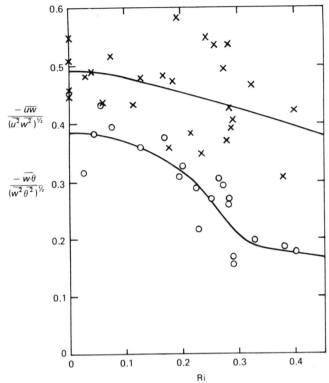

$\dfrac{-\overline{uw}}{(\overline{u^2}\,\overline{w^2})^{\frac{1}{2}}}$

$\dfrac{-\overline{w\theta}}{(\overline{w^2}\,\overline{\theta^2})^{\frac{1}{2}}}$

FIG. 21.24   Variation of normalized Reynolds stress ($\times$) and heat transfer ($\bigcirc$) with Richardson number in a stratified shear flow. From Ref. [407].

relatively less efficient as a heat transfer mechanism than as a momentum transfer mechanism. This can be understood physically in this way: a fluid particle that is displaced but then falls back to its original position without any mixing with its new environment does not transfer any heat, but it can transfer momentum through the action of pressure forces. At the higher values of the Richardson number, the turbulence may be thought of as random superposition of internal waves (Section 15.4).

In other situations, complexities may arise from variations in the Richardson number, as defined above, from place to place. As an example of the consequences of this we consider flow close to a horizontal upward-facing heated or cooled surface—sufficiently close that the velocity scale is determined entirely by the wall stress and is u_τ as discussed in Section 21.5. The temperature scale is then similarly determined by the vertical heat transfer, which, like the stress, varies little in the region under consideration. If H is the rate of heat transfer per unit area (taken as positive if the transfer is upwards, for example from hot ground to cooler air), this scale is

$$\theta_H = H/\rho C_p u_\tau. \tag{21.26}$$

u_τ enters this expression because the temperature variations needed to produce a given heat transfer are smaller when the turbulence is more vigorous. Specification of the problem in terms of u_τ and θ_H is particularly useful in application to the atmospheric boundary layer, as these are often known when the dynamics of the whole system are not.

If stratification is having little effect on the dynamics of the turbulence, we know that the velocity gradient in the region under consideration is given by eqn (21.17),†

$$\partial U/\partial z = u_\tau/Kz \tag{21.27}$$

Similar considerations applied to the temperature field lead to‡

$$\partial T/\partial z = -\theta_H/Kz. \tag{21.28}$$

Whether the turbulence is indeed unaffected by the stratification is indicated by the Richardson number

$$\mathrm{Ri} = \frac{g\alpha \, dT/dz}{(dU/dz)^2} = -\frac{g\alpha K\theta_H z}{u_\tau^2} = \frac{g\alpha KHz}{\rho C_p u_\tau^3}. \tag{21.29}$$

The Richardson number and thus the relative importance of the stratification increase with height. It is useful to introduce the height $|L|$ at which $|\mathrm{Ri}| = 1$,

$$L = -\rho C_p u_\tau^3/g\alpha KH. \tag{21.30}$$

This is the Monin–Obukhov length; like the Richardson number it is defined to be positive when the stratification is stabilizing and negative when it is destabilizing.

$|L|$ is more precisely the height at which $|\mathrm{Ri}|$ would equal unity if the profiles were unchanged by the stratification. However, L can be used as a length scale in the analysis of the changes due to stratification. Equation (21.27) may be re-stated by defining

$$\phi = \frac{Kz}{u_\tau} \frac{\partial U}{\partial z} \tag{21.31}$$

and saying that $\phi = 1$ when $\mathrm{Ri} = 0$ (or $L = \infty$). It may then be extended to

† Notation: it is conventional to denote the direction normal to the wall by y except when this is explicitly vertical, and then to denote it by z.
‡ It is a moot point whether the constant here should be the same as that in eqn (21.27), but the two are found to be sufficiently close that we need not make any distinction for the present purpose.

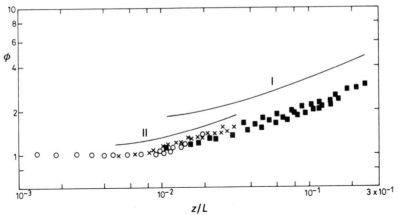

FIG. 21.25   Values of ϕ (eqn (21.31)) estimated from measurements in a wind-tunnel with a cooled floor. z for all points is such that ϕ would be 1 in the absence of stratification. Points are calculated with actual u_τ; lines with u_τ for corresponding unstratified flow (see later in Section). Values of overall Richardson number $(g\alpha\Delta T\delta/U_0^2)$: ■ and I, 0.098; × and II, 0.025; ○ 0.0105. Data from Ref. [67] (and Ref. [66]).

non-zero Ri by saying

$$\phi = f(z/L) \qquad (21.32)$$

provided that the conditions required for (21.27) are still fulfilled (i.e. z is within the wall layer but outside the viscous sub-layer). Figure 21.25 shows some observations in stably stratified flow that confirm (21.32).

Equation (21.29) implies that, when z is sufficiently small compared with $|L|$, the effect of stratification is slight. Provided z is still large enough compared with v/u_τ, the logarithmic profile will be observed. This expectation is confirmed by the region in Fig. 21.25 for which $\phi = 1$. However, departures from this occur at lower values of z/L than one might guess on the basis of eqn (21.29); they are detectable for z/L above about 0.01 and very substantial when $z/L \sim 0.1$. Generally, the effects of stable stratification tend to be underestimated by simple order-of-magnitude considerations (cf. Fig. 21.23).

Moreover, analysis of the data in the above way disguises a significant aspect of the differences between a stratified boundary layer and a corresponding unstratified one. u_τ/U_0, where U_0 is the free-stream speed, is changed by the stratification. (We are here considering primarily laboratory experiments where the comparison may be made directly. In meteorological studies, the information for a comparison is unlikely to exist, and indeed U_0 may not be a well-defined quantity whilst u_τ is.) The reason is the higher Richardson number in the outer part of the boundary layer.

Consider for example the case when Ri becomes large enough for the turbulence to be suppressed in the outer region, so that the Reynolds stress becomes small there. It is this stress that keeps the fluid closer to the wall moving; in its absence this fluid is slowed down by viscous friction at the wall. Experimental observations showing this effect clearly were made in the boundary layer under the top wall of a wind-tunnel, one section of which was heated [289]. Turbulence established before the heated section decayed in the stably stratified region. Figure 21.26 shows velocity profiles at two distances downstream from the start of the heated section. The deceleration of the fluid close to the wall is apparent. (It is this effect which is responsible for the common observation of the wind 'dropping' on a clear evening as the ground cools by radiation.)

Figure 21.25 related to the flow far downstream of the start of wall cooling (the wall being a bottom one). The effect of change in the outer layer on the inner is illustrated by the lines on the figure. These show

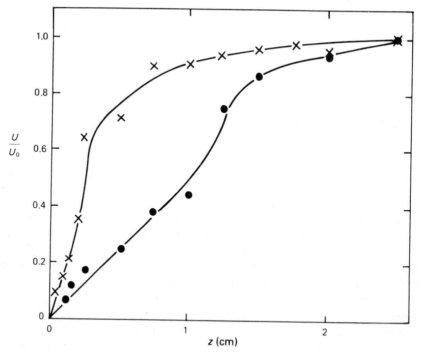

FIG. 21.26 Velocity profiles of boundary layer below heated surface at 24 cm (×) and 74 cm (●) downstream from start of heated section. Data from Ref. [289].

how the data trend is shifted if ϕ is calculated using the value of u_τ that would exist in the absence of stratification instead of the actual value. (For the third set of data the shift is negligible.) This illustrates better the total effect of stratification on the boundary layer (but the reduction of all cases to a single curve is lost).

We have looked at the effects of stratification on boundary layers primarily in terms of data for stabilizing stratification. Similar principles apply to destabilizing. Quantitatively the effects are somewhat smaller; the departure—now a decrease—of ϕ from unity at any given $z/|L|$ is less [66, 167].

When the destabilizing stratification is strong, $z/(-L)$ becomes large in the outer part of the boundary layer and buoyancy is the main source of turbulence generation there. A principal feature of the flow as a whole may then be the following [384]: large eddies of the eruption type originate close to the wall as described in Section 21.6; as they move away from the wall they become increasingly influenced by buoyancy, until, at large heights, they resemble the free convection plumes to be described in Section 22.7.

21.8 Reverse transition [287]

In the last section, we saw an example of turbulence suppression when stabilizing stratification was imposed. Strong enough stratification can force the flow to revert fully to laminar motion. Such a process, known as reverse transition or relaminarization, can occur in a variety of configurations [287]; other examples include pipe and channel flows in which the Reynolds number is reduced by a change of geometry; boundary layers that enter a region of strongly favourable pressure gradient; boundary layers on convex walls; and shear flows in rotating fluids.

Reverse transition involves its own characteristic mechanisms; it is not just a matter of the turbulence energy production being insufficient to sustain the turbulence against increased viscous dissipation or some other process (e.g. buoyancy in a stratified flow). The production mechanism is itself affected. As an example, we may consider observations [70] made in a channel of which the dimensions gradually change with distance downstream from 12.7×76 mm to 12.7×228 mm; the consequent drop in speed reduces the Reynolds number (based on the smaller dimension) from a value at which the flow is turbulent to one at which reverse transition occurs. Figure 21.27 shows maximum values, with respect to position across the channel, of the intensity of longitudinal velocity fluctuations and of the correlation coefficient that provides the Reynolds stress, as functions of the distance downstream from the end of the

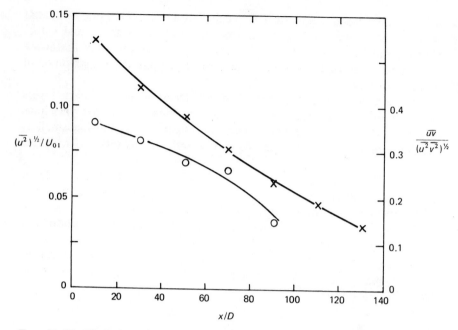

FIG. 21.27 Variation of longitudinal intensity ($\times$) and normalized Reynolds stress ($\bigcirc$) with distance downstream during reverse transition in channel flow. Both quantities plotted are maxima with respect to position across channel. ($Re_{2a} = 1730$). Data from Ref. [70].

cross-section change. One sees that, as the turbulence decays, the velocity components become less correlated. The structure of the turbulence is thus changing in a way that produces a faster approach to laminar motion, perhaps because the generation of eruptions (Section 21.6) is suppressed [287].

22

CONVECTION IN HORIZONTAL LAYERS

22.1 Introduction [47, 105, 106, 292]

We had a preliminary glance at this topic in Section 4.4; but it was only a glimpse of a very large topic. The reason that it has received so much attention is not primarily that it is of practical importance (although it has applications, e.g. Sections 26.2, 26.5, 26.6), but rather that it has become a context for the development of ideas about the consequences of instability and evolution towards turbulent motion. As a situation in which nothing whatsoever of interest would happen but for instabilities, it is a natural candidate for this role.

The configuration with which we are concerned is that of Fig. 4.1(b): a layer of fluid is bounded by two horizontal rigid planes a distance d apart and at different constant and uniform temperatures T_1 and T_2; the lower plate is hotter than the upper ($T_2 > T_1$). Adopting current usage rather than historical accuracy (Sections 4.4, 4.5), this is called the Bénard (or Rayleigh–Bénard) configuration. Variations of the theme, e.g. with different boundary conditions or with internal heating replacing heating from below, have been of some interest; but all the specific results in this chapter will relate to the conditions above.

The aspect ratio of the layer is defined as the ratio of the (smallest) horizontal dimension b to d. In this chapter we suppose that b/d is sufficiently large that one may think in terms of a horizontally infinite layer, although we shall mention lower aspect ratio layers briefly and return to them in Section 24.7. Horizontally restricted layers are generally considered to come into the category of Bénard convection provided that the boundary conditions on the end-walls are such that there is still an equilibrium solution (stable or unstable) of the equations with the fluid at rest.

We suppose that the Boussinesq approximation may be made and so the discussion of dynamical and thermal similarity in Section 14.5 applies. For a reason that will become apparent in the next section, it is usual to specify Bénard convection in terms of the Rayleigh number

$$\text{Ra} = \text{GrPr} = g\alpha(T_2 - T_1)d^3/\nu\kappa \tag{22.1}$$

and Prandtl number

$$\mathrm{Pr} = v/\kappa \qquad (22.2)$$

rather than the Grashof number Gr and Prandtl number.

The Nusselt number

$$\mathrm{Nu} = Hd/k(T_2 - T_1) \qquad (22.3)$$

has a particularly simple and useful interpretation when the length scale is the thickness d of a layer with temperature difference $(T_2 - T_1)$ across it. It is the ratio of the actual heat transfer to the heat transfer that would occur by conduction alone if the fluid remained at rest. Onset of convection is thus marked by Nu increasing above unity.

22.2   Onset of convection [42]

The key point in any discussion of Bénard convection is that there is an equilibrium solution with the fluid at rest. If motion occurs, it does so as a consequence of instability of this equilibrium. Hence stability theory is needed from the outset.

For an infinite expanse of fluid with the density increasing upwards, the physically evident instability is contained in the set of eqns (15.25)–(15.27) for small disturbances in a stratified fluid. The square of the Brunt–Väisälä frequency

$$N^2 = -\frac{g}{\rho}\frac{d\rho_0}{dz} = g\alpha\frac{dT_0}{dz} \qquad (22.4)$$

is negative in this case. Hence the frequency ω of eqn (15.38) is an imaginary quantity

$$\omega = i\sigma \qquad (22.5)$$

and the amplitude of the disturbance is proportional to $\exp(\pm\sigma t)$, with σ real and positive. The case $\exp(+\sigma t)$ corresponds to instability. (As in Section 17.3 instability occurs whenever one amplifying solution exists. The existence of a second decaying solution, which would occur on its own only if one had an initially large disturbance of just the right configuration, is irrelevant.)

For a layer of finite depth, the instability is modified by viscosity and thermal conductivity, together with the boundary conditions. The previously amplifying solution may now become a decaying one (whilst the previously decaying one always remains so). A full stability analysis is needed to discover when convection will occur. The principles and general procedures of such analyses have been discussed in Chapter

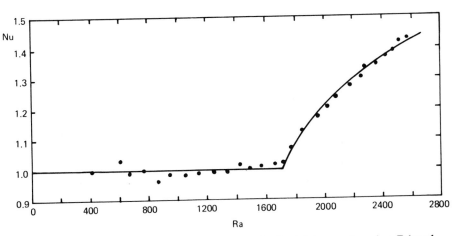

FIG. 22.1 Variation of Nusselt number with Rayleigh number for Bénard convection, showing increase of heat transfer with onset of motion. Data obtained by C. W. Titman.

17—with Couette flow stability (Section 17.5) being the most closely analogous case—so we can consider the theory in outline.

Using Cartesian coordinates with z vertical, a velocity field of the form

$$\mathbf{u} = \mathbf{U}(z) \exp[i(k_x x + k_y y) + \sigma t] \qquad (22.6)$$

together with temperature and pressure perturbations of similar form are superimposed on the equilibrium configuration. The equations of motion are used to investigate the sign of σ (which turns out to be always real). The results are found to depend only on the total wavenumber

$$k = (k_x^2 + k_y^2)^{1/2} \qquad (22.7)$$

and not on k_x and k_y individually. Consequently the stability loop is the locus of $\sigma = 0$ on a plot of non-dimensional wave number

$$\xi = kd \qquad (22.8)$$

versus the Rayleigh number. This locus is independent of the Prandtl number†, and resembles‡ the corresponding locus for Couette flow in Fig. 17.10.

---

† This result could have been anticipated. On the locus, $\partial \mathbf{u}/\partial t$ is zero. $\mathbf{u} \cdot \nabla \mathbf{u}$ is also zero, because there is no mean velocity field and second-order terms in the perturbation velocity field are neglected. Hence, the whole inertia term $D\mathbf{u}/Dt$ is zero. Dynamical similarity is then governed by the Rayleigh number alone. The only balance of terms to be considered is that in the thermal equation, as indicated by (14.44) (no longer restricted to low Grashof number when the inertia term is identically zero).

‡ Most closely if the Rayleigh number replaces the square of the Reynolds number [42].

The most important quantity is, of course, the Rayleigh number below which $\sigma < 0$ for all ξ and above which $\sigma > 0$ for some ξ. This comes out to be

$$\text{Ra}_{\text{crit}} = 1708. \qquad (22.9)$$

(Since the wavenumber has been allowed to take any value the layer is implicitly unlimited horizontally. Also the value 1708 relates specifically to the boundary conditions—rigid, isothermal walls—to which we are confining attention.)

This is another of the successes of linear stability theory. Whether or not motion occurs is most precisely determined experimentally by its effect on the heat transfer. Figure 22.1 shows an example of the observed variation of Nusselt number with Rayleigh number. Nu is equal to 1 when the fluid is at rest and greater than 1 when it is convecting (Section 22.1), and it is apparent that the onset of motion occurs close to the predicted value of Ra.

22.3 Convection just after onset

What is the consequence of the instability discussed above? In other words, what sort of flow pattern will be found when the Rayleigh number is above, but not too much above, its critical value? As emphasized in Section 17.4, linear stability theory does not in principle predict the type of motion that eventually occurs as a result of this growth—non-linear processes must always come into play before that stage. Nevertheless we saw in the context of rotating Couette flow (Section 17.5) that there are some cases (summarized by curve A of Fig. 17.6) for which there are marked similarities between the unstable modes and the observed motion when the critical condition is just exceeded. Bénard convection is in fact another such case. Thus the fact that, for Rayleigh numbers only a little above critical, there is only a limited range of unstable wavenumber k might be thought adequate information for a knowledge of the flow. However, restriction to a single wavenumber does not uniquely specify the flow pattern, for reasons connected with eqn (22.7). Not only may k_x take a whole range of values for a given k, but also linear combinations of solutions of the form (22.6) with different k_x but the same k are single wavenumber patterns. In consequence, there are many single wavenumber patterns, some of them of a complexity that makes them not instantly recognizable as such [357]; for example, one of them is a hexagonal pattern rather like Fig. 4.9. All such patterns occur with equal probability on linear stability theory.

Consequently, the statement that the developed convection pattern is broadly similar to a single wavenumber pattern still allows a variety of possibilities. It is not surprising that, as noted in Section 4.4, the observed flow can depend markedly on departures from the simple specification of Bénard convection, notably temperature variation of viscosity (e.g. Fig. 4.9). In this chapter, however, we are focussing attention on the behaviour when the effect of such departures can be assumed to be slight. Then the pattern actually adopted consists of long parallel rolls; i.e. the single wavenumber pattern that it resembles may be taken as having (cf. (22.7))

$$k = k_x, \quad k_y = 0. \tag{22.10}$$

This, of course, involves an appropriate choice of axes; and, since there is no preferential horizontal direction, the choice must be made after, not before, the rolls have formed.

The knowledge that this is what occurs comes primarily from stability theory—applied not to the rest configuration as in Section 22.2, but to the various possible convective structures [105]. This indicates that, were one of the other patterns to arise somehow, it would undergo a transition to the roll pattern.

We have seen an experimental realization of this type of convection in Fig. 4.7. Figure 22.2 is a sketch of another pattern from a similar experiment [308] (the lines correspond to bright lines in the shadowgraph). In both cases there is a very clear roll pattern, but with conspicuous defects in it. Such defects commonly occur in Bénard convection experiments, as may be understood in the following way. Suppose that the temperature difference across a Bénard layer of very large horizontal extent is increased so that the Rayleigh number passes through its critical value. Convection will be initiated in various parts of the layer, giving rise to the roll pattern discussed above. The orientation of the rolls will depend on whatever small perturbations happen to be present as the temperature difference is raised. In fact rolls of different orientation may be generated simultaneously in different parts of the

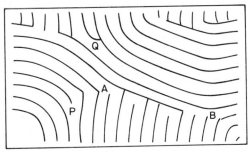

FIG. 22.2 Sketch of convection pattern (Ref. [308]) at $Ra = 3.2 \times 10^3$, $Pr = 70$, with defects identified.

layer sufficiently far apart that they are not initially interacting. There is a striking analogy here with crystal growth, except that it happens in two dimensions, not three. When a crystal is formed, for example by cooling of a melt, crystal grains of different lattice orientation are formed in different places; as the whole material solidifies these meet on grain boundaries—the counterpart of the line from A to B in Fig. 22.2. (Grain boundaries are also conspicuous in Fig. 4.7, particularly one running across the picture just above the middle.) Counterparts of other types of crystal defect may also be found in a 'convection lattice'; dislocations are to be seen at P and Q in Fig. 22.2.

22.4   Evolution with increasing Rayleigh number: general remarks

The fact that a convection pattern, occurring when Ra is a little above its critical value, commonly contains defects has very important consequences for the ways in which the convection may evolve as Ra is further increased. The 'grain boundaries' and dislocations can have a strong effect on subsequent developments—as we shall consider in Section 22.6.

However, there is also much interest, particularly in connection with how stability theory and transition to turbulence are related, in knowing how a convection pattern without these defects evolves. Just as in crystallography, if one wants a 'perfect crystal' one has to adopt special procedures to obtain it. This is the situation to which we shall give fullest consideration (Section 22.5).

Although the critical Rayleigh number is independent of the Prandtl number, subsequent developments are not. In general, corresponding developments tend to occur at a lower value of Rayleigh number when the Prandtl number is lower. As an extreme example, in mercury, with Pr = 0.025, it is difficult to observe any form of convection other than turbulent. However, there are not only quantitative but also qualitative differences in the evolution at different values of Pr. We thus have to consider a typical evolution rather than attempt to cover all possibilities.

That remark applies even though we are confining attention to cases in which 'non-ideal' features such as temperature variation of viscosity are supposed negligible. Of course such features are never totally absent in experimental work. However, one can do experiments in which it is reasonable to suppose, on the basis of a synthesis of experimental and theoretical results, that they are not grossly modifying the flow.

Much the same comment may be made about the effect of side-walls. In this chapter we discuss the layer as if it were horizontally infinite, and thus consider experiments with large aspect ratio—accumulation of observational experience and theoretical results being needed to decide what is large enough. There is, however, some interest in the opposite extreme when the side-walls are very definitely constraining the flow. Observations with apparatuses of relatively small aspect ratio have led to interesting comparisons with theories to be discussed in Chapter 24. In this situation the flow is affected not only by the velocity boundary conditions imposed by the vertical walls but also by the way in which these walls affect the temperature field. Consequently, there can again be a burgeoning of subdivisions. What is important, however, is that in any particular apparatus one can observe a repeatable sequence of developments (e.g. as Ra is increased). Quantitatively this is liable to be peculiar to the apparatus, but, at least sometimes, such developments occur in ways that seem to have very general significance. That remark may sound paradoxical; clarification must await Section 24.7.

22.5 Evolution of defect-free convection

The principle of the experiments considered in this section has been mentioned in Section 22.4. Very regular patterns of convection are deliberately induced. This may be done for example by heating the layer radiatively through a grid of the appropriate geometry [122]. This heating is imposed before the vertical temperature difference and, of course, produces weak convection. Then when the vertical difference is introduced making the Rayleigh number supercritical, the Bénard convection also adopts this pattern. The horizontally varying heating may then be switched off, and the evolution of pure Bénard convection with this pattern investigated. Controlling the flow in this way is obviously comparable with the introduction of periodic disturbances to control the development of shear flow instabilities (Sections 17.7, 18.1). There is, however, the difference that in those cases the control is used to investigate the first instability whereas here it is needed only for the later stages.

This procedure, called pattern control, enables one to investigate, experimentally in addition to theoretically, what happens to a 'perfect crystal' of rolls (Fig. 22.3) as the Rayleigh number is increased. We will suppose that the wavenumber of the rolls is similar to that that occurs without pattern control; i.e. the control serves only to remove the defects, not to impose an unusual wavenumber. (More precisely, we suppose that the imposed wavenumber is within a range that exists stably in a Rayleigh number range a little above critical [105]. Experiments have been done with imposed wavenumbers outside this range, and some interesting instabilities observed by which the wavenumber adjusts to a stable value [105, 108, 122]. However, these will not be described here.)

FIG. 22.3 Regular roll pattern generated in Bénard convection. Ra = 2.0 × 10⁴, Pr = 100. From Ref. [108]. (All pictures in Figs. 22.3 to 22.8 are shadowgraphs, Section 25.4.)

The sequence of pictures in Figs. 22.3–22.8 shows convection patterns occurring at Rayleigh numbers ranging upwards from the critical value. The pictures are shadowgraphs (see Section 25.4); the bright regions correspond to depth-averaged temperature being high and the dark regions to its being low.

Considered as a whole the pictures provide a nice illustration of transition to turbulence. The first and last pictures contrast highly ordered motion at one end of the sequence with turbulent motion at the other end, and it is seen that there is a series of intermediate stages by which transition from one to the other takes place. Incidentally it should be remembered that the first picture (Fig. 22.3) is actually the second stage of the sequence, because this pattern is itself the consequence of an instability that occurs only if the Rayleigh number is high enough (Sections 22.2, 22.3); however the first stage could only be represented by a picture which was 'a perfect and absolute blank'.†

It should also be noted that there is one difference between the earlier and later pictures of the sequence not apparent by looking at them. The motion changes from being steady to unsteady, as discussed below. Hence, Figs. 22.3 and 22.4 would look identical if they had been taken

---

† Like the Bellman's ideal map in Lewis Carroll's 'Hunting of the Snark'.

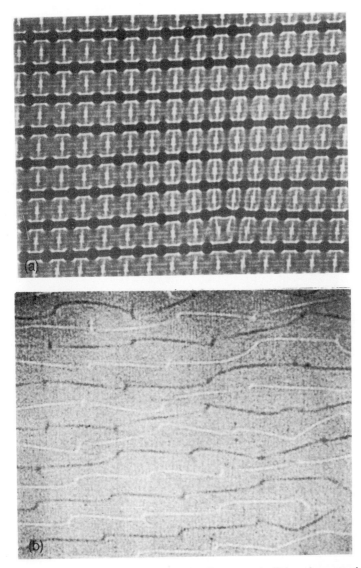

FIG. 22.4 Consequences of instability of roll pattern in Bénard convection: (a) $Ra \approx 5 \times 10^4$, $Pr = 100$; (b) $Ra = 1.1 \times 10^4$, $Pr = 3.7$. From Refs. [107, 108].

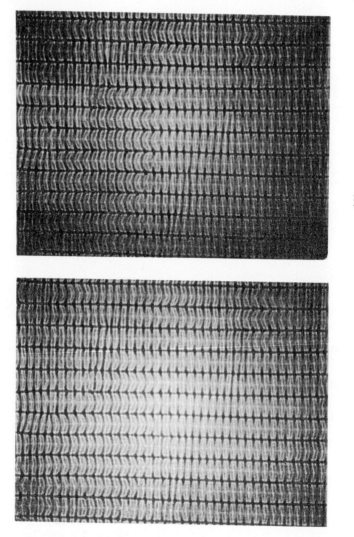

FIG. 22.5 Two phases of periodic oscillation in Bénard convection at Ra = 6.1×10^4, Pr = 16. From Ref. [109].

earlier or later, but the other pictures would not—they would give the same overall impression, but details would be different. In particular, the spatial irregularity apparent in Fig. 22.8 reflects irregular fluctuations in time.

As one goes through the sequence from ordered to turbulent motion, the proportions of information provided respectively by theory and by experiment alter. The earlier stages can be discovered almost entirely by

FIG. 22.6   Bénard convection at Ra $= 4.8 \times 10^4$, Pr $= 7$. From Ref. [106].

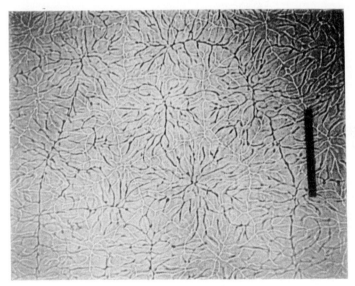

FIG. 22.7   Bénard convection at Ra $= 1.3 \times 10^5$, Pr $= 7$. From Ref. [106].

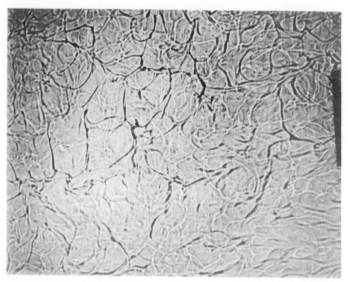

FIG. 22.8   Bénard convection at $Ra = 1.7 \times 10^7$, $Pr = 7$. From Ref. [106].

theoretical analysis, with experiments serving largely to confirm the predictions (and incidentally to provide illustrative material such as that being used here). At somewhat higher Rayleigh number theory still provides much understanding of the observations, but the experiments are less peripheral; e.g. they may indicate just what flow patterns should be included in the theoretical analysis. The later stages, at still higher Rayleigh number, have not yet yielded to theoretical analysis and our knowledge of them comes entirely from experimental observation.

Let us look in more detail at the intermediate stages of the sequence. A set of pictures, such as Figs. 22.3–22.8, at various values of the Prandtl number (determined, of course, by the availability of good illustrations†) does not strictly represent a unified sequence. However, so long as the figures are seen as illustrations of typical developments and not as a catalogue of all possible types of flow, this may not matter much. It has already been remarked that corresponding developments occur at lower Rayleigh number when the Prandtl number is lower. Consequently a development described as arising from an increase in Ra may sometimes be illustrated by a picture taken at lower Ra but also lower Pr.

† As a practical point, high Rayleigh number experiments are more readily performed at lower values of Pr. Pr is varied by using fluids with large differences in kinematic viscosity and relatively small differences in thermal diffusivity. Consequently, achieving a high Rayleigh number with a high Prandtl number, along with a large aspect ratio, requires an impracticably large apparatus.

The first development as Ra is increased from a value a little above critical is a change in the geometry of steady convection. This is illustrated by Fig. 22.4. The two parts of this figure are for different values of the Prandtl number. Although, for the most part, we are not considering all alternatives, there is such a striking difference in this development for Pr greater than or less than about 10 that a single illustration would be inadequate. In all of Figs. 22.3, 22.4(a) and 22.4(b), a uniform array of convection rolls has been established by the method described above. At low enough Rayleigh number this pattern persists indefinitely (Fig. 22.3). At higher Rayleigh number spontaneous changes occur. These may generate rolls perpendicular to the original ones and of smaller wavelength, producing a rectangular pattern like that in Fig. 22.4(a) (known as a bimodal convection); or, at lower Pr, they may lead, via an instability illustrated by Fig. 22.4(b) to a new roll pattern of larger wavelength.

In either case, the new flow, once established, is a new pattern of steady convection. The transition has not introduced any fluctuations in time.

The next development with increasing Rayleigh number does produce such fluctuations—periodic ones. Figure 22.5 illustrates this, choosing the case in which the previous stage had produced bimodal convection. A standing wave pattern has spontaneously arisen, causing the longer boundaries of the rectangle to oscillate. The two pictures in Fig. 22.5 show the same flow at times separated by half an oscillation period. A probe measuring the temperature or velocity at a fixed point within this flow would indicate a periodic variation (cf. Fig. 24.9(A) for similar behaviour in a small aspect ratio apparatus).

Both the transition from one steady pattern to another and the onset of periodic unsteadiness can be understood as instability of the pre-existing type of convection—and indeed have been successfully analysed from that point of view.

Hence, the increasing complexity of the flow can be interpreted as the result of a sequence of instabilities, each giving rise to a new pattern which is stable for some Rayleigh number range but which itself becomes unstable at higher Ra. The further developments seen in Figs. 22.6–22.8 can probably be interpreted in a similar way. However, from this point on, there is no full theory; the description is almost entirely observational and so the interpretation involves more guesswork.

The later stages of evolution are dominated by changes in the pattern of fluctuation with respect to time—from the very regular periodicity considered above towards the irregular fluctuations characteristic of turbulence. Associated with this, however, are changes in the spatial structure of the flow, leading first to replacement of the roll dominated patterns by different patterns (e.g. Fig. 22.6) and ultimately to the disappearance of any permanent spatial structure. (Figs. 22.7, 22.8).

In Fig. 22.6, the large polygonal features are permanent structures; that is to say, once the flow has been established the polygonal pattern remains unchanging so long as the Rayleigh number is maintained constant, although the details of just where the polygons locate themselves will be different each time the experiment is performed. There is thus a steady mean circulation in any one experiment. However, there are fluctuations superimposed on this; i.e. the flow within the mean circulation is unsteady and probably essentially turbulent. These fluctuations are responsible for the 'spoke-like' pattern of bright and dark patches to be seen within the polygons of Fig. 22.6, although, because of the unsteadiness, the behaviour is not fully illustrated by a single picture of this sort.

The permanent spatial structure is not present at the highest Rayleigh numbers at which observations have been made (Fig. 22.8). The motion can then be described as fully turbulent. This does not, however, imply that the fluctuations are 'featureless'. We shall look briefly at the features of this motion in Section 22.7.

22.6 Evolution of convection with defects [183]

However, we must first remember that the above account of evolution with increasing Rayleigh number derived from experiments with pattern control. What happens without it (in what might be considered the 'natural' situation)? Then the cellular pattern arising when Ra is a little above critical is likely to contain dislocations and grain boundaries (Section 22.3). These have a marked effect on the development with increasing Ra. In particular, unsteadiness may set in at much lower Ra than it would in their absence. It is likely that instabilities similar to those occurring in 'perfect crystal' motion again occur, but promoted and modified by the spatial imperfections. One can again observe initially periodic oscillations being superseded, with increasing Ra, by aperiodic ones; the details are plainly much more sensitive to small inevitable peccadillos of the individual apparatus than when one has pattern control. Figure 22.9 illustrates the onset of unsteadiness in uncontrolled flow in a special way. This picture was obtained by having the fluid illuminated along only one line and moving the camera so that the image of this line moved across the film. The result is essentially a space–time representation of the flow, and the 'ribbed' pattern indicates the occurrence of periodic fluctuations. The appearance of this pattern in some places and not in others presumably relates to the detailed spatial structure in two dimensions.

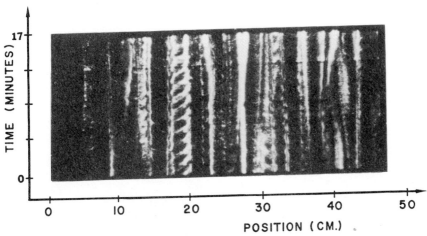

FIG. 22.9 Space–time flow visualization (see text) of Bénard convection at Ra $= 3.4 \times 10^5$, Pr $= 57$. From Ref. [234]

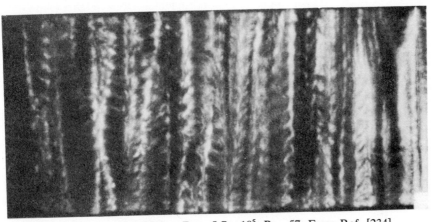

FIG. 22.10 As Fig. 22.9 at Ra $= 5.7 \times 10^5$, Pr $= 57$. From Ref. [234].

Figure 22.10 shows another space–time representation of the same type at a higher Rayleigh number. One sees that the fluctuations are now occurring throughout but with a trend towards less regularity. It is likely that the later stages of development and particularly the structure of the turbulent motion at high Ra are essentially the same whether or not the experiment has involved pattern control.

22.7 High Rayleigh number convection and turbulence

To understand the processes occurring at high Rayleigh number, it is helpful to look at the mean temperature distribution across the layer.

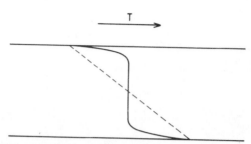

FIG. 22.11 Typical mean temperature distribution in Bénard convection at high Rayleigh number (compared with corresponding conduction profile).

(The averaging is as discussed in Section 19.2 and may be either over time or over horizontal planes.) This takes the form shown schematically in Fig. 22.11, with large temperature gradients close to the boundaries and a nearly isothermal region in the interior. The reason is analogous to the reason for large velocity gradients close to the walls in shear flows (Section 21.5) but with heat transport replacing stress in the argument. The total vertical heat transport is the sum of that transported by the motion and that conducted down the temperature gradient

$$H = \rho C_p(\overline{w\theta} - \kappa\, \partial T/\partial z)\qquad\qquad (22.11)$$

(cf. also Section 21.7). H must be independent of z, the vertical distance across the layer. At high Rayleigh number the heat transport is large, in the sense that Nu $\gg 1$. In the centre of the layer this transport is brought about by the first term of eqn (22.11): essentially just the fact that rising currents in the convection are typically hotter than falling ones. Close to each boundary, however, $\overline{w\theta}$ becomes much smaller—the boundary conditions require it to be zero right at the walls—and the second term of (22.11) must become much larger. Incidentally, the Nusselt number is given by the ratio of the temperature gradient at either wall to the temperature gradient that would exist throughout the layer in the absence of convection (dotted line in Fig. 22.11).

 Throughout either of the sequence of developments with increasing Rayleigh number described in Sections 22.5 and 22.6, the temperature profile tends away from the linear form that occurs when the fluid is at rest and towards the form of Fig. 22.11. This fact is actually of some importance in more detailed analysis of some of the instabilities described in Section 22.5. But it is in the fully turbulent regime that it is most central to the discussion. By this stage, the large gradient regions occupy only small parts of the layer.

 In consequence, the most strongly unstable regions are those close to the boundaries and the dominant features of the flow—the counterpart of the large eddies of shear flows described in Sections 21.4 and 21.6—originate there. We consider the hot bottom wall; similar processes occur close to the cold top wall, with the roles of hot and cold fluid reversed. There are intermittent eruptions of hot fluid away from the bottom boundary. Colder fluid moves close to the wall to replace the fluid in an eruption. This is gradually warmed by conduction from the wall until there is a thick enough layer of hot fluid for another eruption to be initiated. The process then repeats itself [162].

Each eruption gives rise to one or sometimes more columns of hot fluid rising through the interior region. These features are sometimes called thermals, sometimes plumes. Over a range of Rayleigh number (probably dependent on Prandtl number), the thermals penetrate right across the layer, generating transient stable blobs of fluid close to the opposite boundary [123].

Ultimately, however, at the highest Rayleigh numbers, the thermals lose their identity before reaching the opposite boundary. The processes at each boundary now occur independently of those at the other. They can be investigated by studying the motion above a single heated plate at the bottom of a large expanse of fluid (Figs. 22.12–22.14), corresponding in a sense to infinite Rayleigh number.

A tendency has sometimes been observed for several successive eruptions to occur with noticeable periodicity and for successive thermals to be in the same places each time [354, 363]. However, this does not persist over a long time scale and is often not to be observed at all. In general, in a system of large enough horizontal extent, one expects to

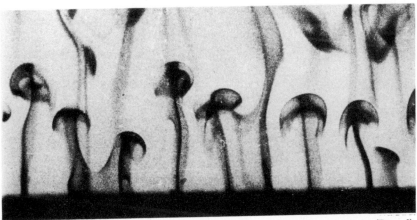

FIG. 22.12 Thermals rising from a heated horizontal surface in water (Pr = 6); dye produced *uniformly* at surface by pH technique (see Section 25.4). From Ref. [354].

FIG. 22.13 Thermals rising from a heated horizontal surface in air $(Pr = 0.7)$; flow visualization using recondensed water vapour evaporated at moist surface. From Ref. [317].

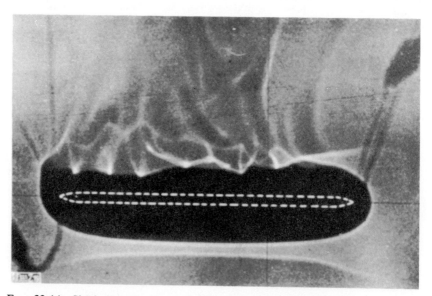

FIG. 22.14 Shadowgraph of convection around a heated horizontal plate; cf. Fig. 14.5. From E. Schmidt, Forschung auf dem Gebiete des Ingenieurwesens, Band 3, Seite 181 (1932), by kind permission of VDI-Verlag GmbH, Düsseldorf.

find a spatially random distribution of thermals with some of them just forming, some in vigorous convection and some fading away, with the interval between formation and fading away being very variable. Figure 22.14 shows the random character of the instantaneous temperature field.

Thermal formation and persistence appears to be the dominant feature of convection regardless of Prandtl number (at least from values rather less than 1 to very much greater than 1; experimental evidence for $Pr \ll 1$ is limited). There are, however, differences in the motion at different Pr. Unless $Pr \gg 1$, there is vigorous disorganized small-scale motion as well as the rather coherent large-scale motion of the thermals. The former occurs both within the thermals and in the slowly downward drifting cold fluid between the thermals [382]. At very high Pr, this is absent and individual thermals are essentially laminar flow structures [391]. The flow as a whole is then rather different from flows typically indicated by the name turbulent, but the randomness in space and time of the occurrence of the thermals makes the name still applicable on the point of view of Section 19.1.

23

DOUBLE DIFFUSIVE FREE CONVECTION

23.1 Introduction [37, 206, 395]

We have seen in Chapter 14 that the density variations that drive free convection may be introduced into a fluid through either temperature variations or concentration variations, and that the two are closely analogous. In this chapter we consider what happens when the two are simultaneously present. It might be thought that the analogy is readily extended to cover this situation, but in fact a whole range of new phenomena arise. Some of these are particularly striking because they occur in circumstances where at first sight one would expect nothing interesting to happen.

The most immediate examples of this are ones in which the imposed density variations are entirely vertical; as in Chapter 22, this implies that any motion is the consequence of instability of the rest configuration. Consider, in particular, the case in which one of the two causes of density variation produces an increase with height and the other a decrease. It is found that the former can give rise to convection even when the latter is sufficiently strong that the net effect of the two is a decrease; i.e. lighter fluid overlies heavier.

There are evident applications to oceanography. For example, solar radiation may warm the surface layer of the sea but this may also give a high evaporation rate increasing the salt concentration; thus hot, salty water often overlies colder, less salty water. The converse situation can arise when cold fresh water (e.g. from a river formed by melting ice) flows into the sea; if it is sufficiently cold it will initially remain close to the sea-bed and thus underlie warmer, salt water.

Partly because of these applications, the most widely considered combination is that of heat and dissolved common salt. This combination led to the older name for the phenomena we are considering: thermo-haline convection. However, it is only one example of a more general situation—for instance one might have no temperature variations but two different solutes—and the usual name now is double diffusive convection.

The crucial feature necessary for the occurrence of the special phenomena of double diffusive convection is that the two components should have different diffusivities,

$$\kappa_1 \neq \kappa_2. \tag{23.1}$$

If this were not the case, then the two components would necessarily remain coupled in a way that would allow little difference from convection due to just one of them. When the two components are heat and common salt,

$$\kappa_1 = \kappa, \quad \kappa_2 = \kappa_c \tag{23.2}$$

in the notation of Sections 14.2 and 14.3, and

$$\kappa_c \approx 10^{-2} \kappa. \tag{23.3}$$

However, a much smaller contrast between the two will still produce the effects under discussion. In the following, the discussion is presented in terms of the heat–salt combination, because this is easily envisaged. The more general case is given by replacing 'heat' by 'the component with higher diffusivity' and 'salt' by 'the component with lower diffusivity'.

In this chapter we consider primarily the configuration mentioned above—a fluid layer with the net density decreasing upwards as a result of opposing contributions from heat and salt. What happens depends on which is the destabilizing and which the stabilizing component; the two cases constitute Sections 23.2 and 23.3. There are other configurations of significance (one is mentioned in Section 23.4) but we can illustrate the essential features of the topic with this one.

The obvious relevance to oceanography has already been mentioned, but it is far from being the only field of application. Other places where the importance of double diffusive effects has been particularly recognized are materials science (e.g. crystal growth, Section 26.11) and geology (e.g. the relationship of such effects in magma chambers to observed layering of igneous rocks). Further applications are discussed in Ref. [395].

23.2 Salt-driven convection

We start our discussion of a fluid layer with vertical density variations with the case in which the salt is the destabilizing component; i.e. variations in its concentration are the cause of the convection. Thus if z is vertically upwards, the imposed conditions have

$$\mathrm{d}c_0/\mathrm{d}z > 0, \quad \mathrm{d}T_0/\mathrm{d}z > 0, \tag{23.4}$$

where the basic notation is that introduced in Chapter 14 and the suffix 0 indicates, as in Chapter 15, the imposed vertical variations. We denote the density changes due to salt concentration and temperature changes by $\Delta\rho_c$ and $\Delta\rho_T$; thus

$$\rho = \rho_r + \Delta\rho; \quad \Delta\rho = \Delta\rho_c + \Delta\rho_T; \quad \Delta\rho_c = \rho_r\alpha_c c; \quad \Delta\rho_T = -\rho_r\alpha\Delta T$$

$$(23.5)$$

where ρ_r is a fixed reference density, and

$$d\rho_{co}/dz = \rho_r\alpha_c dc_0/dz; \quad d\rho_{T0}/dz = -\rho_r\alpha\, dT_0/dz. \quad (23.6)$$

Hence, we are considering

$$d\rho_{co}/dz > 0; \quad d\rho_{T0}/dz < 0; \quad d\rho_0/dz < 0. \quad (23.7)$$

In this situation the predominant phenomenon is that known as salt fingering, shown schematically in Fig. 23.1. An array of long thin fingers of salty water descends, interspersed with a similar array of rising fresh water. The motion is driven by the higher density of the salty fingers. Why is it not inhibited by the fact that the water entering these is also hotter than that entering the upgoing fingers? The answer lies in the difference in diffusivities. It is essential to the dynamics that thermal diffusivity produces significant heat transfer between the downgoing and upgoing fingers. Consequently, the temperature contrast between them at any level is smaller than the temperture difference between top and bottom. There is not much temperature-produced density contrast at a

Hot salty water

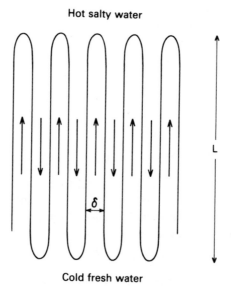

Cold fresh water

FIG. 23.1 Salt fingers (schematic).

given level; thus this does not much affect the buoyancy force. The heat transfer can occur without a corresponding diffusion of salt because of the lower diffusivity of the latter. The salt-produced density contrast of fluid entering the downgoing and upgoing fingers can be sustained over their height and thus produce a relative buoyancy force at every level.

This argument can be formulated more quantitatively in a way that both helps with the understanding of it and shows what determines the width of the fingers. We suppose that $\kappa \gg \kappa_c$, not just $> \kappa_c$. This is not in fact necessary for salt fingering to occur but it simplifies the presentation. When $\kappa \gg \kappa_c$, it is possible for the horizontal concentration variations to be effectively the same as the vertical ones whilst horizontal temperature variations are much reduced from the vertical. The former requires

$$\boldsymbol{u} \cdot \nabla c \gg \kappa_c \nabla^2 c \qquad (23.8)$$

so that the concentration is advected into the salt fingers little changed by diffusion. In order of magnitude

$$Wc/L \gg \kappa_c c/\delta^2 \qquad (23.9)$$

where W is a typical vertical velocity in the fingers, L is the depth over which the convection is occurring (its exact definition depending on just how the system is set up) and δ is the width of the fingers (Fig. 23.1). Thus

$$W/L \gg \kappa_c/\delta^2. \qquad (23.10)$$

On the other hand, thermal diffusivity is changing the temperature as fluid moves up and down the fingers. Hence,

$$\boldsymbol{u} \cdot \nabla T \sim \kappa \nabla^2 T \qquad (23.11)$$

i.e.

$$W \, \Delta T_V/L \sim \kappa \Delta T_H/\delta^2 \qquad (23.12)$$

where ΔT_V is a typical temperature difference between top and bottom and ΔT_H that between downgoing and upgoing fluid. Moreover, we are supposing that heat conduction makes

$$\Delta T_H \ll \Delta T_V \qquad (23.13)$$

so we require

$$W/L \ll \kappa/\delta^2. \qquad (23.14)$$

In order to use (23.10) and (23.14) to constrain δ, we require an estimate of W from the dynamical balance between the buoyancy force and one (or both) of the inertia force and the viscous force. The buoyancy force is provided by $\Delta \rho_c$ (which is, in the first place, the difference between upgoing and downgoing fingers, but, in view of the above discussion, may be taken as the difference between top and bottom). Hence,

$$g\Delta\rho_c \sim |\rho u \cdot \nabla u| \quad \text{or} \quad g\Delta\rho_c \sim |\mu \nabla^2 u| \tag{23.15}$$

i.e.

$$g\Delta\rho_c/\rho_r \sim W^2/L \quad \text{or} \quad g\Delta\rho_c/\rho_r \sim vW/\delta^2 \tag{23.16}$$

(where the choice of length scales follows the principles introduced in Chapter 11). However, in view of (23.14)

$$W^2/L \ll vW/\delta^2 \tag{23.17}$$

unless the Prandtl number v/κ is very small ($\text{Pr} \approx 6$ for water). Thus the buoyancy force/viscous force balance applies and

$$W \sim g\Delta\rho_c\delta^2/\rho_r v. \tag{23.18}$$

This can be substituted into (23.10) and (23.14) to indicate the possible range of δ. It is convenient to introduce a concentration Rayleigh number[†]

$$\text{Ra}_c = g\Delta\rho_c L^3/\rho v \kappa_c. \tag{23.19}$$

Then

$$1/\text{Ra}_c \ll (\delta/L)^4 \ll \kappa/\kappa_c \, \text{Ra}_c. \tag{23.20}$$

The appearance of the fourth power here implies that the finger widths are quite closely determined by the dual requirements of large heat transfer (right-hand inequality in (23.20)) and small salt transfer (left-hand inequality) even when κ/κ_c is large. The deduction has implicitly assumed that δ/L is small; since the instability producing the fingers occurs at high Rayleigh number (cf. Section 22.2) this is internally consistent.

As mentioned above, one does not actually require κ/κ_c to be much greater than 1 for the occurrence of fingers. The same mechanism can operate when κ/κ_c is only a little greater than 1. It is then a matter of smaller and larger transfer of the two components, rather than very small and very large. Obviously the size of the negative $d\rho_{T0}/dz$ that can be overcome by a given positive $d\rho_{c0}/dz$ decreases as κ/κ_c decreases.

So far the type of situation in which salt fingers occur has been specified in only rather general terms. They certainly occur in a variety of detailed configurations and may well be a universal feature of convection driven by the lower diffusivity component. Sometimes, however, they are only part of the full picture, as we shall see in examples below.

[†] Equation (23.19) is the obvious definition of such a Rayleigh number, but it should be noted, when referring to other texts, that $\kappa_c \, \text{Ra}_c/\kappa$ is often used instead.

Linear stability theory applied to a configuration of the Bénard sort with appropriate concentration and temperature differences across the layer indicates motions of a salt finger type [73]. The wavenumbers of the unstable modes are determined by the considerations leading to (23.20) rather than being directly related to the layer depth (as they are in ordinary Bénard convection, Section 22.2).

Experimentally one cannot set up just this configuration because of the difficulty in imposing constant concentration boundary conditions (Section 14.3). In general, the most readily practicable experiments are ones in which an initial density distribution is set up and there is then some evolution of the configuration during the course of the experiment. A particularly straightforward procedure in principle and yet a very informative one is to add carefully a layer of hot salty water on top of one of cold fresh water (or the counterpart of this in which two solutes are used instead of heat and one solute). The instability originates at the interface, which then thickens into a region containing salt fingers. Figures 23.2 and 23.3 show vertical and horizontal sections of fingers generated in this way. The layer containing the fingers does not continue to thicken until it fills the fluid region. A quasi-steady state is reached with the original layers still existing above and below this layer. (Flow visualization gives a picture like Fig. 23.4 below, but with only three layers.) The top and bottom layers are, of course, initially homogeneous

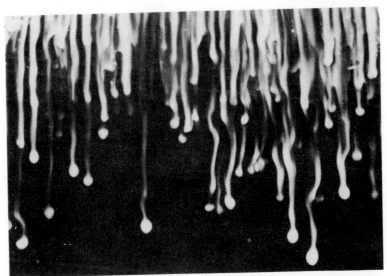

FIG. 23.2 Salt fingers formed by setting up a stable temperature gradient and pouring a little salt solution on top. Visualization by adding fluorescein to the salt and illuminating through a slit from below. From Ref. [206].

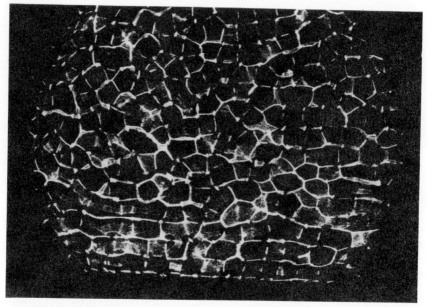

FIG. 23.3 Shadowgraph of cross-section of salt fingers. From Ref. [345].

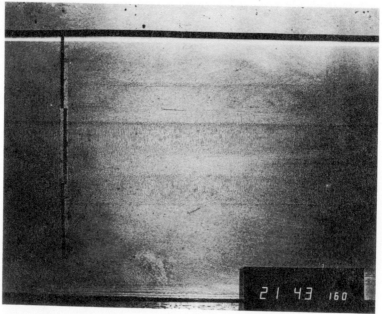

FIG. 23.4 Five layers, two containing salt fingers arising as described in text. From Ref. [252].

in concentration and temperature and therefore inactive. However, once the salt fingers are formed, they are pumping salty fluid into the top of the bottom layer and less salty fluid into the bottom of the top layer. (The associated reduction in potential energy is the energy source for the whole motion.) Because of the diffusion in the fingers, the temperature fields are not correspondingly affected to a great extent. Hence, concentration driven convection occurs in these layers. Overall the system has three length scales, only one of which is externally imposed: the total depth of the fluid, the thickness L of the salt finger layer, and the width δ of individual fingers.

The tendency for L to find its own maximum value leads to interesting developments if the initial density distribution is set up in a way that makes it initially larger than this maximum [252]. The finger-containing layer can then split, as shown in Fig. 23.4. In this example the end result is five layers, two containing salt fingers and three with isothermal concentration-driven convection. This observation is relevant to the interpretation of much larger numbers of layers observed in the ocean, and may be seen in the context of the very general tendency, discussed in Section 23.4, for layering to arise in double diffusive systems.

23.3 Heat-driven convection

We turn now to the reverse case in which the temperature distribution is the driving component. Thus, instead of (23.4) and (23.7), we have

$$dT_0/dz < 0, \quad dc_0/dz < 0 \tag{23.21}$$

giving

$$d\rho_{T0}/dz > 0, \quad d\rho_{c0}/dz < 0, \quad d\rho_0/dz < 0. \tag{23.22}$$

Unlike the previous case, the behaviour of a marginally unstable system and that of a vigorously convecting one do not have an obvious common feature. The development from one to the other is somewhat obscure. One can, however, understand the way in which the difference in diffusivities enables each type of motion to occur despite the overall stable stratification.

The counterpart for the present case of the linear stability theory mentioned above predicts overstability (defined in Section 17.4). This prediction has been confirmed experimentally (Fig. 23.5), by gradually increasing the temperature difference across a salt-stratified layer so that the Rayleigh number passes through its critical value.

The way in which double diffusive processes produce amplified oscillations may be understood through a simple displaced particle

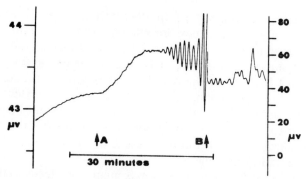

FIG. 23.5 Temperature versus time record in salt stratified thermal convection near marginal stability. Signal is differential output of two thermocouples at different heights. Heating rate was increased slightly at A. Left-hand scale applies up to B, right-hand one thereafter. From Ref. [344].

argument. Consider a fluid particle that is displaced upwards, say, from its equilibrium position. Such a particle is heavier than its new surroundings and thus tends to fall back to its original level and then overshoot. In the absence of any diffusion whatsoever its behaviour is identical to that involved in the theory of the Brunt–Väisälä frequency (Section 15.4). If, however, thermal diffusion occurs, without a corresponding concentration diffusion, the particle will be cooled when it is above its equilibrium position and thus surrounded by cooler fluid. It thus becomes even heavier relative to its surroundings, making the overshoot larger than the original displacement. Similarly, at the bottom, warming gives it buoyancy that makes the next oscillation still larger.

This argument may be made clearer by formulating it algebraically, extending the theory in Section 15.4. As there, we denote the density at $\Delta z = 0$ by $\rho_0(0)$. Then the density of undisturbed fluid at height Δz is

$$\rho_a = \rho_0(0)\left[1 + \alpha_c \frac{dc_0}{dz}\Delta z - \alpha \frac{dT_0}{dz}\Delta z\right] \qquad (23.23)$$

(the suffix a signifying 'ambient'), which may also be expressed

$$\rho_a = \rho_0(0) + \frac{d\rho_0}{dz}\Delta z = \rho_0(0)(1 - N^2 \Delta z/g) \qquad (23.24)$$

where N is the Brunt–Väisälä frequency. A displaced particle originating at $\Delta z = 0$ has density

$$\rho = \rho_0(0)(1 + \alpha_c \Delta c - \alpha \Delta T). \qquad (23.25)$$

We suppose, however, that the concentration diffusivity is so small that the particle conserves its concentration,

$$\Delta c = 0. \tag{23.26}$$

Its temperature, on the other hand, does change, and we assume that its rate of change is proportional to the difference between ambient temperature and its temperature

$$dT/dt = K(T_a - T) \tag{23.27}$$

which may be rewritten

$$\frac{d\Delta T}{dt} = K\left[\frac{dT_0}{dz}\Delta z - \Delta T\right] = -J\Delta z - K\Delta T \tag{23.28}$$

where J and K are positive constants. The dynamical equation, a modification of (15.21), is

$$\rho \, d^2\Delta z/dt^2 = g(\rho_a - \rho) = \rho_0(0)(-N^2\Delta z + g\alpha\Delta T) \tag{23.29}$$

which, since we are making the Boussinesq approximation, becomes

$$d^2\Delta z/dt^2 = -N^2\Delta z + g\alpha\Delta T. \tag{23.30}$$

Equations (23.28) and (23.30) form a pair of differential equations with respect to time for the displacement Δz of the particle and its temperature change ΔT in its displaced position. We look for solutions in the form

$$\Delta z = Z \exp(i\omega t), \quad \Delta T = \Theta \exp(i\omega t) \tag{23.31}$$

(where, of course, the real parts correspond to the physical quantities). Substituting in (23.28) and (23.30)

$$-\omega^2 Z = -N^2 Z + g\alpha\Theta \tag{23.32}$$

$$i\omega\Theta = -JZ - K\Theta. \tag{23.33}$$

Eliminating Θ/Z,

$$(\omega^2 - N^2)(i\omega + K) = g\alpha J. \tag{23.34}$$

It simplifies the algebra to confine attention to the case in which the unstable thermal stratification is weak compared with the stable salt stratification so that the oscillations are only slightly modified from those described by eqns (15.21) and (15.22). (Since we have taken the concentration diffusivity to be effectively zero, this is sufficient to give the effect we are looking for). Hence, we put

$$\omega = N + \omega', \quad \omega' \ll N. \tag{23.35}$$

We also take

$$\omega' \ll K \tag{23.36}$$

(since $K \sim N$ when the thermal diffusion occurs on a time scale comparable with the oscillations). Then (23.34) approximates to

$$2\omega' N(iN + K) = g\alpha J. \tag{23.37}$$

Since N, K, and J are all real, ω' is complex:

$$\omega' = \omega_r' + i\omega_i'. \tag{23.38}$$

Substitution in (23.37) leads to

$$\omega_r' = Kg\alpha J/2N(N^2 + K^2), \quad \omega_i' = -g\alpha J/2(N^2 + K^2). \tag{23.39}$$

The crucial point is that ω_i' is negative. Since†

$$z = Z \exp[(iN + i\omega_r' - \omega_i')t] \tag{23.40}$$

$$T = \Theta \exp[(iN + i\omega_r' - \omega_i')t], \tag{23.41}$$

this implies that the modified Brunt–Väisälä oscillations gradually grow in time; i.e. there is overstability.

Two points may be noted about this analysis. Firstly, the result only applies when the concentration and temperature gradients have the appropriate signs; positive dc_0/dz would make N imaginary and positive dT_0/dz would make J negative. Secondly, substitution of the results for ω back into (23.32) or (23.33) shows that Θ/Z is complex. This implies that Δz and ΔT are not oscillating in phase. (Taking real parts of (23.31), it implies that if we put $\Delta z \propto \cos \omega t$, then $-\Delta T \propto \cos(\omega t - \phi)$.) This was to be expected; it is an essential part of the mechanism that at any given value of Δz, the particle should be hotter when it is ascending than when it is descending. As for ordinary convection, this is what drives the motion.

The above argument is, of course, a gross simplification of any full

† In relation to the notation introduced in Chapter 17,

$$i\omega = \sigma; \quad \omega_i = -\sigma_r; \quad \omega_r = \sigma_i.$$

That notation is not used here because we are looking specifically at a modification to a sinusoidal oscillation.

fluid dynamical situation. Also, as with our discussion of salt fingers in Section 23.2, although the double diffusive processes are conveniently understood in the context of large contrast between κ and κ_c, a small contrast is sufficient to give rise to these processes. Nevertheless the argument is an effective way of understanding why linear stability analysis of the full situation predicts overstability and thus what is happening in Fig. 23.5.

That in turn is important in understanding why motion occurs at all, but, as noted earlier, relates directly to observations only when the Rayleigh number is only a little above its critical value. What happens when it is far above that value is known primarily from experimental work, although the dynamical processes involved can be interpreted. Figure 23.6 shows typical temperature and concentration distributions in these circumstances. The convection divides into a series of layers; nearly all the variation of each quantity occurs close to the edges of the layers.

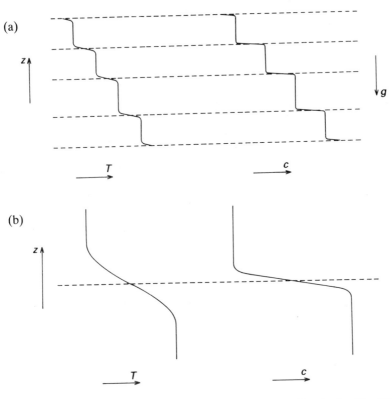

FIG. 23.6 Schematic temperature and concentration profiles in double diffusive convection: (a) overall pattern; (b) around an interface.

Within each layer convection is occurring, but the density difference between the layers associated with the concentration difference prevents this convection penetrating from one layer to another. Heat and salt are both being transported vertically, by advection within the layers and by diffusion down the gradients between them. Within each convecting layer, the upgoing fluid must be lighter than the downgoing, and this

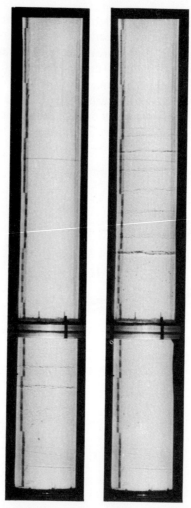

FIG. 23.7 Two stages of development of layers in tank of salt stratified water heated from below. Visualization by shadowgraph method (Section 25.4). (Notes: refraction through cylindrical tank gives a misleading perspective—tank width is actually about $\frac{1}{3}$ its height; marks common to both pictures are due to imperfections in the Perspex.) From Ref. [205].

implies that the heat transport is dominant over the salt transport. The temperature distribution within one layer resembles that in Fig. 22.11, the rising hot fluid being supplied from the hot bottom boundary layer. The corresponding concentration boundary layer is thinner (Fig. 23.6(b)). Consequently the density reduction due to the high temperature of the fluid going into an upflow is not cancelled out by a density increase due to high concentration. This behaviour is possible only because of the difference in diffusivities. Without it, the high concentration gradient between layers, compared with the temperature gradient, would give a higher salt transport than heat transport—the opposite way round from the relationship needed to sustain the convection. (To prevent the argument from being too cumbersome, there is some loose usage above. For example, it is strictly meaningless to 'compare a concentration gradient with a temperature gradient', because the two are dimensionally different. Such improprieties would be removed by replacing 'temperature (concentration)' by 'density change due to temperature (concentration)' with similar changes for 'heat' and 'salt'.)

The convection within one layer is not very different from Bénard convection. However, we have seen (Chapter 22) that this takes a variety of forms. There are thus various regimes of the double diffusive convection in which different detailed patterns occur within the same overall picture, but we shall not go into these details here.

The ways in which the experiments leading to the above description have been done are varied (although again constrained by the boundary conditions of some ideal configurations being impracticable). As an example, we look at the case in which an initially isothermal column of liquid is set up with a stable salt gradient in it; the base of the column is then heated. Layers form sequentially from the bottom, as the effect of the heating spreads upwards and more of the fluid starts to convect. Figure 23.7 shows two stages of such a sequence.

23.4 Layering

The tendency to form layers seems to be the most pervasive feature of double diffusive convection, arising in one way or another in widely different experimental configurations. We have seen it occurring in both types of double diffusive convection arising from vertical density gradients, e.g. Figs. 23.4 and 23.7. These two patterns of behaviour both involve layers in which there is convection due to the driving component little modified by the presence of the other component. There is, however, the difference that when heat is the driving component then layers are simply separated by narrow regions with steep gradients, but

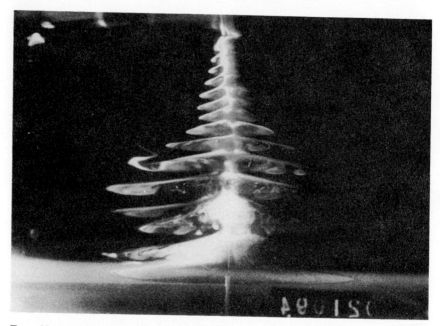

FIG. 23.8 Layers formed by convection above a point heat source in salt stratified water. Visualization by combination of fluorescent and coloured dyes. From Ref. [392].

when salt concentration is the driving component they must be separated by other dynamically active layers—the ones containing salt fingers.

We take one further example, although without any details. This has two purposes, firstly to illustrate the above assertion about the pervasiveness of layering and secondly to act as a reminder that there are important double diffusive flows that do not arise from instability of vertical gradients. Again a vertical stable salt concentration gradient is set up, and a heat source introduced. In contrast with Fig. 23.7, the heat source is now a localized one. In the absence of the salt stratification this would produce a plume like that in Fig. 14.2 or 14.7(a). The behaviour with salt stratification is shown in Fig. 23.8. There is again a plume, but additionally horizontal layers of double diffusive convection spread out around this [120, 392]. The tendency for horizontal influence to extend into regions that are quite unaffected in the absence of stratification is loosely analogous with the same tendency in other types of stratified flow as discussed in Section 15.1.

24

DYNAMICAL CHAOS

24.1 Introduction [88, 134, 340, 361, 378]

Turbulent flows have often been regarded as the most important yet least understood set of phenomena of fluid dynamics. They have also been regarded as a topic distinctive to fluid dynamics. In recent years it has been realized that many non-fluid systems can exhibit behaviour which shares important characteristics with turbulent motion. (More precisely this fact was, at least in a general way, appreciated by some people in the second half of the nineteenth century. It is only recently, however, that there has been widespread realization of the implications for the study either of turbulence or of other systems.) Some of these systems are far simpler than any fluid flow; we shall see examples in Sections 24.3 and 24.4. Knowledge and understanding of them can thus be helpful in our efforts to understand what is happening when flows become turbulent (although as we shall see in Sections 24.6 and 24.7, the extent to which the new ideas 'solve the problem of turbulence' can be—and has been—overstated.) There are also some other fluid dynamical applications of these ideas, as we shall see in Section 24.5. Hence, fluid dynamicists need an understanding of these developments.

The word 'turbulence' is specifically associated with fluid flow. A name is needed for the more general set of phenomena, and the word 'chaos' has taken on a scientific meaning for this purpose. Also, as we shall see in the next two sections, an important aspect is that the chaotic behaviour is exhibited by classical, deterministic systems; hence, the name 'deterministic chaos' is often used.

Chaotic behaviour can arise only when the governing equations are non-linear. The fundamental advances have come theoretically through discoveries about the properties of various non-linear difference and differential equations. We shall not go into any of the general theory here [88, 134, 340, 378]. However, we shall use the particular case of the Lorenz equations to give a fuller explanation of the concept of deterministic chaos (Section 24.2).

Although the rigorous justification for supposing that deterministic chaos is a basic property of some non-linear systems (and not just a matter of inadequate control) comes from theory, its nature may be best illustrated by an actual physical system. In Section 24.3, therefore, we digress from fluid dynamics to look at a non-fluid mechanical system. We shall also use an electrical system, in Section 24.4, to illustrate the topic of 'routes to chaos'. Both these systems are essentially demonstrations; they do not have obvious applications. It should, therefore, be noted that chaotic behaviour of non-fluid dynamical systems is of importance in a whole range of applications; examples are solar system dynamics; non-linear laser optics; electric circuits with non-linear feedback; chemical reaction rates; dynamics of species populations.

Because some of the systems that exhibit chaotic behaviour are so simple, it seems strange that the realization that they can behave in this way is only recent. It is most unlikely that such behaviour had never previously been observed with comparably simple systems. However, it would almost always be undesirable behaviour in any practical system; presumably people reacted to it by saying 'something has gone wrong' and making adjustments to get rid of it. Fluid flow turbulence, in contrast, is far too pervasive ever to have been regarded in this way.

24.2 The Lorenz equations [259, 353]

Amongst the equations or sets of equations that have been particularly significant in the development of ideas about chaos are the Lorenz equations, introduced in Section 17.3. We remember in particular from that section that there are conditions in which none of the steady state solutions is stable; we left open the question of what happens then.

The Lorenz equations are actually a slightly more general set than we considered previously:

$$dX/dt = -PX + PY \tag{24.1}$$

$$dY/dt = -Y + rX - XZ \tag{24.2}$$

$$dZ/dt = -bZ + XY. \tag{24.3}$$

These differ from the previous set by allowing $b \neq 1$; the case $b = 8/3$ has received much attention. The equations were originally developed as a highly simplified model (mathematically a very severe truncation) of the equations of Bénard convection (Chapter 22) [259]. That is perhaps more of historical than of current scientific interest. If you wish to picture solutions of the equations, it is probably best to think of convection in a loop (Figs. 17.4, 17.5) even though that really applies only for $b = 1$.

Equations (24.1)–(24.3) give the rates of change of X, Y, and Z in terms of their present values. It is significant, in the context of properties to be seen below, that they are thus 'deterministic' equations. That is to say, if X, Y, and Z are known *precisely* at some initial instant, then their evolution for all subsequent time is in principle determined.

The principal results in Section 17.3 go over to the more general case with only minor modifications. In particular, eqn (17.43) becomes

$$r_c = P(P + b + 3)/(P - b - 1). \qquad (24.4)$$

For $r > r_c$, all three steady solutions are unstable. As r passes through r_c, the two steady solutions involving circulation become unstable in the form of amplifying oscillations ($\sigma_r > 0$, $\sigma_i \neq 0$). (r_c exists only when $P > (b + 1)$; for $P < (b + 1)$, the two solutions are always stable, but this case is consequently of less interest.)

What happens when $r > r_c$? X, Y, and Z must vary continuously in time. It is the form of these variations that is of special interest. This is not indicated by the analysis in Section 17.3, for the reason given there that the non-linear perturbation terms become significant. The principal way in which the behaviour has been elucidated is by numerical solutions of the equations; one starts with some initial condition for (X, Y, Z) and integrates forward in t. There is now a body of more formal theory that provides understanding of why the numerical solutions behave as they do [353], but we confine attention here to empirically observed features of the solutions.

Before we consider results for $r > r_c$, it is worth mentioning briefly what numerical solutions show for $r < r_c$. For the most part, such solutions ultimately tend to one of the stable steady state solutions—thus confirming, for example, that the consequence of the instability of the rest configuration when $r > 1$ is a transition to one of the two steady solutions with circulation. However, there is a range in which different initial conditions can lead to different ultimate behaviour; for example, for the frequently investigated case of $P = 10$, $b = 8/3$ this range is [353, 416]

$$24.06 < r < 24.74 (= r_c). \qquad (24.5)$$

In this range, solutions for some initial conditions tend to one of the steady state solutions whereas those for others exhibit indefinitely the sort of behaviour that is found for all initial conditions when $r > r_c$.

Figures 24.1–24.3 show three examples of solutions for $r > r_c$. They are actually all for $r/r_c = 1.13$ and $b = 8/3$ but various values of P. In each case the integration was started at a time sufficiently earlier than the start of the trace that the choice of initial conditions is not having a large direct effect. Our discussion will refer particularly to the case in Fig. 24.2. The other two cases are included primarily to emphasize that a wide variety of phenomena is contained within the Lorenz equations, although, since yet further variety can be obtained by varying r/r_c and b, they can do so only by example.

In the solution in Fig. 24.1, X, Y, and Z vary periodically (although

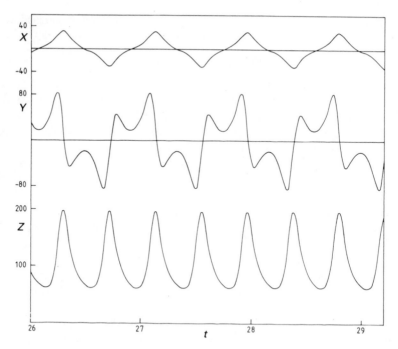

FIG. 24.1 Section of numerical solution of Lorenz equations for $r = 131$, $b = 8/3$, $P = 4$. Computations by R. Toogood.

far from sinusoidally). This is perhaps the most obvious possibility, given that solutions with steady circulation become unstable with respect to growing oscillations when $r > r_c$. It is significant that it does indeed sometimes happen; we return to the question of when at the end of this section. But our main interest is in the cases when the fluctuations are non-periodic, as in Figs. 24.2 and 24.3.

Figure 24.2 is in two complementary parts, one showing a short time stretch in detail and the other a much longer stretch without detail. In the former, all of X, Y, and Z are shown—the relationships between their changes are illustrated and one could estimate from it the ways in which the various terms in the equations are bringing about the changes. Figure 24.2(b) shows only X, and all details of variations between minima and maxima of this are omitted.

One sees that X (and Y) alternates between intervals during which it oscillates about a positive level and ones during which it oscillates about a negative level. In one sense, therefore, this is a highly patterned behaviour (quite different from velocity fluctuations in a turbulent flow, Fig. 2.6). It is nevertheless chaotic, as is seen when one considers the duration of each interval or equivalently the number of oscillations within

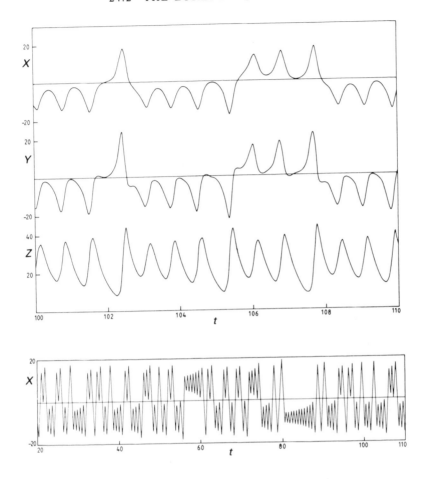

FIG. 24.2 Sections of numerical solution of Lorenz equations for $r = 28$, $b = 8/3$, $P = 10$. (a) Short section showing detailed variation of X, Y, and Z. (b) Long section showing broad features of variation of X. Computations by R. Toogood.

it. It is apparent from Fig. 24.2(b) that this number varies widely and without obvious pattern. Examination of long stretches of numerical data, together with other methods of analysis, have shown that it is indeed varying chaotically; that is to say, no matter how long the fluctuations continue, one will never find repetition of the detailed pattern.

An important aspect of such chaotic behaviour is sensitivity to initial conditions. A small change in these leads to a solution that ultimately diverges from the original one. The smaller the change the longer is the

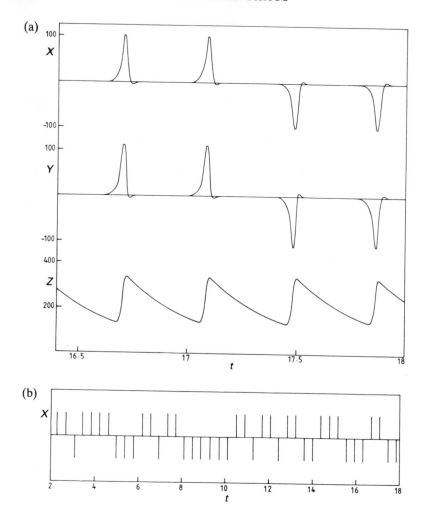

FIG. 24.3   As Fig. 24.2 for $r = 237$, $b = 8/3$, $P = 200$. Note that in (a) X and Y do not become identically zero between peaks, but much smaller than can be shown on this scale; and that in (b) the sign of peaks in X is shown by a sequence of similar lines although peaks are not identical in form.

time for which the two solutions remain close, but they never do so indefinitely. If two such solutions were displayed as in Fig. 24.2(b), the overall impression given would be the same; however, sufficiently long after initiation, the times at which the changes of level occurred in the two solutions would be totally uncorrelated.

This has the implication that, although in principle exact prescription of the initial conditions determines the solution throughout subsequent time, one has no way of actually finding this solution. Any procedure of numerical integration will involve approximations that have the same effect as unknown small changes in the initial conditions (although they are introduced continuously, not just at an initial instant). The relationship of a computed solution, such as that in Fig. 24.2, to the unknown exact solution is similar to that between two solutions with slightly different initial conditions.

More generally the concept of sensitivity to initial conditions has important implications for the predictability of systems governed by equations with this property. In any real system there is always some lack of exactitude in one's knowledge of initial conditions. Even if this is small, it means that the detailed behaviour becomes totally unforecastable in the long run. Determinism in principle does not necessarily imply predictability in practice. (In Section 24.3 we will look at an actual physical system that shows the implications of that statement rather effectively.)

It is only the detailed form of the fluctuations that is unforecastable. Average quantities (e.g. the mean value of Z or the root mean square value of the fluctuations in X) will take the same values in different realizations of the same conditions; i.e. different computations, perhaps using different numerical procedures, for different initial conditions but the same values of r, P, and b. This matter is considered more fully in the appendix to this chapter, as an illustration of the closure problem (Section 20.2).

Figure 24.3 gives an example of behaviour at high P, again in the form of a short detailed section plus a long schematic one. X and Y remain small for much of the time but have a series of peaks. These are almost equally spaced with the result that Z is almost periodic (but not exactly so—there are detectable variations in, for example, the minimum value of Z). However, the peaks can be either positive or negative: and this is another example of chaotic behaviour; a sequence like that in Fig. 24.3(b) might be obtained by the toss of a coin—heads for a positive peak, tails for a negative. (See Ref. [163] for more about the Lorenz equations at high P.)

The contrast between the periodic behaviour of Fig. 24.1 and the chaotic of Figs. 24.2, 24.3 was obtained by increasing P. It should not, however, be inferred that this is a simple trend. The question of when solutions of the Lorenz equations are periodic and when non-periodic is in fact a matter of some complexity. In particular, examples have been found [353] of periodic behaviour occurring at higher r than chaotic for the same values of b and P. Changes are not necessarily in the direction of greater randomness the further one goes from the first instability of steady solutions. We shall return to this point in Section 24.7.

24.3 Chaotic motion of a forced spherical pendulum

The concept of chaotic motion may seem somewhat abstract. We therefore now look at one particular dynamical (although not fluid dynamical) arrangement that exhibits it. This demonstration also forcibly makes the point that chaotic motion may occur even in remarkably simple systems.

The following description is based on observations with a demonstration apparatus [390]. The demonstration was motivated by theoretical work [273, 274] which provides the real justification for the interpretation given. We will return to this point after the description of the demonstration.

We consider a pendulum that is free to swing equally in any direction, so that the bob can move over a spherical surface—an arrangement known as a spherical pendulum or a conical pendulum. The point of suspension is oscillated sinusoidally in a straight line (the x-direction) with an amplitude small compared with the pendulum length. This, of course, causes the pendulum to swing in the same direction. If the driving frequency (f) is close to the pendulum's own natural frequency (f_0), the swing amplitude becomes large (obviously); the motion may then become unstable with respect to perturbations in the perpendicular (y) direction (far from obviously). As a result, there is a range of driving frequencies (for quantitative details see Refs. [273, 274, 390]) in which the pendulum bob orbits within the spherical surface, rather than just oscillating on an arc of the surface.

It is this orbital motion that exhibits the phenomena with which we are concerned. There is a surprisingly complex sequence of changes in this motion as one traverses the driving frequency through the main range. However, we can illustrate the most important features by considering just two types of motion, each of which occurs over a significant sub-range. (To a first approximation, one is likely to encounter the first type if f is a little higher than f_0 and the second type if f is a little lower than f_0.) Only the second type is a chaotic motion, but the first is worth mentioning for comparison.

Within one sub-range the pendulum bob moves on an elliptical orbit, with principal axes in the x- and y-directions, with the same frequency as the drive. The motion is thus a highly ordered one, although different from what one might anticipate.

It is interesting to consider the predictability of this behaviour; as discussed in Section 19.1, we expect the fact that the motion is the consequence of instability of a different type of motion to imply some loss of predictability. Suppose that one set the pendulum in motion, waited long enough for transients to die away, and then attempted to predict (on the basis of previous experience) what one would observe if one looked at the pendulum. One could in fact make a nearly completely correct prediction. The only feature that one could not predict would be whether the pendulum was orbiting clockwise or anticlockwise. There is no asymmetry in the system that makes one sense more probable than the other. (An analogy may be made with the discussion of a vortex street in Section 19.1. In both cases the behaviour is not completely predictable, but observation for only a short time makes it so for all subsequent times.)

We turn now to the other sub-range that we are considering. (With the particular apparatus on which this description is based, this involves a reduction in the driving frequency of typically 3%.) The motion is chaotic on a time scale long compared with the oscillation period. If one observes the horizontal projection of the motion for just one or two periods then one sees a quite well-defined orbit; one can say whether the bob is moving on a circular, elliptical or linear path, how big this is, and how, for an ellipse or line, it is oriented. If, however, one observes the motion for a longer time, one finds that the size, ellipticity and orientation of the orbit are continuously changing. Moreover there is no discernible pattern to the changes. Figure 24.4 gives an example (drawn from trajectories on the screen of a videotape of the motion) of how one might see the bob moving if one glanced at it at a succession of equally spaced times—although one would not expect to see just this sequence ever again.

An attempt at prediction, like that outlined above, would now have little chance of success. Accumulation of observations would allow one to specify statistical properties of the motion, such as the average kinetic and potential energy of the pendulum over a large number of periods (cf. the statistical specification of turbulent motion, Section 19.2). But it would not allow one to predict the detailed sequence of changes in any run. Sensitivity to initial conditions (explained in Section 24.2) makes this different every time.

From the demonstration alone, it is difficult to be certain that the motion is chaotic, in the specific sense that the word is now used, and not just very complicated. The fact that no pattern to the changes can be discerned does not guarantee that there is none. The justification for the interpretation is theoretical analysis of the forced spherical pendulum. Despite some unavoidable differences between the theoretical and actual pendulums, the phenomena exhibited by them show good correspondence. The theoretical results can be subjected to specific mathematical tests (concerning the spectrum of the motion, cf. Sections 19.5, 24.7) to establish their chaotic character. There is every reason to suppose that changes such as those in Fig. 24.4 are intrinsic to the dynamics, not something that could be eliminated if only one had better control of the apparatus.

We have noted that chaotic behaviour can arise only for systems

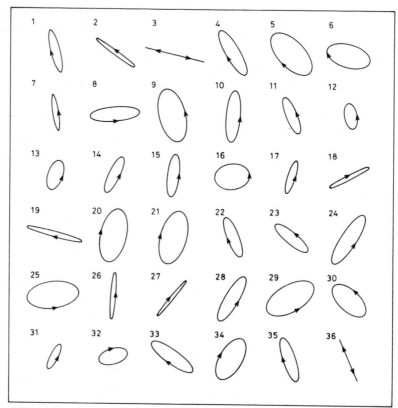

FIG. 24.4   Orbits of pendulum bob in chaotic motion (see text) at intervals of $10\,T_0$, where T_0 is the natural period. Orbits are drawn as ellipses but are actually not quite closed. (Ratio of drive frequency to natural frequency $= 0.996$; cf. Ref. [390]).

governed by non-linear equations. Non-linearity is in fact also needed to provide the coupling between the x- and y-motions that produces the latter. Why is the pendulum a non-linear system? It is a consequence of the large amplitudes to which the pendulum swings. In the obvious notation, the restoring torque on the pendulum is $mgl\sin\theta$. The theory works to the approximation

$$\sin\theta = \theta - \theta^3/6. \qquad (24.6)$$

If one were to make the linear approximation of $\sin\theta = \theta$, none of the effects under consideration would be obtained. (This is the only non-linearity in the theory. With the demonstration apparatus there are other non-linear effects present, particularly in the damping; the bob is a ball moving through the air at a Reynolds number $\sim 10^3$, cf. Sections 3.4 and 7.3. But this does not seem to make much difference to the behaviour.)

24.4 Routes to chaos

It is evident from the mathematical and physical examples described in the foregoing sections that systems exhibiting chaotic behaviour also exhibit ordered behaviour for different values of the governing parameters. They thus make a 'transition to chaos' as the parameters are varied. The ways in which this transition can occur form a topic central to theoretical ideas about deterministic chaos. In the present context they are important for attempts to relate transition to turbulence to transition to chaos, a matter we shall be considering in Section 24.7.

The topic of transition to chaos or 'routes to chaos' is remarkable for—what at first sounds paradoxical—its simultaneous universality and diversity. Universality refers to the fact that totally different systems can exhibit the same route to chaos—the same not just in that they show the same broad features, but very closely the same including quantitative details. On the other hand, there are various quite distinct routes, and a single mathematical or physical system may exhibit more than one of them; a change in some parameter can lead to a qualitative change in the route adopted as a second parameter is varied.

We cannot give an adequate treatment of this matter here. The theoretical structure is large and in places complicated. Also any account could well become rather rapidly out of date. What we need is, firstly, a more specific indication of what is meant by a route to chaos and, secondly, some ideas on which we can draw in Sections 24.5 and 24.7.

For these purposes we will look at experimental observations with a particular system that provides effective illustrations. Some might think this a rather idiosyncratic approach to a primarily theoretical topic, but it seems appropriate to this book. Readers wanting a definitive account of routes to chaos should turn to, e.g., Refs. [88, 134, 340, 378].

The system we consider is the electrical circuit shown in Fig. 24.5.† A

---

† This circuit is a convenient one for exploring phenomena associated with chaos: it is very simple to set up. Various investigations have been reported [101, 117, 213, 227]. In these the diode used was a varactor diode. However, it has been found that, with appropriate values of the other components and the frequency, various types of diode can be used including 'common-or-garden' rectifier diodes. One is then making use of non-standard features of the diode which will vary between nominally similar diodes. The voltage values at which the developments occur, or even the detailed nature of the developments, may change if one switches to a different diode of the same type. The sequence described here applies to a set of observations with one particular diode. Incidentally, in this experiment, the self-capacitance of the coil was not completely negligible; this influenced the detailed form of the pictures, but was almost certainly not an essential property for the main developments.

The non-linearity essential for chaotic behaviour is contained in the diode characteristics. There is, however, some question about just what properties are involved. Indeed, theoretical models that seem to correspond to some of the observations have been developed around two different diode properties—variable capacitance in reverse bias [227] and the persistence of charge carriers for some time after bias reversal [323].

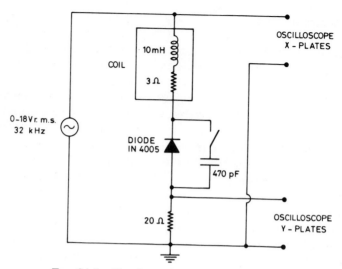

FIG. 24.5   Circuit used to produce Fig. 24.6.

coil, diode, and resistance in series are connected across an alternating voltage source; the capacitor shown in parallel with the diode is initially absent: we consider it later. Changes in the component values, the particular diode, and the source frequency can give a rich variety of observations. For our purposes it is sufficient to display the results of one particular experiment.

Figure 24.6 shows a sequence of photographs of an oscilloscope screen, on which the current through the circuit is displayed in the y-direction versus the voltage across it in the x-direction. Through the sequence the source voltage is being increased with all else held constant; the x- and y-scales change between pictures, not necessarily in proportion.

In picture (a) the circuit is behaving as one might guess—the single loop on the screen indicates that the current varies in the same way for every cycle of the source. However, as the voltage is increased much less obvious behaviour ensues. In (b) there is a double loop, indicating that the pattern of current variation is different for alternate cycles of the source. A particular voltage can be identified at which a transition from the behaviour of (a) to that of (b) occurs. Further voltage increase gives another transition: the pattern of current variation repeats on only every fourth cycle of the source (picture (c)). In (d) and (e) the current variations have become chaotic; the pattern never repeats itself exactly. In (e) the current variation during one period of the source may be such that the path on the oscilloscope screen (the 'phase space trajectory') may lie anywhere within a certain region. At the lower voltage of (d) there are gaps in the picture; although the behaviour is now chaotic, the trajectories remain fairly close to ones that were followed when it was periodic. As the voltage is increased the 'permitted' regions gradually spread until the gaps are filled in.

Further increase of the voltage from that of (e) leads to a reversion to ordered behaviour. Now the pattern of current variation repeats exactly over every three periods of the source (picture (g)). Between (e) and (g) there is a range of behaviour not fully illustrated by a still picture such as (f), which looks like a superposition of (e) and (g): if one watches the oscilloscope one sees the picture alternating randomly between the chaotic pattern and the ordered one. The typical duration that it spends in either mode is very long compared with the source period.

Pictures (h), (i), and (j) continue the story as the voltage is increased. The repetition period of the current changes from three times to six times the basic period. Then chaotic behaviour recurs, with the phase space trajectories being initially confined to bands and subsequently filling a whole region. There is, however, a significant difference from the changes between pictures (d) and (e). Now, instead of the bands gradually spreading to fill the gaps, there is a particular voltage at which the gaps are suddenly filled (a type of change known as a 'crisis').

One variant of the experiment is worth mentioning. A small capacitance was introduced in parallel with the diode (Fig. 24.5). The sequence of observations was broadly the same with one significant exception. No region of alternation between pictures (e) and (g) occurred. Instead the pattern jumped suddenly from one picture to the other as the voltage was varied. However, there was marked hysteresis in the voltage at which this occurred; there was a range of voltages in which the behaviour could be either permanently chaotic or permanently ordered, depending on the side from which the range had been entered.

The above observations involve a variety of routes to chaos. The developments leading from the simple behaviour of Fig. 24.6(a) to the first chaotic zone (d) are an example of a widely discussed route, the period-doubling sequence. The interpretation of these observations, based on theory [88, 340], is that there is actually an infinite number of changes like that from (a) to (b) or from (b) to (c). At each step the period of the current variations doubles. One thus gets a sequence of periods: $1, 2, 4, 8, 16, \ldots, \infty$. Although the number of steps is infinite, the whole sequence occurs within a finite range of the input variable, in our case the voltage. The period thus reaches ∞ at a particular voltage. This means that the pattern of variation never repeats itself. It has thus become chaotic.

This description, of course, implies that the range for period 2^n gets

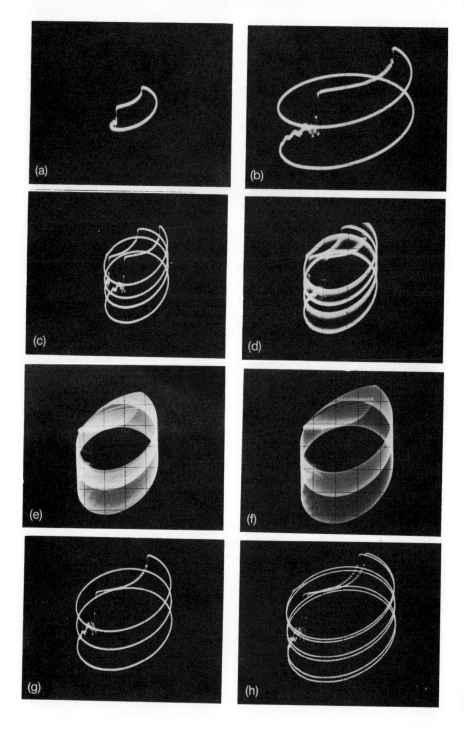

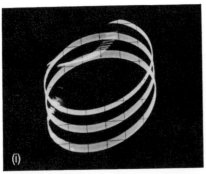
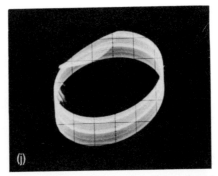

FIG. 24.6 Photographs of oscilloscope screen with circuit in Fig. 24.5. Input r.m.s. voltages: (a) 1.5, (b) 3.5, (c) 5.2 (d) 5.3, (e) 7.0, (f) 7.3, (g) 8.2, (h) 9.4, (i) 10.2, (j) 18.6.

rapidly narrower as n increases. In any real experiment one will be able to detect only a limited number of stages.

It is worth emphasizing that the sequence is one of *period*-doubling, not of frequency doubling. Each stage introduces a new *sub*-harmonic, not a new harmonic. (Harmonics are present even for period 1, because the fluctuations are highly non-sinusoidal.) The process is thus not the long-known one of the production of harmonics by non-linearity.

The observations with the coil-doide circuit included a second period-doubling sequence after the window of ordered behaviour. This time one may suppose that the sequence is: 3, 6, 12, 24, . . . , ∞. It is essentially the same route to chaos. We have, however, already noted that there is a difference in subsequent developments within the chaotic zone.

One may also consider what happens when one leaves the ordered window on the other side. One is starting from an ordered behaviour, with period 3, and observing a transition to chaos as the voltage is *decreased*. The route is now quite different. It is in fact the intermittency route, effectively defined by the above observations. In the transiton zone the behaviour alternates randomly between the behaviours on either side of it. The intermittency factor, defined as the fraction of the time that the fluctuations are chaotic, varies continuously across the transition zone from 0 at the ordered side to 1 at the fully chaotic side. (Fig. 24.9 will illustrate this in a different context.)

It should be noted that the causes of this type of intermittency are different from those of intermittency arising from alternate laminar and turbulent motion, as in Figs. 2.10 and 21.4.

Finally, the experiment with the additional capacitance has been mentioned to illustrate that the route can be changed by a small modification of the system. Now, on any given occasion (decreasing or

increasing the voltage), the switch between ordered and chaotic behaviour takes place catastrophically, without an intervening zone of
intermittency. However, because of the hysteresis, there is still an
intermediate zone—one in which either behaviour may occur
permanently.

One other route to chaos, the quasi-periodicity route, not illustrated
by the coil–diode circuit will be mentioned in Section 24.7.

24.5 Chaos in fluid dynamics

Certainly the primary reason why ideas about chaos have been seen as
important for fluid dynamics is their potential relevance to turbulence.
We discuss this in the next two sections. However, there are also quite
different ways in which flows exhibit chaotic behaviour—with the
connection to the general theory of chaos being perhaps more definitely
established. It seems likely that examples of such flows will multiply in
coming years. Thus, although some of the topics are outside the main
scope of this book, we look briefly at this matter in this section.

Firstly, it is worth noting that if the Lorenz equations (Section 24.2)
are considered as modelling convection in a loop (Figs. 17.4, 17.5), then
the chaotic solutions of the equations do not correspond to turbulent flow
in the loop. They correspond to chaotic fluctuations of the whole
circulation. The question of whether the flow becomes turbulent (as
Poiseuille flow becomes turbulent, Sections 2.6, 18.3) is distinct. One
could envisage any of the following: nonchaotic laminar flow; chaotic
laminar flow; non-chaotic turbulent flow; and chaotic turbulent flow.
Relating this to actual flows, is restricted by the severe simplifications in
deriving the equations, although there have been some experiments
showing interesting behaviour [133, 184].

Cases where the connection is more immediate concern various flows
with free surfaces. It has, for example, been shown theoretically that
non-linear forced surface waves in a circular cylinder are governed by
equations closely similar to those for the pendulum described in Section
24.3 [275]. It has also been found experimentally that the columns
formed by a layer of viscous liquid falling over an edge can move around
chaotically. [313].

As an example to illustrate the matter a little more fully we consider a
dripping tap [263]. The interval between successive drips sometimes
varies chaotically. It is unclear just when this happens—one is unlikely to
be able to observe it in a casual experiment at the kitchen sink—but Fig.
24.7 shows observations made by varying the pressure behind a suitably
shaped nozzle. Like the system described in the previous section,

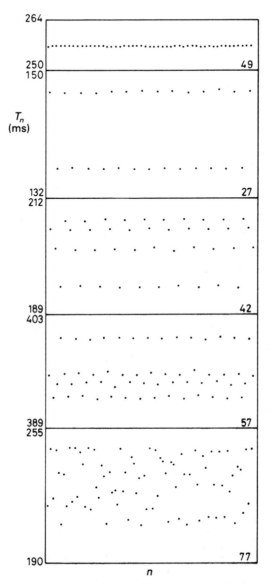

Fig. 24.7 Five examples of variation of time interval between falling drops plotted against interval number (measurements made by drops intercepting a laser beam directed onto a photocell.) Plots provided by M. G. Priestley, N. Proudlove, and S. Whitby, Department of Physics, University of Bristol.

considerable complexity is found; there is more than one transition to
and reversion from chaos. The full picture has not yet been elucidated.
Figure 24.7 thus does not represent a single sequence. However, it shows
cases in which every interval is the same, every alternate interval is the
same (period doubled), every fourth interval is the same (period
quadrupled—two examples with different detailed sequences), and the
intervals vary chaotically.

24.6 Implications for turbulent flow

The implications of modern ideas about chaos for fully turbulent motion
(as opposed to transition to turbulence to be considered in Section 24.7)
are primarily concerned with our general understanding of the word
'turbulence'. Turbulent motion can now be seen as an example of
deterministic chaos. In a sense that has long been the case, since it has
generally been supposed that turbulent flows are contained within a
deterministic set of equations—the Navier–Stokes and continuity equations. However, in the absence of a formal derivation, it has been
possible to question whether this is really the case. This matter can now
be considered resolved. Since far simpler deterministic systems exhibit
chaotic behaviour and since we understand how this comes about, the
view that the occurrence of turbulence implies a failure of deterministic
equations is no longer tenable.

The new ideas may also help with the formulation of a more precise
definition of turbulence, as we shall discuss at the end of the next section.

Beyond this, the implications for fully turbulent flow are slight. Some
of the simpler systems provide a context in which certain ideas about
turbulent flow may be elucidated; e.g. the analysis of the closure problem
in the appendix to this chapter. But attempts to understand particular
turbulent flows, as outlined in Chapters 19–21, have not been significantly modified or aided by the new developments. It is necessary to
stress this rather negative fact mainly because of excessive claims that
have been made for the new ideas. It has been said that 'the problem of
turbulence has been solved'. Aside from the question of what is 'the
problem of turbulence' (there are many), this can give a quite false
impression.

24.7 Implications for transition to turbulence

If we turn to the processes by which flows become turbulent, rather than
their structure when fully turbulent, more positive statements can be

made. We have seen in Chapters 17, 18, and 22 that these processes are diverse. Studies of non-fluid systems undergoing transition to chaos and developments in the mathematical theory of this transition are also revealing a varied phenomenology. It is of interest to see whether and where the two share common ideas. In particular one may ask whether the established routes to chaos (Section 24.4) are to be found during transition to turbulence. Since there is a body of theory associated with these routes, some of it of considerable generality, identification of one of the routes implies that the transition is at least partially understood.

The short answer is that several of the routes have been identified in fluid transition experiments, but that the significance of this for transition in general is currently very uncertain. Quite a number of experiments leading to such identification have been with small aspect ratio Bénard convection (Sections 22.1, 22.4) [87, 181, 246]; small in this context means not much greater than unity. Another fruitful configuration has been rotating Couette flow (Section 17.5) [158, 402]. We illustrate the point by presenting briefly the results of experiments in which two of the routes introduced in Section 24.4, the period-doubling route [177, 182, 246] and the intermittency route [86, 150, 269], were found.

Figure 24.8 shows a sequence of spectra for increasing Rayleigh number measured in a Bénard apparatus. The appearance of sub-

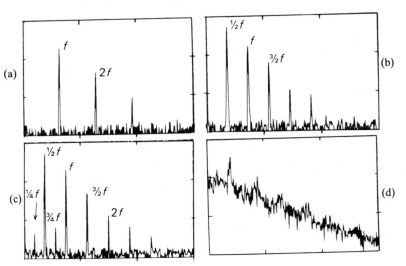

FIG. 24.8   Power spectra of local velocity in small aspect ratio Bénard convection, showing period doubling. $Pr = 2.5$. Values of Ra/Ra_{crit}: (a) 21.0, (b) 26.0, (c) 27.0, (d) 36.9. In each part the abscissa is linear in frequency and the ordinate is logarithmic in power with a factor of 10^6 between bottom and top. From Ref. [182] by permission of the New York Academy of Science.

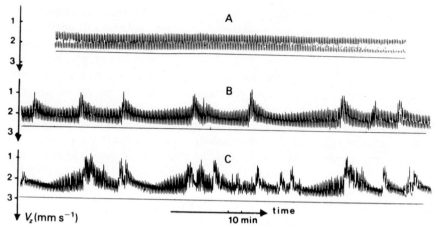

FIG. 24.9 Oscillograms of vertical velocity (increasing downwards) in small aspect ratio Bénard convection, showing intermittency. $Pr = 130$. Values of Ra/Ra_{crit}: (A) 270; (B) 300; (C) 335. From Ref. [86].

harmonics $(f/2, f/4)$ of a previously present frequency (f) reveals period-doubling. The continuous spectrum of the last diagram indicates that periodic fluctuations have been replaced by chaotic ones.

Figure 24.9 shows vertical velocity fluctuations in a different Bénard experiment. Consecutive oscillograms correspond to increasing Rayleigh number, and one sees a change from periodic fluctuations to irregular ones through an increasing intermittency factor.

One other route to chaos, not illustrated by the experiment described in Section 24.4, has received particular attention in the context of transition to turbulence. It is therefore illustrated in Fig. 24.10, although

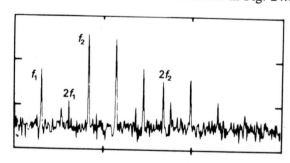

FIG. 24.10 Power spectrum of local velocity in small aspect ratio Bénard convection, plotted as in Fig. 24.8 at $Pr = 5$, $Ra/Ra_{crit} = 35.0$. Peaks can be identified at sequentially f_1, $f_2 - f_1$, $2f_1$, f_2, $f_2 + f_1$, $2f_2 - f_1$, $f_2 + 2f_1$, $2f_2$, $f_2 + 3f_1$, $2f_2 + f_1$, $2f_2 + 2f_1$. From Ref. [181].

a full discussion would take too much space. The figure is a spectrum similar to those in Fig. 24.8 and shows that there are two basic frequencies (f_1 and f_2) present together with their harmonics and integer combinations ($mf_2 \pm nf_1$). This occurs as an intermediate stage between a single frequency spectrum (similar to Fig. 24.8(a)) and a continuous one representing chaotic fluctuations (similar to Fig. 24.8(d)). There are several variations of the route, in which other intermediate stages, if any, differ [158, 181, 247, 268, 269]. Fluctuations with two or more incommensurate frequencies (i.e. frequencies of which the ratio is not a ratio of integers) are called quasi-periodic. Hence, the route is known as the quasi-periodicity route.

An obvious question is posed by this diversity of routes: when does each occur? The answer is not known. In fact quite small changes in an apparatus can lead to an unexpected switch to a different route.

A more general open question concerns the extent to which these observations are typical of transition. Will it be possible to relate transition, for example, in large aspect ratio Bénard convection or in shear flows to the above ideas? The links with the observations described in earlier chapters are certainly not obvious. On the other hand, the experiments in which particular routes to chaos have been identified required very precise experimental techniques. Any of three outcomes seems possible at present: (i) known routes to chaos will be identified in other flows by more refined experiments; (ii) new routes to chaos will be found and related to these flows; or (iii) transition to turbulence is often inherently different from transition to chaos in simple systems (i.e. systems with few degrees of freedom are not a good guide to those with very many).

Two further general points are of interest. Firstly, we noted in Section 24.2 that solutions of the Lorenz equations may change from chaotic to periodic as r is increased; also the experiment described in Section 24.4 showed a chaotic zone with ordered ones on either side of it. Reversion to order can occur; systems do not necessarily get progressively more disordered as one goes further beyond a first instability. Does this have implications for transition to turbulence? We can be confident that, for practical purposes, the general notion that high Reynolds number (or its counterpart) implies turbulence is valid. But there do sometimes seem to be trend reversals on the route to this. Reappearance of periodic motion has been observed in rotating Couette flow [402]. In Sections 3.3 and 17.8, we saw that the vortex street frequency has ranges of decreased regularity; although these changes can be related to developments in the flow, there are tentative suggestions that they might also be viewed in the present context [355].

Secondly the standard routes to chaos imply the existence of a well-defined onset of chaotic behaviour. This is relevant to the question raised in Section 19.1 as to whether the demarcation between non-turbulent and turbulent motion is sharp, or whether there is a 'grey area'. In experiments described above (e.g. Figs. 24.8 and 24.10), the onset of turbulence was specified clearly by the stage at which the spectrum changed from a discrete to a broadband structure. However, the generality of this is uncertain. We saw, for example, in Section 18.4 that in transiton in jets and other free shear flows randomness appears to develop continuously; it would be difficult to assign a stage in Fig. 18.11 or 18.12 at which the motion is first turbulent, and it may be impossible. The fact that this is an open question is obviously related to our earlier open question: can a standard route to chaos always be identified in transition to turbulence?

Appendix: The Lorenz equations and the closure problem

The Lorenz equations (Section 24.2) provide a convenient context in which to illustrate the closure problem (Section 20.2). The formulation is much simpler than for a turbulent flow but the point of principle is the same.

The average properties of a chaotic solution of the Lorenz equations are presumably well defined (where the comments in Section 19.2 about the alternatives of ensemble and time averaging again apply). One might hope to determine these average properties from averaged forms of the equations. We will see that one cannot in fact do so without some additional statistical hypothesis.

The approach in general is to formulate expressions for the rate of change with time of various averaged quantities, and then set this rate of change to zero. We suppose from the outset that certain averages are zero from symmetry considerations; in particular that

$$\bar{X} = \bar{Y} = 0 \qquad\qquad (24.7)$$

(the overbar denoting averaging as in Section 19.2). This is reasonable because the Lorenz equations have the property that if (X, Y, Z) is a solution, so is $(-X, -Y, Z)$; hence a solution that can reach negative X and Y from positive is likely to be statistically symmetrical about $X = 0$, $Y = 0$. This is fairly evidently a property of Figs. 24.1 to 24.3. (There can be solutions to which this argument does not apply because they do not pass from positive to negative X and Y, but we need not complicate the discussion by including these.) The same symmetry argument implies that various other mean quantities, e.g. $\overline{XZ}$, are zero.

With this simplification, only one of the three Lorenz equations gives non-trivial information by direct averaging. Putting

$$d\bar{Z}/dt = 0 \qquad (24.8)$$

in eqn (24.3) gives

$$-b\bar{Z} + \overline{XY} = 0. \qquad (24.9)$$

This is the counterpart of the interaction between the mean flow and the turbulence discussed after eqn (19.13). One would like to use it to determine $\bar{Z}$, so one next formulates equations for the fluctuations in the hope of determining $\overline{XY}$.

(To maintain the closest analogy with the usual procedure for turbulent flows, we should divide Z into mean and fluctuating parts

$$Z = \bar{Z} + z. \qquad (24.10)$$

We do not in fact do so because it adds to the algebra without affecting the point of principle. Since

$$\overline{Z^2} = \bar{Z}^2 + \overline{z^2} \qquad (24.11)$$

no real difference is involved. The matter does not arise for X and Y because of eqn (24.7).)

Multiplying eqn (24.1) throughout by X and averaging gives

$$\tfrac{1}{2}\,d\overline{X^2}/dt = -P\overline{X^2} + P\overline{XY} \qquad (24.12)$$

and thus, with mean quantities unchanging,

$$-P\overline{X^2} + P\overline{XY} = 0. \qquad (24.13)$$

Similarly, from eqns (24.2) and (24.3)

$$\tfrac{1}{2}d\overline{Y^2}/dt = -\overline{Y^2} + r\overline{XY} - \overline{XYZ} = 0 \qquad (24.14)$$

$$\tfrac{1}{2}\,d\overline{Z^2}/dt = -b\overline{Z^2} + \overline{XYZ} = 0. \qquad (24.15)$$

Multiplying (24.1) by Y and (24.2) by X, adding, and averaging gives

$$d(\overline{XY})/dt = -(P+1)\overline{XY} + r\overline{X^2} + P\overline{Y^2} - \overline{X^2Z} = 0. \qquad (24.16)$$

(The other permutations of this procedure give equations of which every term is zero by symmetry.) Equations (24.13)–(24.16) are the set of four equations for $\overline{X^2}$, $\overline{Y^2}$, $\overline{Z^2}$, and $\overline{XY}$. They give relationships between them,

$$\overline{XY} = \overline{X^2} \qquad (24.17)$$

and

$$r\overline{X^2} = \overline{Y^2} + b\overline{Z^2} \qquad (24.18)$$

but they do not provide values (for any given P, b, and r) of the four quantities. In particular they do not give a value of $\overline{XY}$ to put back in eqn (24.9) and so determine $\bar{Z}$.

This is because of the appearance of the averaged triple products $\overline{XYZ}$ in (24.14) and (24.15) and $\overline{X^2Z}$ in (24.16). The problem is now becoming clear. If one formulates equations for the averaged triple products (as one can do by, for example, substituting (24.1) to (24.3) into

$$\mathrm{d}(XYZ)/\mathrm{d}t = XY\,\mathrm{d}Z/\mathrm{d}t + YZ\,\mathrm{d}X/\mathrm{d}t + ZX\,\mathrm{d}Y/\mathrm{d}t \qquad (24.19)$$

and averaging), one finds that these involve averaged quadruple products such as $\overline{X^2Y^2}$ and $\overline{X^3Y}$. At every stage, new averaged quantities of higher order are introduced. One never gets a closed set of equations.

We leave the matter with that rather negative conclusion. If one's aim were to predict averages for the Lorenz system, one would next need to introduce some assumption about the higher-order terms to terminate the above process. Our purpose, however, has been merely to illustrate the principle of the problem because it is a crucial reason why turbulent flows cannot simply be 'solved'.

25

EXPERIMENTAL METHODS

25.1 General aspects of experimental fluid dynamics

We do not have space for a full description of all the experimental techniques used in obtaining the results discussed in this book. This chapter can give only a general survey, intended to place the various methods in some perspective. The techniques appropriate to different branches of fluid mechanics—and to different experiments within one branch—are diverse, and decisions about those to be used in a particular project require detailed consideration of the successes and failures of those used in previous related projects. One must add to this that the limitations to an experiment often lie in the performance of transducers, and the experimentalist should always be on the alert for new possibilities.

The purpose of an experiment in fluid mechanics may range from direct verification of a theory (as in Figs. 11.2, 15.12 and 17.19) to a general exploration of the phenomena that occur in a given situation. Most of the experiments that have contributed to the ideas described in this book fall somewhere between these two extremes, although the proportions contributed by theory and by experiment to the final story are very variable. Most often one is dealing with a situation for which mathematical difficulties preclude a full theory, but in which it is still useful to refer to the equations of motion in deciding what measurements to make and how to interpret the results. Studies of the energy balance in turbulent flow, as in Figs. 21.8 and 21.16, provide a straightforward example of this. Even in topics for which there is a wholly adequate theory, primarily exploratory experiments may have played an equally important role. For example, hydrodynamic stability is now one of the more highly developed theoretical branches of the subject but the need for this type of treatment of the equations of motion would not have been apparent without experimental observations of instabilities.

The design of an experiment involves careful attention to the requirements of dynamical similarity. One must ask what range of phenomena one wishes to study and thus what values the relevant non-dimensional parameters should have. Then one must consider how these can be achieved, or, if they fall outside the practicable range, what departures are least unacceptable.

In this process of design, there are three questions that frequently arise: (i) will the work consist primarily of quantitative measurements or of observations of flow patterns? (ii) will the work be done with an existing installation or with specially built apparatus? (iii) what fluid will be used?

The distinction implied by the first question—transducing versus flow visualization—is not complete. One can, for example, measure velocities by timing the movement of dye. However, it does provide a useful general classification of experiments. The decision depends partly on the previous state of knowledge of the topic under investigation—one is most likely to opt for flow visualization in a preliminary exploration—and partly on the efficacy of available transducers in the particular situation. However, the two approaches are often complementary. No purely qualitative study is likely to answer all questions about a flow. On the other hand, measurements are often difficult to interpret without the assistance of flow visualization. Within this book, rather frequent use of the results of visualization experiments has been made because the photographs provide ready illustrations of the flows under discussion. The reader should not conclude that experimental fluid mechanics is primarily a matter of 'look and see'; one always aims to express results as quantitatively as possible.

There is a third category of experiment, quantitative but not involving detailed probing of the flow. In these experiments, some bulk quantity associated with the flow is measured. Examples are the mass flux in pipe flow (Sections 2.3 and 2.7), the torques acting on the cylinders in rotating Couette flow (Sections 9.3 and 17.5), and the heat transfer in convection experiments (Sections 14.4, 14.8 and 22.2—particularly Fig. 22.1). The experimental methods involved depend very much on the particular experiment, so they will not be discussed further in this chapter.

Some fluid mechanics laboratories consist mainly of standard flow systems into which different experiments can be introduced. Others consist mainly of equipment that has been built for particular experiments. This depends primarily on the branch of fluid mechanics being studied. The investigation of the flow past obstacles or of boundary layers requires a uniform flow with minimal velocity fluctuations. To obtain this is in itself a complicated matter. Hence, such experiments are normally carried out in a laboratory permanently equipped with a wind-tunnel (the name for any system providing a working air stream), a water flume or channel (similar systems with water), or a towing tank (a large tank of stationary water through which an obstacle can be moved). A few of the many examples of results obtained with such equipment are Figs. 11.2, 12.9, 21.7 and 21.19. Readers interested in more details of such techniques are referred to Refs. [22, 29]. In contrast, many experiments, for example in convection, are difficult to fit into standard systems, and every experiment involves the construction of a special piece of apparatus. Examples of observations made in such experiments include Figs. 15.2, 17.11, 22.1 and 22.3–22.8.

The choice of working fluid, unless dictated by the available standard equipment, is usually influenced by two considerations, the achievement of the desired values of the governing non-dimensional parameters and the performance of flow transducers. Whenever possible, of course, either air or water is used. The choice between these two often depends on the type of experiment; more successful techniques for velocity measurement have been developed for air than for water, whilst flow visualization is generally more successful in water than in air. Other considerations may, however, enter. For example, one might face a situation in which the fact that water more readily gave the desired values of the governing parameters (its lower kinematic viscosity is often advantageous in this respect) had to be set against the practical difficulties of containing it in an arrangement with movable probes. Other fluids are sometimes used because they give better values of the governing parameters or because they have properties particularly appropriate to an experiment. For example, silicone oils are widely used in convection experiments: they can be obtained with different viscosities, but otherwise similar properties, which provides a convenient method of varying the Prandtl number; and the viscosity varies with temperature much less than for many fluids.

25.2 Velocity measurement [51]

Obviously the most important quantity to be measured in most flows is the fluid velocity. Here we shall look at the principles by which various methods of measuring velocity work; the reader interested in the arrangements of a full working system and the procedures for operating it should follow up the references. The general name for any instrument for measuring fluid velocity is 'anemometer'.

There are three principal instruments for velocity measurement—that is to say, instruments that have found application in many different types of experiment. These are the Pitot tube, the hot-wire anemometer (and similar devices) and the laser-Doppler anemometer. The remaining techniques have not given rise to general-purpose instruments but have proved useful in particular experiments.

The Pitot tube [51] is illustrated in Fig. 25.1. The inner tube with a hole at the nose, S, of the instrument is entirely sealed from the outer tube. Several holes (typically five), of which two are shown in the figure, round the periphery of the tube all lead into the outer annulus. The

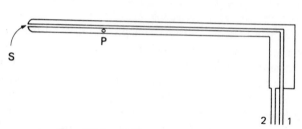

FIG. 25.1 A Pitot tube.

theory of the operation of a Pitot tube is contained essentially in eqn (10.19), Bernoulli's equation applied to the streamline that ends at the forward stagnation point of an obstacle placed in a stream. If the Pitot tube points into the flow, the pressure p_S will occur at the point S and will be measured by a manometer connected to outlet 1 (the manometer must, of course, block the tube, so that there is no flow through it). Calculation of the speed, u, from the relationship, $p_S = \frac{1}{2}\rho u^2 + p$, requires information about the pressure p. This is provided by the peripheral holes, P. These are some distance downstream from the nose so as to be beyond the region of pressure variation. Since the pressure difference across the boundary layer is negligible, the pressure at these holes is the static pressure p. The average pressure from several holes is observed, as the pressure at a single hole would be more sensitive to misalignment of the tube than the p_S-reading. A manometer connected between outlets 1 and 2 reads the pressure $\frac{1}{2}\rho u^2$, thus providing a direct measurement of u. Provided that the Pitot tube is small compared with the length scale of variations of the flow, u and p can be interpreted as the speed and pressure that would exist at the position of the Pitot tube in its absence.

  Pitot tubes have been most widely used in air, where they can measure speeds from about $1\ \mathrm{m\,s^{-1}}$ upwards. With greater difficulty they can be used in water [311] to measure speeds down to about $3\ \mathrm{cm\,s^{-1}}$. Since one applies inviscid theory to determine the velocity, it is essential for the Reynolds number of a Pitot tube to be high. However, except for specially made tiny instruments, this requirement is fulfilled at all measurable speeds. The main limitations to the use of Pitot tubes are their size and the slowness of their response; they cannot be used in flows of very small length scale, nor to measure rapid velocity fluctuations. In most turbulent flows, for example, only the mean velocity can be measured with a Pitot tube.

  The principle of operation of a hot-wire anemometer is virtually indicated by its name. An electrically heated wire is cooled by the flow, the rate of cooling depending on the velocity. In the simplest mode of operation, the current through the wire is maintained constant; its temperature and thus its resistance, measured by the voltage across the wire, depend on the velocity. In an alternative, widely used, mode, a feedback circuit maintains the wire at a constant resistance and so at constant temperature; the current needed to do this is a measure of the fluid velocity.

Hot-wire anemometers have been most widely and successfully used in gas flows. In general, hot-wires are more sensitive at low speeds than high; however, if the speed is too low, free-convection heat transfer takes over from forced convection, making the cooling insensitive to velocity. Hot-wire anemometers can be used in air readily down to about $30 \, \text{cm s}^{-1}$; below this, measurements are more difficult although not impossible.

There are other probes that operate on exactly the same principle but are geometrically different: an important one is the hot-film anemometer; the heated element consists of a thin metallic film on the surface of a wedge-shaped thermally and electrically insulating base. Because of their greater robustness, such probes are mostly used in liquids. Use in water, or other electrically conducting liquid, also requires a very thin layer of insulation over the metallic film. In some other devices semiconductor beads are used instead of metallic sensors.

Hot-wire (and hot-film) anemometers operate best in just those conditions where Pitot tubes fail. They are small and rapidly responding. They have thus become the principal instruments for studying fluctuating flows, in particular the phenomena of transition and turbulence. They have the additional advantage of giving the information in electrical form, which can be processed electronically to give quantities such as intensities, correlations, and spectra. Nearly all the measurements presented in Chapters 18 to 21 were made with hot-wire anemometers.

Although a hot-wire anemometer is simple in principle, its actual use is a matter of some complexity. A large body of information [32, 44, 257, 300]—and a certain amount of 'folklore'—has grown up around hot-wire anemometry. This covers topics such as: the calibration of hot-wires (this is always necessary—hot-wires are not absolute instruments); the geometrical combinations of hot-wires needed for measuring different components of velocity fluctuations (a particularly important arrangement being two wires in the form of an X as shown in Fig. 25.2); the particular problems in achieving accuracy that arise when the velocity fluctuations are not small compared with the mean velocity; the problems associated with velocity measurements in the presence of temperature variations, to which a hot-wire probe is also sensitive; and much else.

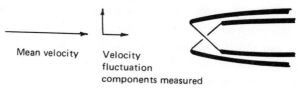

Mean velocity   Velocity
fluctuation
components measured

FIG. 25.2 An 'X-wire', which, when arranged so that the mean velocity is in the plane of the X, can be used to measure both components of velocity fluctuation in that plane.

The laser-Doppler anemometer measures velocity by measuring the Doppler shift of light scattered within the moving fluid. The scattering centres are tiny particles of dust, small enough that they are always moving effectively with the instantaneous fluid velocity. Usually there are enough such particles already present in any liquid or gas; if not, some are introduced. The light source must be a laser for the beam to be sufficiently monochromatic. Then the range of speeds that can be measured is very wide; fractional Doppler shifts as small as 10^{-15} can be measured and thus speeds down to less than $1\ \mu m\ s^{-1}$—far lower than those normally encountered in fluid dynamics—though not all systems are capable of this.

Figure 25.3 shows the principle of operation. The incident and scattered beams are inclined to one another so that the point of intersection locates the point at which the velocity is being measured. The scattered beam interferes at the photomultiplier with an attenuated beam direct from the laser to give beats, the frequency of which is a measure of the Doppler shift. Figure 25.3 is, of course, highly schematic; the full optical systems are much more complex than this, and take a

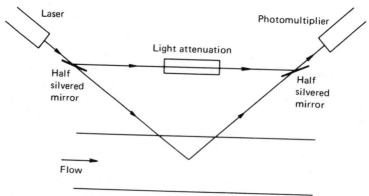

FIG. 25.3 Schematic diagram of laser-Doppler anemometer.

variety of forms. There are also variations in the way in which the signal is detected and processed.

Sometimes the light is carried to and from the point of observation by optical fibres. This has the advantages of not requiring high-quality optical windows on the apparatus, of making it easier to locate the measurement point, and sometimes of avoiding problems of refraction of the light by density variations. It does, however, lose what is often one of the main advantages of laser-Doppler anemometry, the fact that there is no probe to disturb the flow.

Like hot-wire anemometers, laser-Doppler ones respond to rapid fluctuations in the velocity and thus can be used to study the details of transitional and turbulent flows. They have, for example, been particularly useful in the studies of the details of transition that have been related to 'routes to chaos', as discussed in Section 24.7.

The practical difficulties of laser-Doppler anemometry obviously vary from flow to flow; sometimes it may be easy to mount a precision optical system around the flow, other times far from easy. There are also various sources of inaccuracy, not all of which are obvious. Again there is an extensive body of experience to be drawn upon [149, 151, 152].

The principle of using a Doppler shift to indicate flow velocity can be applied with sound waves instead of light waves. This has not been developed into such a standard technique but has been used to investigate both laminar and turbulent flows [170]. One actually uses ultrasound, i.e. the frequency is far above the audible range.

Almost every effect of fluid motion must at some time have been tried as a means of measuring velocities. Many experiments have used purpose-built probes rather than one of the standard techniques. The topic is too diverse for a complete survey to be attempted here, but a few comments may be made.

Methods that are simple in principle are not to be scorned, though they do not always turn out to be simple in practice. One obvious such method is the timing of the passage of marked fluid from one point to another, and many flow-visualization experiments have been made quantitative in this way. The development of methods for the automatic scanning of flow photographs [161, 215] enables far more information to be obtained and is likely to become increasingly important.

Mechanical methods are also an obvious possibility, making use of the force or torque exerted by a flow on an obstacle. They are widely used on the large-scale, a familiar example being the cup anemometer used in everyday meteorological applications. They are more difficult to adapt to the small scale of laboratory work, but are sometimes effective. Examples are observation through a telescope of the deflection of the free end of a cantilevered quartz fibre [233], and mounting a vane on the extended arm of a moving-coil milliammeter [194]. Such arrangements can give high accuracy in low-speed laminar flows.

Finally, we mention two methods of which the main use is probably the calibration of a wind-tunnel or water-channel below the Pitot tube range. This is important because many devices for measuring low flow speeds in turn require calibration. The first is essentially the marked fluid method again, but with heat instead of a visible marker. A fine wire is stretched across a wind-tunnel to be calibrated, an alternating current put through it, and the wavelength of the resulting temperature variations measured. A probe—a thermocouple, for instance—is traversed along the tunnel, and the phase variations between the input to the heated wire and the output from the probe are observed. This is tedious but it does provide an absolute calibration on which other devices can be based. Probes for local velocity measurement, based on the timing of the passage of a heat pulse from a source wire to a detector wire, have also been developed [346].

Secondly, the frequency of a vortex street behind a cylinder (Sections 3.3, 17.8) is a measure of the velocity, once the relationship between the Reynolds number and the Strouhal number has been established. The fluctuations can be detected with an uncalibrated hot-wire anemometer and so their frequency measured. Again this technique has also been used in a special probe [407].

25.3 Pressure and temperature measurement

We need not consider at any length the methods used for measuring flow parameters other than the velocity, since these methods do not differ significantly from those used in other branches of physics.

There are various ways of measuring the pressure at a point on a wall. The simplest is to link a small hole in the wall, similar to the ones in a Pitot tube, to a manometer. This is widely used for steady or mean pressure measurements but cannot respond to rapid fluctuations. Some form of pressure transducer is needed to measure these. Examples are a piezoelectric crystal or a small diaphragm whose movement changes a capacitance, but the choice of devices is quite wide [352].

The measurement of pressure in the interior of a flow has been considered above in connection with Pitot tubes. There is at present no satisfactory device for measuring rapid pressure fluctuations within a flow.

An important aspect of pressure measurements, as well as of velocity measurements with Pitot tubes, is the design of sensitive manometers. The various types are described in Ref. [29].

It should be added that, whilst pressure measurements are often very useful in wind-tunnel and similar studies, the pressure variations occurring in many experiments, for example in the field of free convection, are too small to be measured.

In experiments on convection, particularly free convection, measurements of the temperature field [233] are usually both easier and more accurate than measurements of the velocity field. The former, therefore, often provide the main source of information about the structure of the flow. However, we can consider the methods used very briefly, since they are, at least in principle, wholly straightforward. The transducers most widely used are resistance thermometers and thermocouples [91, 316]. A resistance thermometer often conveniently takes the form of a hot-wire anemometer probe with a very low current through it. Thermistor beads are also often used as the sensitive elements of resistance thermometers. The junction of a thermocouple can also be made very small, so all these devices can be made to respond rapidly.

Optical systems can also sometimes give quantitative information about temperature fields; however, these will be considered under the heading of flow visualization.

25.4 Flow visualization [25, 36, 68, 409]

The variety of methods of flow visualization is illustrated by the photographs throughout this book. Here we collect the different methods into a systematic list, referring back to the other chapters for examples. Again no claim is made that the list is complete.

Conceptually the simplest procedure is the introduction of smoke or dye. The introduction of smoke into a gas flow without unduly disturbing it is rather difficult except in wind-tunnels specially designed for the purpose. Also, the time scale is often too rapid for the eye to follow the phenomena; high-speed photography or some other special technique may be needed. Hence, dye-in-liquid experiments are often preferred. Nevertheless very good results can be obtained with smoke [29], as is illustrated by Figs. 14.6, 15.1, 17.18, 18.9, and 21.18.

A widely used method of smoke generation is to evaporate paraffin, or a similar oil, and to allow it to recondense as fine droplets. Usually this is done in a special generator and the smoke is then injected into the flow, often in a purpose-built smoke tunnel. However, the technique is also used in the form of a heated 'smoke wire' onto which the oil has been previously coated or is allowed to trickle continuously [68]; this has some of the versatility of the hydrogen bubble method in water described below.

A particularly effective dense white smoke is produced by exposing titanium tetrachloride to the atmosphere, but this is both toxic and corrosive. A convenient way of producing 'instant smoke' for simple exploratory experiments is the reaction between hydrochloric acid and ammonia vapours, but again its extensive use has its dangers.

Many of the photographs in this book illustrate the use of dye; a selection, illustrating different dyes and different ways of using them, consists of Figs. 3.4, 3.5, 13.6, 15.7, 16.4, 16.9, 17.14 and 21.6. Convenient and effective dyes for introducing into water flows include potassium permanganate, gentian violet, and methyl blue. The first has the advantage of producing particularly strong contrast, but at the cost of a sometimes inconveniently large density increase.

As alternatives to introducing dye, there are several ways in which marker can be produced electrically—the hydrogen bubble technique, the tellurium technique, the electrolytic precipitation technique, and the pH technique. The first works best at high speeds, such as those in many water-channel experiments; the second and third at intermediate speeds; and the last at the low speeds likely to be encountered in convection and rotating fluid experiments.

In the hydrogen bubble method a wire stretched across the flow is the cathode of an electrolytic system, the anode being, for example, a wall of the channel. Typically 10 V are needed with tap water, larger voltages with deionized water. Provided the wire is very fine (typically 10–100 μm), the hydrogen bubbles produced at it are small and numerous enough to move with the flow and give the appearance of a white dye. By insulating portions of the wire and/or pulsing the voltage, patches of dye can be produced, giving most effective pictures like those in Fig. 21.19.

A cathode made of, or coated with, metallic tellurium releases a dense brown dye, which is effective for flow visualization [413]. Care is needed to avoid toxic effects.

In a similar way a white colloidal cloud is produced at an anode made of certain materials [367]. Those containing tin (in water containing sodium carbonate) are particularly effective; solder is thus a convenient material. This method, known as the electrolytic precipitation method, requires rather larger voltages than the others, typically 30–50 V.

The last of these electrical methods [75] makes use of the fact that there is a change in pH (the measure of acidity or alkalinity) in an electrolyte in the vicinity of an electrode. Hence, if the working fluid contains an indicator and is titrated to be close to the end-point of that indicator, the application of a voltage between a fine wire and some other point in the flow can produce a local colour change at the wire. This method has two marked advantages: the production of the dye produces no density change or displacement of the flow; and the dye gradually reverts to its previous state, so the system can be run continuously without becoming filled with dye. It has the disadvantage that it can be used only in closed systems, not in flows where fluid is continuously entering and leaving. The most usual indicator is thymol blue; this changes from amber (neutral) to dark blue (basic) at a cathode and so gives a clearly visible marker. Figures 16.2 and 22.12 were obtained in this way. With both this method and the tellurium method, it is important that hydrogen bubbles should *not* be produced. For both methods, voltages in the range 1–10 V are satisfactory, and this gives no trouble in deionized water.

Variations in pH and thus in the colour of an indicator can be introduced in other ways. For example, we saw in Fig. 15.9 how injection of very small quantities of acid and alkali produced regions of different colours (appearing as different brightnesses in a black-and-white picture) thus showing the main features of the flow pattern.

Other flow-visualization experiments make use of suspended particles (e.g. Figs. 4.4, 12.1, 16.17). Of the various particles available, poly-styrene beads are particularly useful in water, as their density is only slightly greater than that of water. Particles in the form of small flakes take up different orientations in different regions of a flow [333]; when light enters from one direction and is viewed from another, bright and dark patches reveal the structure of the flow (e.g. Figs. 4.8, 16.11, 17.8). Aluminium powder has been extensively used in this way ('silver' paint is a convenient source of sufficiently fine particles), particularly in silicone oils and other highly viscous liquids. Other types of flake, including ones originating as fish-scales, are also available.

In all flow-visualization experiments it is necessary to give some thought as to what features of the flow are shown up. Smoke and dye experiments usually show particle paths or streaklines (Section 6.1). The electrical methods allow other possibilities; one can, for example, release dye all along a wire at a given instant and obtain information from its subsequent locus. Similarly, suspended particles may be used in various ways. Photographing them for an interval gives portions of particle paths, which, if short enough, can be used to synthesize a streamline pattern even in unsteady flow (e.g. Fig. 3.7). Other techniques, such as using reflecting flakes, give overall information about the structure of the flow.

The last group of flow-visualization methods to be considered consists of optical techniques making use of refractive index variations in the fluid [200]. Such variations occur as a result of density variations associated with a temperature or concentration field. They are thus particularly appropriate for studies of convection and stratified flow. However, even in nominally constant-density flows it may be possible to introduce density differences which are large enough to allow the use of one of these techniques whilst keeping the internal Froude number high enough for the flow to be unaffected. All these techniques work best in two-dimensional configurations, as the observed pattern is the integrated effect of the passage of the light through the fluid.

In the shadowgraph method, parallel light enters the fluid and is deflected where there are refractive index variations. If the second spatial derivative of the refractive index is non-zero the amount of deflection varies giving a pattern of bright and dark regions related to the flow structure. Shadowgraphs appear in various places in this book, including Figs. 14.2, 14.5, 18.10 and 22.3 to 22.8.

In the schlieren method, parallel light is again used, but is brought to a focus after passing through the fluid. A knife-edge or other stop at this focus blocks off some of the light. If the light has been deflected in the fluid the amount passing this stop will be increased or decreased. Changes in intensity thus relate to the first spatial derivative of the refractive index. By having an optical arrangement in which the stop is at a conjugate point of the source whilst the observing screen is at a conjugate point of the fluid, one can obtain an image of the flow pattern with these intensity variations. There are many variations in the details of the arrangement [200]. Amongst the schlieren photographs in this book are Figs. 3.12, 15.11, 21.2 and 26.22. (It should be noted that the name 'schlieren' is occasionally used to refer to optical methods in general rather than this specific one, which is then known as the Töpler–schlieren method.)

In an interferometer [371], light that has passed through the fluid interferes with light from the same source that has not passed through the fluid. Whether this interference is constructive or destructive depends on the optical path difference, and so the pattern of bright and dark bands reflects the pattern of constant density surfaces (in a two-dimensional flow configuration). Again there are various optical systems. Figure 17.19 shows two interferograms.

A development of interferometry is the use of holographic techniques. The initial interferogram is itself used as a diffraction grating to reconstruct an image. There are various ways in which this principle can be applied [138, 241]. In the one example in this book, Fig. 15.5, the net effect is similar to that of ordinary interferometry; the light and dark lines represent constant-density surfaces. The advantage over direct interferometry is that the demands on the quality of the optical system are less severe.

25.5 Numerical experiments

Since the cases in which the equations of fluid motion can be solved algebraically are limited, numerical integration plays an important role.

This was the case even before the advent of electronic computers; a 'classical' example is the integration of eqn (11.28) to give the Blasius profile, Fig. 11.2. However, modern computers obviously enable problems of much greater variety and complexity to be tackled. In such work the proportion of numerical to algebraic analysis varies greatly. The traditional approach, essential before the days of high-powered computers, was to proceed algebraically as far as possible and resort to numerical integration only when there was no alternative; the Blasius profile integration again provides an example. This approach is still often adopted, as the analysis may provide insight into the significant dynamical processes. However, it is not now the only possible approach. At the opposite extreme a numerical approach is adopted from the outset; numerical analysis is applied to the full Navier–Stokes and continuity equations with appropriate boundary and initial conditions. One then says that one is doing a 'numerical experiment'. The name implies an analogy with a laboratory experiment. In either case the aim is to discover what processes and phenomena result from the physical laws governing the dynamics and expressed by the equations. The mathematical properties of the equations producing the results are not directly elucidated (although one will, of course, often attempt to interpret the observations in terms of such properties). For example, supposing that the flow investigated becomes more complicated than might be expected because of an instability, this fact would be discovered not by a stability analysis but by the observation, in the laboratory or on the computer, that a complicated flow occurred.

Numerical experiments, like laboratory experiments, must be planned with attention to the requirements of dynamical similarity (Chapter 7). In laboratory work one may find that the desired values of the non-dimensional parameters are not actually achievable. This is much less likely to be a problem in numerical work, which is therefore particularly important for studies of regimes inaccessible in the laboratory (e.g. convection in conditions where non-Boussinesq effects—see the Appendix to Chapter 14—are significant, perhaps in connection with an application such as solar granulation, Section 26.6).

However, numerical experimentation has, of course, its own problems, lack of awareness of which can easily lead to spurious conclusions. We have seen in various places in this book that flows often develop length scales quite different from the imposed ones. Care is thus needed with any numerical procedure, on the one hand, that its grid size is small enough to resolve, for example, a boundary layer (Sections 8.3, 11.2–11.4, 12.4) and, on the other hand, that it extends into all regions of importance, such as a wake (Section 11.5), upstream wake (Section 15.2) or Taylor column (Section 16.4).

Extensive numerical work has been done on turbulent flows [322]. The interpretation of laboratory experiments has been elusive (Sections 21.4, 21.6) and the aim of numerical experiments is to aid this. Of course, these are not necessarily easier to interpret—techniques such as conditional sampling (Section 19.6) may again be invoked—but they can illuminate different aspects. The very wide range of scales present in a turbulent flow (Sections 20.3, 21.3) produces particular problems for numerical modelling. Indeed, except for flows at untypically low Reynolds number, even the most powerful current computers cannot handle all the data needed for a full numerical analysis of turbulent motion. Various ways of circumventing this difficulty have been developed, usually including some representation of the effect of smaller eddies in an analysis of the larger eddies, but we cannot go into details here.

In this book only occasional direct reference has been made to the results of numerical modelling. Most specifically, Fig. 21.22 is an example of results for a turbulent flow as discussed above. Figures 3.3 and 6.2 show computed flow patterns past a circular cylinder; some algebraic analysis of the equations was involved as well as numerical analysis. The derivation of Figs. 6.3–6.6 from Fig. 6.2 can be regarded as a sort of numerical experiment, although a very untypical one since all the dynamical results were contained in Fig. 6.2 and the experiment concerned only kinematic transformations of these results.

26

APPLICATIONS OF FLUID DYNAMICS

26.1 Introduction

The main body of this book has been concerned with developing a basic understanding of the phenomena of fluid motion. Applications have been ignored, apart from passing references to illustrate a particular point. One does not capture the full flavour of the subject without some emphasis on the fact that moving fluids occur in a wide variety of practical situations. Some of these applications have had a profound effect on the directions of advance of the basic studies.

It would not be possible in the space—nor perhaps very interesting— to give a systematic survey of applied fluid mechanics. Instead, we look at, and discuss briefly, a selection of particular topics. The selection has been made with two purposes in mind: to illustrate the variety of branches of applied science in which fluid dynamics arises; and to show applications of different topics in the main body of the book. The applications discussed are thus not intended to be those of greatest importance or topicality. They are, however, all topics of current or recent research.

26.2 Cloud patterns

The very varied patterns formed by clouds reflect the variety of dynamical processes that occur in the atmosphere and are often a useful immediate indication of these processes. Pictures from the Earth's surface [305, 341] have in recent years been supplemented in an important way by satellite images [342]; the latter often reveal meteorological phenomena on a scale too large for them to be seen from the ground or an aeroplane but too small to be revealed by synoptic methods.

Clouds form when moist air is cooled so that the saturation vapour pressure falls below the actual vapour pressure. Although cooling can occur in a variety of ways, the most common is the cooling associated with expansion as air rises; that is, cooling associated with the adiabatic temperature gradient (Section 14.6). One may thus observe situations in which cloud is forming in rising air and evaporating in descending air to give a pattern of cloudy and clear patches related to the flow.

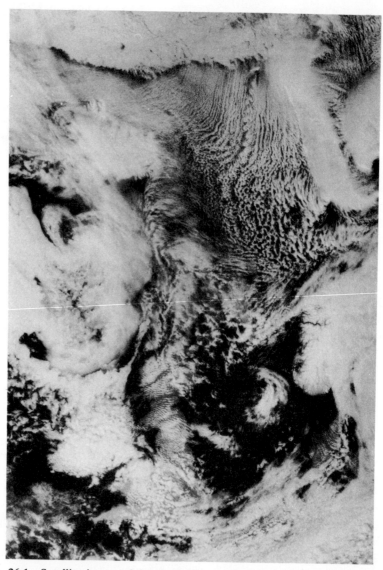

FIG. 26.1 Satellite image of Arctic Ocean and Britain. Visible light; 27 March 1985, 13.36 GMT. Image provided by Department of Electrical Engineering, University of Dundee.

Of the many possible illustrations of this, we choose 'cloud streets'. One can fairly frequently observe clouds in long parallel lines, quite evenly spaced. When they can be seen from the ground, these are usually either convection rolls aligned by the mean flow as considered in Section 15.1 (cf. Fig. 15.1), or billows generated by shear instability in stably stratified air as considered in Section 17.6 (cf. Fig. 17.14). The two processes can be distinguished, even if the stratification is unknown, by the orientation of the cloud lines with respect to the wind direction; convection cells are aligned along the wind, billows across it.

The top part of the satellite image in Fig. 26.1 shows cloud streets on a larger scale. Formation of such streets can be found especially during so called cold air outbreaks, where very cold and dry air is flowing over relatively warm ocean water. The instability producing the pattern is more complex than any we have considered in this book; it involves both unstable stratification and instability associated with the Ekman layer profile (Fig. 16.8) [103, 285]. We have looked at such instabilities separately; thus the overall effect may be seen as an interaction between the types of motion represented by Figs. 15.1 and 16.9.

Figure 26.1 obviously shows a variety of other features. We comment briefly on two. First, the patterns over northern Britain illustrate that cloud alignment may be produced by quite different mechanisms; these are lee waves (Section 15.3) produced by a northerly wind interacting with a rather complicated topography [69]. Second, there is a vortex street within the cloud streets; this originates from Jan Mayen Island, a volcanic island of maximum height 2277 m above sea level, right at the top of the picture. Probably the closest analogy in this book is Fig. 15.3; the presence of the vortex street suggests that there is stably stratified air above the unstably stratified layer producing the cloud streets.

26.3 Waves in the atmospheric circulation

The large-scale circulation of the atmosphere frequently involves wave motions. These are prominent at mid-latitudes in both the northern and southern hemispheres. Figure 26.2 illustrates this. It shows the trajectory of a balloon designed to float in the atmosphere at positions of constant air density; that is effectively at a constant height, of, in this case, about 12 km. The balloon was released in New Zealand, and its position observed through radio signals for 102 days. Wavelike oscillations in latitude as it travels round the earth in the prevailing westerly winds are apparent. Similar features in the northern hemisphere can be seen in the isobar patterns shown in Fig. 19.1, although the development of strong cyclones within the wave pattern makes this situation rather complex.

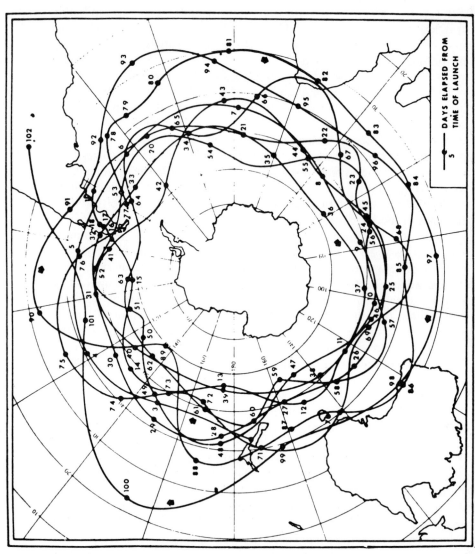

FIG. 26.2. Trajectory over 102 days of a balloon released from New Zealand and circulating at a height of about 12 km. From

The longest waves (three or four wavelengths round the Earth) are, in their essential dynamics, Rossby waves (Section 16.7) [307]. Sometimes they form a fixed pattern arising from disturbance of the general circulation by mountains; i.e. they are a natural counterpart of the flow illustrated by Fig. 16.14. Waves progressing relative to the Earth are also observed [262]; their origin is less apparent.

Shorter waves are probably more often associated with temperature variations between pole and equator—they are baroclinic rather than barotropic waves. These temperature variations are responsible for the general westerly flow on which the waves are superimposed; the Coriolis force turns the meridional circulation that would arise on a non-rotating Earth. This process is modelled by the laboratory experiments described in Section 16.9. The generation of the waves can now be understood as an instability. Despite the difference in geometry, the atmospheric waves may be compared quite closely with some of those observed in the laboratory (Fig. 16.17). The non-linear development mentioned in Section 16.9 has its counterpart in the development of the jet-stream, a similar narrow region in which much of the momentum of the atmospheric circulation is concentrated.

26.4 Formation and movement of sand dunes [71, 191]

Of the many ways in which air and water movement contribute to the formation of the landscape, the development of sand dunes is one of the most apparent. We focus attention on desert dunes (rather than coastal ones). These take various forms but certain characteristic patterns are found in many different places. The type occurring in a particular place is determined by various factors: whether the dunes are part of a continuous sandy desert or whether most of the sand is accumulated into dunes leaving rocky desert between them; the size distribution of the sand grains; whether the wind blows from one or two principal directions or is highly variable; whether there are obstructions to the wind; and the amount of vegetation. Figure 26.3 shows longitudinal or 'seif' dunes. Figure 26.4 shows dunes on the surface of Mars—to illustrate in passing the importance of fluid dynamics in planetary science—but of types also commonly found on Earth; the picture is dominated by linear transverse dunes, but isolated crescent-shaped or 'barchan' dunes can be seen above the crater and towards the bottom left.

The detailed pattern of a dune field is continuously changing and individual dunes can migrate across a desert. An understanding of dune dynamics may not only be of fundamental interest but also help to indicate how dune movement may be controlled—for example, when dunes threaten to engulf a village or irrigated land [405].

FIG. 26.3 Aerial photograph of seif dunes in the Libyan sand sea. From Ref. [306].

There are two interacting general aspects of fluid dynamics determining the shape into which a dune forms and how this varies with time—the movement of sand by the wind and the modification of the wind pattern by the dune itself (together with a non-fluid dynamical question of the maximum slope at which sand can repose).

Sand is moved by wind in three different ways: suspension (which induces dust storms), saltation (jumping), and surface creep associated with saltation. Of these saltation is much the most important in influencing dune morphology. Individual grains are lifted from the ground by the wind and fly through the air to land some distance downwind; heights and lengths of trajectories are typically in the ranges 1–10 cm and 10 cm–1 m, respectively. Theories of the saltation process [191, 295] are complex and cannot be described here, but we note that they involve various topics considered in this book. The dynamics of turbulent boundary layers (Sections 21.5, 21.6), perhaps with allowance for modification by stratification (Section 21.7) are important in determining both whether the stresses on the sand bed initiate saltation and the velocity distribution through which the grains are moving. The trajectories of the grains are then determined by the forces on them due to their motion relative to this distribution; this involves the drag

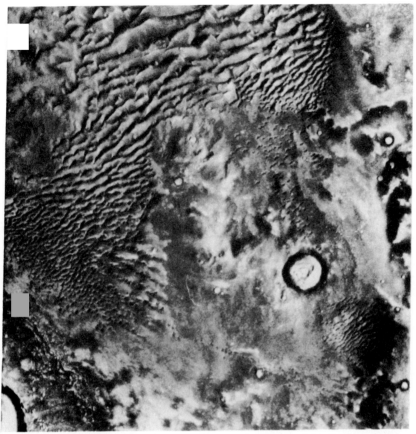

FIG. 26.4 Viking Orbiter image of surface of Mars at 47°S, 340°W. Width of region shown = 57 km. Ref. [115].

coefficient when the Reynolds number is neither very high nor very low (Section 7.3) and, since the grains may spin, the Magnus effect (Section 13.4).

How a dune evolves by means of these sand transport processes is determined by the wind flow around the dune. Over parts of the surface the wind is continuously removing sand. Other parts are sheltered and sand deposited there accumulates. We are now concerned with very high Reynolds number flow past obstacles, though often the boundary layer thickness is comparable with the dune height and so the problem is more complicated than such problems considered in this book. Extensive field observations, supplemented by laboratory modelling, have given a good understanding of the processes involved for some types of dune. As an example, Fig. 26.5 shows the interpretation of a seif dune (Fig. 26.3)

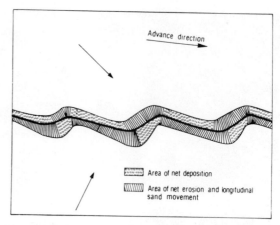

Fɪɢ. 26.5   Schematic diagram of erosion and deposition regions on a seif dune due to two predominant wind directions. From Ref. [393].

which advances gradually from a sand sheet (a 'zibar') under the action of wind from two predominant directions (in summer and in winter). This interpretation is based on detailed studies of a particular dune in the Sinai Desert [393].

26.5   Continental drift and convection in the Earth's mantle

The important role of fluid mechanics in meteorology and oceanography is obvious. Its role in the geophysics of the Earth's interior is somewhat less familiar, and we consider here an example of this.

   The main divisions of the Earth (discovered through the reflection and refraction of elastic waves originating in earthquakes) are a solid crust (the top few tens of kilometres), a solid mantle (extending from the bottom of the crust to a radius a little over one half the Earth's radius), a liquid core, and, probably, a solid inner core. One would expect fluid dynamical problems to arise in the core; motions there generate the Earth's magnetic field. More surprisingly, there are fluid dynamical problems associated with the mantle.

   It is known that solids under stress undergo not only an elastic deformation but also a continuous creep process. If this continues for a very long time, it may result in a complete change of shape of the solid, which is then behaving as a fluid. Thus, over time scales of geological development, it may be appropriate to treat the mantle as a fluid, and indeed possibly as a Newtonian fluid.

   It is now widely accepted that the Earth's crust is undergoing

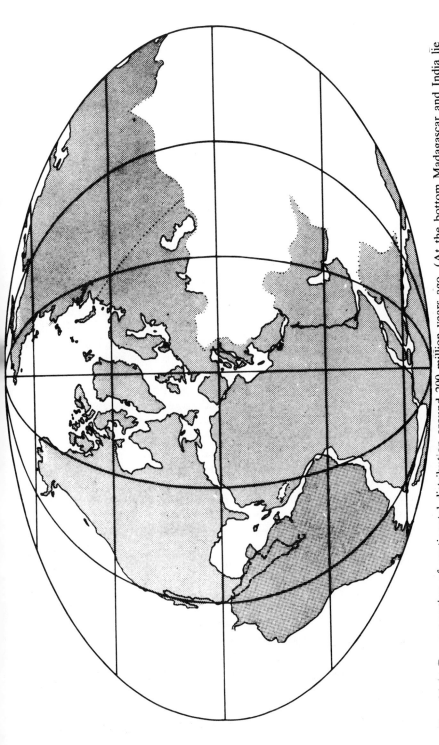

FIG. 26.6 Reconstruction of continental distribution around 200 million years ago. (At the bottom Madagascar and India lie between Africa and Antarctica, and a portion of Australia is seen to the right of Antarctica.) After A. G. Smith, from *Understanding the Earth*, Artemis Press Ltd., 1972.

continuous changes; that, for example, the distribution of the continents over the surface of the Earth has changed [310]. Figure 26.6 shows a suggested reconstruction of this distribution around 200 million years ago [172]. The relative motions of different portions of the crust occur in narrow linear zones, the regions in between moving as rigid plates. This has led to the topic being given the name 'plate tectonics'. The plate movements are responsible not only for continental drift, but also for the formation of mountain ranges, rift valleys, mid-ocean ridges, and ocean trenches and for earthquakes and volcanic activity.

Various suggestions have been made about the cause of the movements; the majority involve some aspect of thermal convection [298, 339]. It is a matter of discussion whether the plates themselves play a basic role in the dynamics or whether they are driven by a flow occurring in the mantle below. It is the latter view that would make concepts from this book more directly applicable. Thermal convection could well occur in the mantle. Convection due to chemical differences, rather than temperature differences, and double diffusive convection associated with both chemical and temperature variations (Chapter 23) are also possibilities. Much of the discussion has been based on our knowledge of Bénard convection; the ideas in Chapter 22 have to be extended to allow for large variations of viscosity with temperature. Flow types considered range from simple cellular flow to the high Prandtl number type of turbulence mentioned at the end of Section 22.7. Although it is almost certain that mantle flow occurs, the details remain very uncertain.

(It is worth noting that mantle convection is an exception to two statements elsewhere in this book that certain ideas are rarely relevant to applications. The exceptions arise because one is concerned with an extremely viscous fluid—the Prandtl number is around 10^{24}. Firstly, compressibility effects, and other non-Boussinesq effects, do enter marginally into the problem. This might seem very surprising when the velocities are tiny, but it is an example of the fact that, at low Reynolds number, relationship (5.61)—not (5.59)—is the criterion for neglect of compressibility. Secondly, the Rayleigh number criterion is often invoked into discussions of whether convection occurs—despite the statement in Section 14.6 about its irrelevance to natural situations.)

26.6 Solar granulation [99, 178]

Observations of the Sun's photosphere, the thin layer close to its surface from which the visible radiation comes, show a granular structure, as illustrated by Fig. 26.7. (This picture has a sunspot at its centre—also a fluid dynamical phenomenon, but one strongly involving electromagnetic

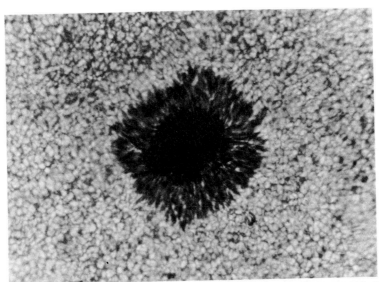

FIG. 26.7 Solar granulation and sunspot. Photograph courtesy of the National Solar Observatory/Sacramento Peak, Sunspot, New Mexico.

processes and thus ideas beyond the scope of this book.) Discrete bright patches are separated by a network of narrower dark regions. The patches are typically 10^3 km across, so that there are of the order of 3×10^6 granules over the surface of the Sun. The pattern is a continuously changing one, the lifetime of an individual granule being typically 10 minutes. Doppler shift measurements of velocities indicate that the bright (hot) regions are ascending and the darker (colder) regions are descending.

The pattern is thus interpreted as a convection pattern. The convection is broadly of the Bénard type in that it is horizontally homogeneous over distances large compared with its scale. Probably the most significant way in which it differs from the convection described in Chapter 22 is that departures from the Boussinesq approximation are strong; the depth of the convecting region is probably of the order of the scale height (see the appendix to Chapter 14). There are other differences, such as the roles of hydrogen ionization and recombination and of radiative transfer in the heat balance [99]. The motion is undoubtedly turbulent—the 10-minute lifetime of a granule is comparable with estimates of the time taken for the gas to make one circulation. This is consistent with the fact that the Rayleigh number is very high. Indeed it is so high that, on the basis of laboratory experiments, one would not expect to observe a cellular structure even as an ever-changing pattern. The fact that one does is perhaps a consequence of the importance of the scale height.

26.7   Effluent dispersal [160, 228]

The problems of air pollution and water pollution evidently involve fluid dynamical considerations amongst their many other aspects. In principle, the type of question posed is usually the same—where is pollutant, emitted from a source, subsequently found and in what concentration? In practice, the flow situations in which pollution may occur are very varied, and different branches of fluid dynamics are involved in answering the question in different cases. We thus consider a particular example.

This is the discharge of sewage or other effluent into the sea or an estuary. The design problem is, evidently, to ensure that the contaminant has become sufficiently dilute to be unobjectionable before reaching places where it could be harmful—an aim that has, of course, not always been fulfilled in the past. Sewage has approximately the same density as fresh water. It is therefore buoyant when discharged into salt water. The problems involved are thus broadly the same as those associated with the discharge of hot water into cold, a matter of practical importance in connection with the return of cooling water from power stations. Excessive heating can be biologically damaging (the problem sometimes known as thermal pollution). Also it is important that the warm water should not return too directly to the power station intake with consequent loss of thermodynamic efficiency.

Fig. 26.8   Model of Tees estuary. Photograph supplied courtesy of Hydraulics Research Ltd., Wallingford, U.K.

A major aspect of the investigation of any system or proposed system is the determination of the pre-existing flow patterns, due, for example, to tides or ocean currents. Field observations are usually supplemented by experiments with models of the site; the model may be either a physical one in a laboratory or a numerical one in a computer. To illustrate laboratory modelling, Fig. 26.8 shows a model of the Tees estuary on the north-east coast of England, and Fig. 26.9 shows a simulation of tidal flows in such a model made during an investigation of the best location for a sewage outfall [277].

Despite the fact that each site has its own characteristics, there are many fluid dynamical topics of general relevance. To illustrate this, we consider the example of buoyant effluent being emitted horizontally into stationary ambient water [278]. (This discharge would be one of a manifold in a large system [160].) The principal features of the flow, which may be expected to be turbulent, are shown schematically in Fig. 26.10. The effluent at first travels horizontally as a jet (Section 11.6). The action of buoyancy causes this jet to curve upwards, and in time its motion becomes nearly vertical, its properties then being essentially those of a thermal plume (Section 14.8). How quickly the upward curving takes place depends on the internal Froude number of the initial jet. Throughout this jet/plume flow the processes of turbulent entrainment and mixing (Section 21.2) are taking place, leading to dilution of the

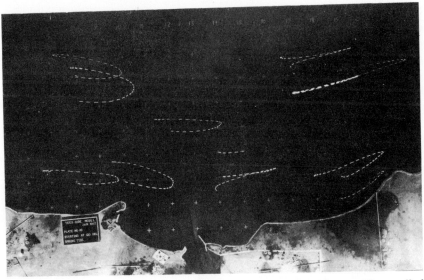

FIG. 26.9 Trajectories of candle floats in Teeside model. Photograph supplied courtesy of Hydraulics Research Ltd. Wallingford, U.K.

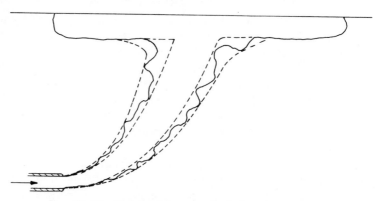

Fig. 26.10   Sketch of buoyant jet below free surface.

effluent [160]. When the plume reaches the surface, the buoyant fluid spreads out in a surface layer. This is stably stratified and so mixing is inhibited (Section 21.7). Further dilution may occur here but is a much less efficient process. Thus the contaminant concentration where the effluent reaches the surface is an important quantity, determined by the characteristics of turbulence in stratified jets and plumes.

26.8   Wind effects on structures

During recent years greatly increased attention has been given to the effect of the wind in the design of buildings and other structures [119, 329]. Both the direct effect of possible wind damage to the structure and the indirect effect of possible undesirable changes to the surrounding wind pattern due to the presence of the structure are matters of importance. In this section we look at two examples of the former. One is an economically trivial example, chosen because the fluid dynamics is straightforward, yet interesting. The other, the collapse of the Ferry-bridge cooling towers, was a major incident that illustrates the complexity of a more typical situation. It should be emphasized that, although in both these examples the trouble was not foreseen, the majority of the work in this field is, of course, concerned with prevention not cure; wind-tunnel experiments of the type mentioned below are now frequently performed at the design stage.

Figure 26.11 shows the 'crowning feature' of the Civic Centre at Newcastle upon Tyne; three 'castles' are mounted on prongs in a form corresponding to the city coat of arms. Some time after erection the two outer castles developed an oscillation, moving in antiphase in the plane of the 'trident'; their supporting prongs were behaving like a tuning fork.

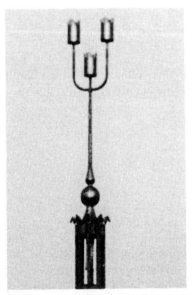

FIG. 26.11   The crowning feature of the Civic Centre, Newcastle upon Tyne.

Cine-film of the oscillation shows the distance between the castles varying by up to 5 per cent, but larger oscillations almost certainly occurred. The wind at the time was roughly perpendicular to the plane of the trident and was not unusually strong (about $15 \, \mathrm{m \, s^{-1}}$). This behaviour resulted from resonance between the frequency of eddy shedding by the cylindrical castles and the natural frequency of the tuning fork mode of oscillation. We saw in Section 3.3 that, because of eddy shedding, the spectrum of the velocity fluctuations and consequently that of the force fluctuations on the cylinder, has a sharp peak at a frequency corresponding to a Strouhal number of about 0.2, even at very high Reynolds number. In this case, the behaviour was sufficiently simple that it could be diagnosed without any experiments, but it illustrates the point that, when instabilities occur, it is not necessarily the strongest winds that are most dangerous. It was cured by adding a load inside the castles to change the resonant frequency and by filling the support tubes with sand to damp the oscillations.

On 1 November 1965, three cooling towers at Ferrybridge, Yorkshire, collapsed in a gale (Fig. 26.12). The three were members of a group of eight in two rows of four. All three were in the leeward row, indicating that the interaction between the towers had been important. Elucidation of the mode of failure was quite a complicated story, but essentially it was a quasi-static tensile failure, i.e. a tensile failure that would have

FIG. 26.12   Collapse of third tower at Ferrybridge. From Ref. [118].

occurred in a steady wind of the highest speed actually reached temporarily, of the shell under the wind loading [118]. There were two main stages in this elucidation. Firstly, wind-tunnel tests on a model of the complex (Fig. 26.13) determined pressure distributions over the towers; it was not possible to achieve the full-scale Reynolds number as is ideally required (Chapter 7), so it was hoped that the highest possible value would serve. Secondly, the results of these were fed into membrane

FIG. 26.13 Wind-tunnel model of Ferrybridge power station. From Ref. [118].

theory computations of the stresses in the shell produced by the combination of the wind pressures and the shell's own weight. The results were very sensitive to the details of the pressure distribution. Even though the total wind force on the leeward towers was less than that on an isolated tower, the redistribution of the forces by the presence of the other towers could lead to failure. In this case, it was not thought that any dynamic effect (such as that in the example above) played a role; the leeward towers will have experienced larger velocity fluctuations due to the presence of the front row (and these might have contributed to the velocity temporarily reaching the critical value), but the frequencies were not appropriate to any resonance. Clearly the subtlety of a situation like this is such that either very extensive model testing is necessary at the design stage or large safety margins must be allowed.

26.9 Boundary layer control: vortex generators

It is apparent that boundary layers play an important role in many applications (see Chapter 11, Sections 12.4, 21.5, 21.6). Whether a boundary layer is laminar or turbulent, whether and where separation occurs may have a marked or even controlling influence on the performance of a whole flow system. Attempts to change or control the character and development of boundary layers may thus be worthwhile.

Such boundary layer control [238] has found application principally in aeronautical engineering. Aims have been as diverse as attempting to keep the entire boundary layer on a wing laminar (so as to minimize the viscous stress) and promoting transition to turbulence right at the leading edge (to prevent separation, for reasons apparent from Sections 12.4 and 12.5).

We cannot here mention all the variety of methods used in boundary layer control, so we look at one example, the use of vortex generators, illustrated by Fig. 26.14. This is not a specially widely used technique in aeronautical engineering. There are, however, also other applications: vortex generators on fairings (in the shape of thick aerofoils) on towing cables and offshore drilling rig riser pipes can reduce drag and vibrations due to vortex shedding [112]; and they have similarly been used to suppress wind-excited oscillation of a bridge [403] (cf. Section 26.8).

Figure 26.14 shows the wing of a Trident 1 aircraft. (The photograph was taken during development flying, and one can see the tufts attached to the wing to show the flow pattern over it.) The vortex generators are the line of specially shaped protuberances close to the leading edge.

In general, vortex generators are introduced to prevent or delay separation in a situation in which the fact that the boundary layer is

FIG. 26.14   Trident wing with vortex generators. Photo provided by Hawker-Siddeley Aviation Ltd.

turbulent is insufficient to do so. Each vortex generator produces a longitudinal vortex extending downstream from it, by a process somewhat analogous to the generation of wing-tip vortices (Fig. 13.8). These enhance the mixing across the boundary layer, bringing rapidly moving fluid from outside the boundary layer in close to the wall. This supplements the mixing due to the turbulence. It is apparent from Fig. 12.5 that the change in velocity profile will inhibit separation.

Vortex generators were introduced on the Trident wing (Fig. 26.14) to cope with a particular problem. The suppression of separation is important during take-off and landing when the leading edge droop is down. The purpose is partly to delay stall (Section 13.2) but more particularly to avoid instability of the aircraft when the angle of attack is close to the stalling angle. Flight and wind-tunnel tests showed that separation started on one part of the wing and spread progressively in such a way as to produce a nose-up pitching moment which tended to further increase the angle of attack. Introduction of the vortex generators at the critical parts of the wing changed this unstable behaviour into a stable one by delaying the outer wing separation until the flow on the inner wing had separated.

26.10 Train aerodynamics

In recent years the fluid dynamical aspects of land transport engineering have assumed increasing importance. Car advertisements which either inform or mystify the public with drag coefficients are an outward manifestation of this. We illustrate this topic primarily with a brief discussion of the aerodynamics of trains, although with some mention of road vehicles to illustrate how the contrasting geometries lead to different primary fluid dynamical considerations.

The combination of increased speeds and the need for fuel economy has resulted in aerodynamic considerations changing from a comparatively peripheral to a major aspect of the design of trains. Its importance is illustrated by the fact, that, for a modern high-speed passenger train, typically 75 per cent of the fuel is used in overcoming air resistance [173]. We shall look briefly at the question of drag reduction below. However, it should first be emphasized that this most obvious aspect of train aerodynamics is far from the only important one.

Other topics that have been the subjects of extensive research include the trackside disturbances produced by trains; the interaction of passing trains; particular problems associated with overhead equipment on electric trains; and the stability of trains of light construction in high winds [173].

A topic of increasing importance concerns trains in (or entering or leaving) tunnels. This involves effects of compressibility and thus basic fluid dynamics outside the scope of this book. It is worth pausing, however, to ask why compressibility is important. The Mach number of a high-speed train is around 0.2; why does the discussion in Section 5.8 not apply? Basically the point is that the time taken for a disturbance travelling at the speed of sound to traverse the length of the tunnel is not usually small compared with the time taken for the train to enter the tunnel. The air throughout the tunnel cannot respond effectively instantaneously to the entry of the train (cf. remarks in Section 10.3, around (10.12)). The inadequacy of the analysis in Section 5.8 derives from the supposition that only a single length scale is involved.

We turn specifically to the question of drag and its reduction [174]. This is, of course, complicated by the wind speed and direction, but we consider only a train moving in still air.

In vehicle dynamics in general it is usual to quote a single value of the drag coefficient C_D (eqn (7.16)†), ignoring any variation with Reynolds number. For a bluff vehicle such as a car or lorry, the justification is essentially the ideas of Section 12.5: the drag is primarily produced by pressure variations over the vehicle body, and the distribution of $(p - p_0)/\frac{1}{2}\rho u_0^2$ relates primarily to the location of separation [83]. For a train, as we shall see in a moment, the direct effects of the turbulent boundary layers are much more important. However, this again implies only weak dependence of C_D on Reynolds number (this being equivalent to weak variation of u_τ/U_0 in the notation of Section 21.5). Constant C_D means that the power required to overcome drag is proportional to the cube of the speed and the total energy so expended in a journey proportional to the square.

Figure 26.15 shows an example of the various contributions to the aerodynamic drag on a train. We see that the main contributions are from the sides, roofs, and undersides of the coaches and from the running gear (bogies) underneath; the length of a train gives rise to a quite different situation from that of a car or a lorry. The most immediately striking difference between modern high-speed trains and earlier designs—the tapered nose—has much more to do with reducing the abruptness of trackside effects than with reducing drag.

Modern trains have a C_D value typically 40 per cent lower than their predecessors. This has been achieved principally by making the train profile as continuous and smooth as possible (Fig. 26.16). Surface finish is important not so much because of the direct forces on excrescences but

---

† In this context it is usual to choose the length scale L so that L^2 is the maximum cross-sectional area perpendicular to the direction of motion.

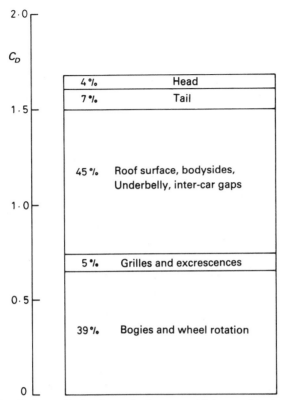

FIG. 26.15 Breakdown of aerodynamic drag on a high speed train (British HST of $1 + 7 + 1$ configuration, length 197 m, in still air). Data provided by British Railways Board, Railway Technical Centre, Derby.

FIG. 26.16 Effect of aerodynamic design of railway coaches: a parcels van of older design and two sleeping cars of modern design.

because these intensify boundary layer turbulence, which in turn increases the shear stress on the coach body. The effect is closely similar to that of vortex generators discussed in Section 26.9. But the aim is just the opposite: there increased shear stress was required because prevention of separation was more important than local force effects; here the latter are all-important. Because of the large contribution in Fig. 26.15 from the bogies, attention to coach undersides has been particularly important. Older practice was to have an assortment of irregularly shaped equipment there; some of this might deflect the air flow onto the bogies. Modern practice is to have a fully faired undersurface.

26.11   Crystal growth [47, 77]

The production of crystals for industrial and technological purposes is both a traditional and a highly modern craft/science. Its importance can be indicated simply by remarking that most semiconductors are crystalline. Crystals are normally grown from a melt, solution, or vapour of the material [77]. In all three cases dynamical processes in the fluid phase can markedly affect the efficiency of the growth process and the quality of the resulting crystal [207, 239]. The range of techniques and the wide range of materials of which crystals are required mean that we cannot even mention all cases. We confine attention primarily to growth from a melt and subsequently to one particular method of such growth.

The fluid dynamical topics arising in the growth of a crystal from a melt are usually in the general field of free, mixed, or forced convection, for the obvious reason that the surface of the growing crystal must be the coolest part of the liquid. In general, the aim is to maintain the temperature over that surface as uniform as possible in both position and time. Temperature variations can have several effects that are undesirable, for example in the high-quality single crystals often needed for semiconductor applications: they can produce non-uniform growth; they can generate stresses within the crystal that greatly increase the density of defects; and they can produce variations in the proportion of any impurity, unavoidable or deliberately introduced, that enters the crystal. There is considerable empiricism involved in growing good crystals, but increasingly people are attempting to gain and apply an understanding of the fluid dynamical processes. For example, one may ask whether success of a technique with one material indicates likely success with another. One consideration is whether the two melts have similar values of Prandtl number. (Semiconductors are metallic when molten and thus have a low Pr, whereas the metal oxides frequently used as substrates for semiconductor devices have $Pr \sim 1$.) The observation (Section 22.4) that, for

Bénard convection, low Pr fluids exhibit unsteadiness even when the Rayleigh number is comparatively low may be relevant despite the more complicated geometry of a crystal-growing system.

Incidentally, the fact that the proportion of an impurity incorporated into the crystal is not usually the same as the proportion in the melt implies that compositional variations will occur within the melt. Consequently, double diffusive phenomena (Chapter 23) are of importance to the crystal grower [179]. However, these are not always present significantly and are ignored in the example chosen for slightly more detailed consideration.

This is the widely used method, known as the Czochralski method, shown in Fig. 26.17. A seed of the crystal is dipped into a melt in a heated crucible. As the crystal grows it is gradually raised so as to keep its surface in contact with the top of the melt. It is also rotated because this improves crystal quality. Sometimes the crucible is also rotated; it is interesting to see two quite different applications—the present one and

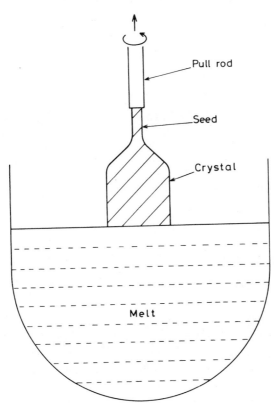

Fig. 26.17 Schematic diagram of the Czochralski method of crystal growth.

454   APPLICATIONS OF FLUID DYNAMICS

geophysical problems such as the general atmospheric circulation—
converging on closely related fluid dynamical topics. However, we
consider mainly the case in which only the crystal is rotated. The
technique is a venerable one, but the high demands of its modern use
necessitate a more detailed understanding of the fluid dynamics. This is
obtained by a combination of observations with actual crystal-growing
systems, laboratory experiments with simpler systems (e.g. with water
replacing the melt), and numerical modelling (Section 25.5) [216, 239].

Let us look briefly at one of the main issues involved. If the crystal did
not rotate, convection driven by the hot crucible walls and the cold
crystal would produce flow radially inwards in the upper part of the
liquid. If the crystal rotated rapidly, this would drive a radially outwards
motion. The two types of flow tend to produce respectively a convex and
a concave interface, instead of the desired flat one. Adjustment of the
rotation rate to give the right balance between the two is thus important.
Figure 26.18 shows a crystal in which the interface changed from convex
to planar at XX as a result of an increase in the rotation rate. There is
also current interest in the instabilities associated with the different types
of flow, as these may lead to unexpected temperature non-uniformities or
fluctuations [207, 216].

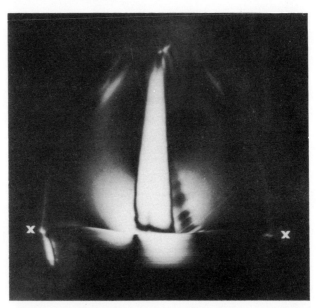

FIG. 26.18 Longitudinal section of gadolinium gallium garnet ($Gd_3Ga_5O_{12}$)
single crystal grown by the Czochralski method with change of rotation rate
during growth. From Ref. [126].

26.12 Undulatory swimming [100, 251]

An obviously important range of biophysical applications of fluid
dynamics concerns the propulsion of animals—the flight of birds and
insects, the swimming of fish and other creatures. There are various types
of propulsion. Here we consider just one, related to topics in earlier
chapters, illustrated by Figs. 26.19 and 26.20. (Both these pictures are of
organisms used for research into this mode of swimming.) Various
species of long thin animals swim by sending waves along their bodies.
Figure 26.19 shows *Amphioxus* (a primitive fish, also known as a

FIG. 26.19 *Amphioxus* swimming, head first from left to right. Photo provided
by J. E. Webb, Westfield College.

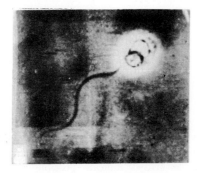

FIG. 26.20 Spore of *Blastocladiella emersonii.* From Ref. [272].

lancelet) swimming in this way; it can do so either head-first or tail-first [406]. Other examples are snakes, leeches, eels, and some species of marine worm. In some cases, there may be several wavelengths within the body length. A very wide variety of microorganisms propel themselves by sending a wave down a flagellum attached to the body [251]; Fig. 26.20 shows the spore of a water mould [272]. The wave is often helical rather than planar, and bacteria propel themselves by rotation of rigid helical flagella. (Also some organisms have many flagella, which are then known as cilia.) The fluid dynamical principles for all cases may, however, be illustrated by considering a planar wave.

The mechanism of undulatory swimming may be understood by considering the force on each portion of the body or flagellum as if it were in isolation. A cylinder moving through a fluid in a direction perpendicular to its axis experiences a larger drag than one moving parallel to its axis. Correspondingly, the drag on a cylinder in oblique motion is not directly opposite to its velocity. Thus it is possible for an animal swimming with every portion of the body moving forward nevertheless to have some portions producing a propulsive force. We may illustrate this by considering an inextensible body in a perfectly sinusoidal wave motion [124]. Figure 26.21 shows two consecutive positions of one wavelength of the body for a case in which the forward

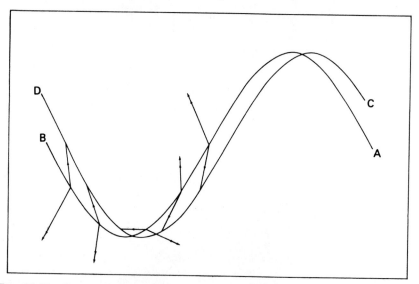

FIG. 26.21 Consecutive positions AB and CD of length of body swimming from right to left by sending wave from left to right. Single arrows indicate motion of portions of body; double arrows indicate possible resulting forces on these portions.

swimming speed is one-third the speed at which the wave travels backwards relative to the body. The short lines joining corresponding points of the body indicate approximately the direction in which each portion is instantaneously moving through the water. The resulting forces on the body may then be in the directions indicated by the double-headed arrows. It is seen that enough of these may have a forward component to overcome the drag in the regions where they have a backward component.

An interesting feature of this swimming mode is that it can be effective over all Reynolds number ranges. For the examples in Figs. 26.19 and 26.20, the Reynolds number (based on body diameter and swimming speed) is respectively around 10^3 and around 10^{-4}. It is clear that the difference in drag associated with different directions of motion is essential to the mechanism. At high Reynolds numbers this difference may be large (a cylinder in transverse motion developing a large drag as discussed in Section 12.5), and so this swimming mode can be quite effective. However, at high Reynolds numbers, there are many other modes available [250], and this one is not usually found in species highly adapted to swimming. At low Reynolds numbers, the drag differences are much smaller, because motion of a body in any direction brings a lot of fluid into motion with it (Section 8.2). (For the same reason the assumption that the drag on each portion of the body is unaffected by the motion of other portions is less satisfactory; a full theory does not require this assumption [251], but it is still useful in interpreting observations of swimming organisms.) Thus a vigorous wave is necessary to produce only slow forward motion. However, since many other swimming modes fail altogether at low Reynolds number, the existence of this mode is important [314].

26.13 Convection from the human body

Because the ambient air is usually cooler than the human body, convection currents are set up around the body. In the absence of other causes of appreciable motion, free convection boundary layers, broadly similar to those described in Section 14.8, surround the body. Figure 26.22 shows a schlieren image (Section 25.4) of the convection currents around a naked standing man. Such convection currents are evidently a significant part of the processes that determine surface and internal body temperature, with a wide variety of applications [125] ranging from questions of general comfort to the problem of hypothermia.

There is another way in which the flow is of considerable medical significance. The concentration of airborne microorganisms is markedly

FIG. 26.22 Composite of schlieren images of convection from human body. Picture provided by R. P. Clark, MRC Clinical Research Centre, Harrow.

higher in the boundary layers than in the surrounding air. In particular, the body is constantly losing skin scales (about 10^{10} per day) through the rubbing actions of limbs and clothes, and many of these scales have microorganisms attached. It has, for example, been suggested [245] that convection may produce the observed connection between skin disease and respiratory disease (e.g. eczema and asthma), by transporting organisms from the skin to the nose; and that it may account for the increase in respiratory infections after a fall in air temperature, which would result in more vigorous convection. Of particular importance is the danger of infection during orthopaedic and other operations. An understanding of the fluid dynamics is thus necessary for the design of air flow systems in operating theatres and of protective clothing. It is involved similarly in the design of infant incubators, industrial clean rooms, and so forth [125].

Figure 26.22 derives from research motivated by these applications, with experiments on human subjects being supplemented by ones with a full-scale heated model of a man [132, 245]. As is evident, the approach was to study the simplest case first. Much further information is now available on the effects of clothing, body movement and wind [125]. Developments of ideas in Chapter 14 on forced and free convection are central to this. For example, in Fig. 26.22, one can see a laminar boundary layer on the back of the legs developing into a fully turbulent one towards the top of the body.

26.14 The flight of a boomerang

The inventors of the boomerang somehow chanced on a design that requires aerofoil theory for an understanding of its performance. Whilst the name boomerang is commonly associated with a device that returns to its thrower, there are also boomerangs which fly almost straight. These were used for hunting food and in war, having the advantage over an ordinary projectile of travelling much further and striking harder; the

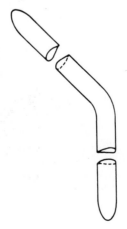

FIG. 26.23 Sketch of boomerang with 'breaks' to show cross-section of arms.

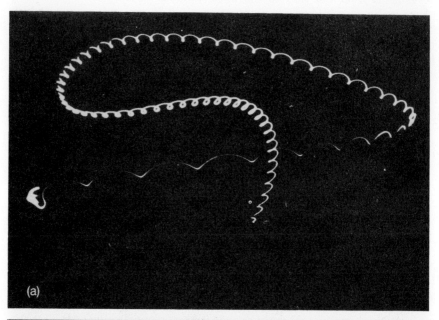

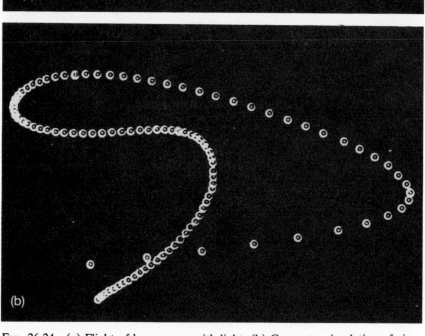

FIG. 26.24   (a) Flight of boomerang with light. (b) Computer simulation of view of boomerang flight. From Ref. [195].

good 'aerodynamic design' gave small viscous energy losses and stability of flight. The differences between returning and non-returning boomerangs are outwardly rather slight; small changes in the detailed shaping have a large effect on the trajectory. The essential features of a boomerang are shown in Fig. 26.23. Each arm has a cross-section of aerofoil shape, with the leading edges on opposite sides of the two arms. There will thus (Section 12.5) be little resistance to rotation of the boomerang in its own plane (anticlockwise in Fig. 26.23). Actual boomerangs are usually rather more complicated than this, often being slightly twisted so that the two arms are not in the same plane. The flight paths can be correspondingly complicated, with changes in the orientation of the boomerang producing different detailed dynamics over different parts of a flight.

However, the basic mechanism by which a boomerang returns can be understood fairly simply in the following way [195]. A returning boomerang has its aerofoil sections shaped or oriented so that a sideways force is generated (as discussed in Sections 13.1 and 13.2). The bend in the middle is not essential to the performance, although it may be important for ease of throwing. Consider a boomerang flying in a vertical plane, so that an observer positioned appropriately sees it travelling from right to left and rotating anticlockwise. Sideways forces act on the two arms, but unequally. The arm instantaneously nearer the top is moving through the air faster than the other, because the linear motion adds to the rotational motion for the upper arm and subtracts from it for the lower. The consequent difference in sideways forces produces a couple on the boomerang. In the absence of rotation, such a couple would twist the boomerang out of the vertical plane. But gyroscopic action associated with the rapid spin of the boomerang leads to it turning about a vertical axis. The boomerang thus travels on a curved path whilst remaining vertically oriented.

Computations based on refinements of this basic idea have provided detailed information about the behaviour of boomerangs of various designs [195]. Figure 26.24 shows a comparison of an actual flight path shown by a boomerang with a light on one tip and the result of such a computation (the programme including a perspective correction so that the pattern corresponds directly to the photograph).

NOTATION

The following list omits some items which appear in only one section of the book and are defined there.

Where a symbol has more than one meaning (either all listed below or some defined only in the text), the use in any place should be apparent from the context.

Dimensions of quantities below are indicated where this provides a useful reminder of the definitions. Primary dimensions are denoted: M = mass; L = length; T = time; Θ = temperature.†

Subscripts: Symbols with a subscript specific to the symbol are listed below. Other subscripts have general uses. Subscripts corresponding to a coordinate indicate the corresponding component of a vector (e.g. u_ϕ = azimuthal component of velocity; k_x = x-component of wavenumber). The subscript 0 indicates a reference or ambient value; a quantity with this subscript is usually a constant, but is occasionally a basic variable specifying a situation (e.g. $u_0(x)$ = velocity outside boundary layer; $\rho_0(z)$ = basic density stratification). Subscripts 1, 2 indicate values at boundaries (except in Sections 17.3 and 23.1, where other definitions are given). Subscripts $_{av}$ and $_{max}$ indicate average and maximum values. In a few places (principally the appendix to Chapter 5 and Sections 19.3 and 20.2) the suffix notation for vector and tensor quantities is introduced; usage is conventional, and is explained briefly in Section 19.3.

Superscripts: An overbar, ‾, indicates an average value. A cap, ˆ, indicates a unit vector (e.g. $\hat{x}$ = unit vector in x-direction; $\hat{n}$ = unit vector normal to surface).

a	half-width of channel
	radius (pipe, cylinder, sphere, liquid column)
	length specifying geometrical similarity
	speed of sound
b	length specifying geometrical similarity
	parameter in Lorenz equations
c	chord of aerofoil
	length specifying geometrical similarity
	concentration

---

† When heat is involved its dimensions (ML^2T^{-2}) are written as a separate group. The reason is that, when interactions between mechanical energy and thermal energy are not physically important (as in the Boussinesq approximation), heat can sometimes be treated as a further primary dimension.

c_g	group velocity
c_p	phase velocity
d	diameter (pipe, cylinder)
	thickness of fluid layer
f	similarity form of stream function (boundary layer, jet)
	vertical component of $2\mathbf{\Omega}$
	isotropic turbulence longitudinal correlation function
$f(\ )$	general function of quantity(ies) in brackets
g	$\lvert \mathbf{g} \rvert$
	similarity form of velocity profile (boundary layer, jet)
	isotropic turbulence transverse correlation function
$\mathbf{g}$	acceleration due to gravity
h	height (vertical slot, Rossby wave layer)
i	$\sqrt{-1}$
k	thermal conductivity $((ML^2T^{-2})L^{-1}T^{-1}\Theta^{-1})$
	wavenumber (L^{-1})
	horizontal wavenumber, $(k_x^2 + k_y^2)^{1/2}$
$\mathbf{k}$	wavenumber (L^{-1})
l	length (channel, plate)
	coordinate along streamline
$d\mathbf{l}$	element of length
m	exponent $(u_{\max} \propto x^m)$
n	coordinate normal to boundary
	frequency
	exponent $(\Delta \propto x^n)$
p	pressure $(ML^{-1}T^{-2})$
	pressure fluctuation in turbulent flow
p_h	hydrostatic pressure
p_S	stagnation pressure
q	$\lvert \mathbf{u} \rvert$
	total velocity fluctuation in turbulent flow
r	polar coordinate
	parameter in Lorenz equations
	space separation (correlations)
$\mathbf{r}$	general position
	space separation (correlations)
s	period of integration
	time separation (correlations)
t	time
u	x-component of velocity
	x-component of velocity fluctuation in turbulent flow
u_τ	wall flow velocity scale
$\mathbf{u}$	velocity
	velocity fluctuation in turbulent flow

v	y-component of velocity
	y-component of velocity fluctuation in turbulent flow
w	z-component of velocity
	z-component of velocity fluctuation in turbulent flow
x	Cartesian coordinate
	distance in main flow direction
x_0	origin of x for self-preserving development
y	Cartesian coordinate
	distance perpendicular to main flow direction
z	Cartesian coordinate
	coordinate perpendicular to planes of two-dimensional motion
	vertical coordinate (positive upwards)
	coordinate parallel to axis of rotation
A	constant of integration
	non-dimensional quantity in theory of Boussinesq approximation
B	constant of integration
	non-dimensional quantity in theory of Boussinesq approximation
$\boldsymbol{B}$	general vector quantity
C	numerical constant of proportionality
	non-dimensional quantity in theory of Boussinesq approximation
C_D	drag coefficient (dimensionless)
C_L	lift coefficient (dimensionless)
C_p	specific heat capacity at constant pressure $((ML^2T^{-2})M^{-1}\Theta^{-1})$ (see note below)
C_V	specific heat capacity at constant volume $((ML^2T^{-2})M^{-1}\Theta^{-1})$ (see note below)
D	drag (MLT^{-2})
	drag per unit length (MT^{-2})
	non-dimensional quantity in theory of Boussinesq approximation
	vertical length scale
D/Dt	substantive derivative
E	rate of kinetic energy transport (ML^2T^{-3} in three dimensions, MLT^{-3} in two)
	energy spectrum (with respect to wave number) (L^3T^{-2})
	internal energy (see note below)
F	spectral transfer function (L^3T^{-3})
$\boldsymbol{F}$	body force per unit volume $(ML^{-2}T^{-2})$
G	pressure gradient $(ML^{-2}T^{-2})$
H	heat transfer per unit area per unit time $((ML^2T^{-2})L^{-2}T^{-1})$

H	heat flux $((ML^2T^{-2})L^{-2}T^{-1})$
J	rate of internal heat generation per unit volume $((ML^2T^{-2})L^{-3}T^{-1})$
K	constant of proportionality
	Kármán constant
L	length scale
	horizontal length scale
	Monin–Obukhov length
	lift (MLT^{-2})
	lift per unit length (MT^{-2})
M	rate of momentum transport $(MLT^{-2}$ in three dimensions, MT^{-2} in two)
N	Brunt–Väisälä angular frequency
P	difference between pressure and hydrostatic pressure
	pressure amplitude (wave or perturbation)
	mean pressure in turbulent flow
	parameter in Lorenz equations
	probability distribution function
Q	heat (in thermodynamic considerations, see note below)
	density amplitude (wave or perturbation)
R	gas constant (see note below)
	correlation coefficient
R_{xx}	correlation coefficient between x-components of velocity at two positions
S	general surface
	cross-sectional area of streamtube
	entropy (see note below)
	$N/2\Omega$ (dimensionless)
$\mathrm{d}S$	element of surface
T	temperature
	mean temperature in turbulent flow
T_a	adiabatic temperature
U	velocity scale
	x-component of velocity amplitude (wave or perturbation)
	x-component of mean velocity in turbulent flow
	bulk velocity superimposed on wave
U_F	velocity of front of turbulent slug
U_R	velocity of rear of turbulent slug
U	velocity of boundary
	velocity amplitude (wave or perturbation)
	mean velocity in turbulent flow
V	transverse velocity scale
	y-component of velocity amplitude (wave or perturbation)

	y-component of mean velocity in turbulent flow
	general volume
	$1/\rho$
dV	element of volume
W	vertical velocity scale
	z-component of velocity amplitude (wave or perturbation)
X	variable in Lorenz equations
Y	variable in Lorenz equations
Z	variable in Lorenz equations
Ek	Ekman number
Fr	Froude number
	internal Froude number
Gr	Grashof number
Ma	Mach number
Nu	Nusselt number
Pe	Péclet number
Pr	Prandtl number
Ra	Rayleigh number
Re	Reynolds number
Re_a	Reynolds number in which the length scale is a
Ri	Richardson number
Ro	Rossby number
Sc	Schmidt number
St	Strouhal number
α	coefficient of expansion (Θ^{-1})
	angle of attack
	angle between boundaries
α_c	coefficient of density variation with concentration
β	compressibility ($M^{-1}LT^2$)
	gradient of vertical component of $2\boldsymbol{\Omega}$ ($L^{-1}T^{-1}$)
γ	ratio of specific heats
	gradient of layer depth
	intermittency factor
δ	transverse length scale
	boundary layer thickness (99 per cent thickness)
δx, etc.	small change in x, etc.
δ_{ij}	Kronecker delta
ϵ	small distance
	turbulent energy dissipation per unit mass (L^2T^{-3})
ϵ_{ijk}	alternating tensor
ζ	z-component of vorticity
η	similarity form of distance across boundary layer or jet
	y-component of vorticity

θ	polar coordinate
	co-latitude
	angle between wave number and vertical
	potential temperature
	temperature fluctuation in turbulent flow
θ_H	wall layer temperature scale
κ	thermal diffusivity (L^2T^{-1})
κ_c	concentration diffusivity (L^2T^{-1})
λ	molecular-mean-free path
	length of blocked region, Taylor column
	second viscosity coefficient $(ML^{-1}T^{-1})$
μ	viscosity $(ML^{-1}T^{-1})$
ν	kinematic viscosity (L^2T^{-1})
ν_T	eddy viscosity
ξ	non-dimensional wave number
	x-component of vorticity
ρ	density (ML^{-3})
ρ_h	hydrostatic density field
σ	force per unit area in main flow direction on surface $(ML^{-1}T^{-2})$
	amplification coefficient in stability theory (T^{-1})
σ_r, σ_i	real and imaginary parts of amplification coefficient
τ	tangential stress in shear flow $(ML^{-1}T^{-2})$
τ_W	value of τ at wall
ϕ	polar coordinate
	velocity potential (L^2T^{-1})
	energy spectrum (with respect to angular frequency) (L^2T^{-1})
	non-dimensional velocity gradient in stratified boundary layer
ψ	stream function (L^2T^{-1})
ω	angular frequency
$\boldsymbol{\omega}$	vorticity (T^{-1})
Γ	circulation (L^2T^{-1})
Δ	length proportional to thickness of boundary layer, jet, etc.
	Ekman layer thickness
Δx, etc.	finite difference in x, etc.
ΔP	pressure difference scale
Θ	temperature difference scale
Λ	transverse pressure variation scale
Π	longitudinal pressure variation scale
Σ	torque per unit length (MLT^{-2})
	turbulence kinetic energy per unit volume $(ML^{-1}T^{-2})$
Φ	gravitational potential (L^2T^{-2})
	energy dissipation per unit volume $(ML^{-1}T^{-3})$

Ψ   time scale
Ω   angular velocity (local)
           angular velocity of reference frame
Ω   angular velocity of reference frame

Note: All extensive thermodynamic quantities are defined to be per unit mass. The practice sometimes adopted of denoting these by lower-case symbols (and reserving capitals for quantities per mole) is not adopted here. (It would introduce clashes with other notation, and we are nowhere concerned with quantities per mole.) The convention used here means in particular that R denotes the universal gas constant divided by the molecular weight.

PROBLEMS

For the purposes of the following problems, air may be taken as a perfect gas with density $1.2 \, \text{kg m}^{-3}$, kinematic viscosity $1.5 \times 10^{-5} \, \text{m}^2 \text{s}^{-1}$, and specific heat at constant pressure $1.0 \times 10^3 \, \text{J kg}^{-1} \text{K}^{-1}$; water may be taken to have density $1.0 \times 10^3 \, \text{kg m}^{-3}$, kinematic viscosity $1.0 \times 10^{-6} \, \text{m}^2 \text{s}^{-1}$, and coefficient of expansion (at 20°C) $2.1 \times 10^{-4} \, \text{K}^{-1}$.

1. A layer of fluid (viscosity μ, density ρ) of depth d flows under the influence of gravity down a wide plane inclined at an angle θ to the horizontal. Assuming that d is constant and that all conditions are steady, find the velocity as a function of the normal distance y from the plane. Determine also the ratio of the average velocity to the maximum velocity, and the mass flux per unit width of the plane. (The free surface supports no viscous stress and the pressure above it may be supposed uniform.)

2. A vertical tube of diameter 2 mm has its upper end open to the atmosphere. At the lower end water is maintained at a pressure of $10^4 \, \text{N m}^{-2}$ above atmospheric. What is the longest length of tube for which this pressure will produce a flow through the tube? (Ignore surface tension.)
   What volume of water will pass through the tube per second if the length is half this? (Assume that the flow is laminar and ignore the entry length.)

3. A two-dimensional channel of constant width d, containing fluid of density ρ and viscosity μ, has one wall at rest and the other moving in its own plane with speed U. The fluid moves under the action of a negative pressure gradient, $-G$, applied parallel to U. In what circumstances will the flow speed exceed U in some region?
   What can you say about the stress on the moving wall when the conditions for this are just not fulfilled?

4. A jet is specified by the diameter of the orifice, the momentum per unit time of the fluid issuing from the orifice, and the density and viscosity of the fluid. What will be the form of the criterion determining whether the jet is laminar or turbulent?

5. For flow of nitrogen in a pipe of diameter 10 mm at a temperature of 300 K, estimate how low the pressure may be without violation of criteria (5.1) and (5.2) for the applicability of continuum mechanics. (Avogadro's constant $= 6 \times 10^{23} \, \text{mol}^{-1}$; collision cross-section of nitrogen molecule $= 6 \times 10^{-19} \, \text{m}^2$.)

6. Show that the continuity equation (5.9) may be written

$$D\rho/Dt + \rho \nabla \cdot \boldsymbol{u} = 0.$$

Derive the equation in this form from first principles, starting with the fact that the mass of fluid in a volume V moving with the flow is conserved; that is, $D(\rho V)/Dt = 0$.

7. (1) Verify relationship (5.61) as the criterion for compressibility effects to be negligible in a low Reynolds number flow.

(2) A body of typical dimension L is accelerated to a speed of order U in a time of order Ψ. What is the criterion for the motion of fluid surrounding the body to be treated as incompressible when Ψ is small compared with L/U (so that $|\partial u/\partial t|$ is larger than $|u \cdot \nabla u|$)? (Find first the criterion for the fractional density change to be small, and then consider whether this is sufficient for the continuity equation

$$\partial \rho/\partial t + u \cdot \nabla \rho + \rho \nabla \cdot u = 0$$

to be approximated by its form for incompressible flow.)

8. The water supply in the apparatus of Fig. 2.1 is suddenly turned off. Supposing that the instantaneous flow pattern throughout the pipe is that of Poiseuille flow, derive an expression for the time in which the height of the water above the pipe entry falls to one half of its original value.

Formulate criteria for the negligibility of the inertial effects associated with (a) the acceleration of the water into the pipe, and (b) the changing velocity within the pipe.

Evaluate the above time for a pipe of diameter 1 mm and length 10 m leading out of a tank of horizontal cross-section $10^{-2}\,\mathrm{m}^2$. Check whether the above criteria are fulfilled if the initial height is 100 mm.

9. Show that the streamlines of the motion

$$u = -\Omega y \quad v = \Omega x \quad w = w_0$$

(where Ω and w_0 are constants) are

$$x^2 + y^2 = a^2 \quad x = a \sin(\Omega z/w_0 + \alpha)$$

(Note: do this by forward integration, not by substitution of the result.)

10. In each of cases (1) and (2) below, a flow is described in a Lagrangian way by giving, as a function of time t, the coordinates (x, y, z) of the fluid particle that is at (x_0, y_0, z_0) at $t = 0$. For each flow, (a) formulate the equations for the path of this particle, (b) derive the Eulerian equations $(u = u(x, y, z, t)$ etc.), (c) consider whether the flow is steady, (d) consider whether the flow satisfies the incompressible continuity equation, and (e) derive a general equation for a streamline. What is the nature of the flows?

(1) $x = x_0 \exp(-2t/s)$ $y = y_0 \exp(t/s)$, $z = z_0 \exp(t/s)$

(2) $x = x_0 \exp(-2t/s)$ $y = y_0(1 + t/s)^2$, $z = z_0 \exp(2t/s)(1 + t/s)^{-2}$

$$(t > 0; \quad s = \text{constant} > 0).$$

11. Derive the general equation for a streamline in the two-dimensional flow (an approximate form of Rossby wave)

$$u = u_0, \quad v = v_0 \cos(kx - \alpha t)$$

where u_0, v_0, k, and α are constants. At $t = 0$, what is the equation for the streamline passing through $x = 0$, $y = 0$?

Derive also the equation for the path of the particle which is at $x = 0$, $y = 0$ at time $t = 0$.

Comment briefly on the comparison of the streamline and the particle path in the two limiting cases, $\alpha = 0$ and $k = 0$.

12. Show geometrically that for cylindrical polar coordinates

$$\partial \hat{r}/\partial \phi = \hat{\phi} \quad \text{and} \quad \partial \hat{\phi}/\partial \phi = -\hat{r}.$$

Hence, by writing

$$\boldsymbol{u} = u_r \hat{r} + u_\phi \hat{\phi} + u_z \hat{z}, \quad \nabla = \hat{r}\frac{\partial}{\partial r} + \frac{\hat{\phi}}{r}\frac{\partial}{\partial \phi} + \hat{z}\frac{\partial}{\partial z}$$

show (1) that the inertia term takes the form given in eqns (5.26), (5.27), and (5.28); and (2) that, for the case when $u_r = u_z = 0$ and u_ϕ depends on r alone, the viscous term takes the form given in eqn (9.3).

Note: the order of operations is indicated by writing $\boldsymbol{u} \cdot \nabla \boldsymbol{u} = (\boldsymbol{u} \cdot \nabla)\boldsymbol{u}$ and $\nabla^2 \boldsymbol{u} = (\nabla \cdot \nabla)\boldsymbol{u}$.

13. (1) Show that, for an unsteady flow in which only the x-component of velocity is non-zero and this varies only in the y-direction, the Navier–Stokes equation reduces to a form analogous to the equation of unsteady one-dimensional heat conduction in a solid.

(2) An effectively infinite flat plate bounding a semi-infinite expanse of fluid oscillates in its own plane with velocity

$$U = U_0 \sin \omega t.$$

Supposing that the induced fluid motion is an oscillation of the same frequency, how do the amplitude and phase vary with distance from the plate?

(3) For a geometry similar to that in (2), the plate is suddenly brought into motion at time $t = 0$ and then moves in its own plane with constant velocity U_0. Both the plate and the fluid were at rest for $t < 0$. Show that, for $t > 0$, the fluid velocity is

$$u = U_0 \operatorname{erfc}[y/2(vt)^{1/2}].$$

Find the force per unit area (as a function of time) needed to produce this motion. Hence, find the total work done after any given time, and determine what proportion of this work has appeared as kinetic energy and what proportion has been dissipated.

$$\left[\operatorname{erfc} x = \frac{2}{\sqrt{\pi}} \int_x^\infty \exp(-x^2)\, dx; \quad \operatorname{erfc} 0 = 1; \quad \int_0^\infty (\operatorname{erfc} x)^2\, dx = \frac{2 - \sqrt{2}}{\sqrt{\pi}} \right].$$

14. Show that for an axisymmetric vorticity distribution $\boldsymbol{\omega} = (0,\ 0,\ \zeta(r))$ in cylindrical polar coordinates $(r,\ \phi,\ z)$, the vorticity equation reduces to

$$\frac{\partial \zeta}{\partial t} = \frac{v}{r}\frac{\partial}{\partial r}\left(r\frac{\partial \zeta}{\partial r}\right).$$

If there is a concentrated line vortex along the z-axis at time $t = 0$ in an otherwise irrotational fluid, show by substitution that the vorticity distribution at any subsequent time is

$$\zeta = \frac{\Phi}{4\pi vt}\exp(-r^2/4vt).$$

What is the corresponding velocity distribution?

15. If, for turbulent flow through a pipe, it is observed that the pressure gradient is proportional to $Q^{7/4}$ (where Q is the volume rate of flow), predict how the pressure gradient would vary with the viscosity and density of the fluid (for fixed Q).

16. The figures below represent observations of the power needed to propel two objects (e.g. toy submarines) through a fluid at various speeds. The objects are geometrically similar, the second being five times as large as the first (linear dimensions). Show that, although the relationship between speed U and power P is different in the two cases, all the figures are consistent and can be combined in an appropriate way. (The other relevant quantities are the density and the viscosity of the fluid.)

Comment on the interpretation of the different relationships between U and P.

Small object ($L = 1$ length unit):

U	1	2	3	5	10	speed units
P	1.0	4.5	12	50	400	power units.

Larger object ($L = 5$ length units):

U	1	2	3	5	10	speed units
P	10	80	270	1250	10000	power units.

17. A circular disc is set spinning with angular velocity Ω in an incompressible viscous fluid. Use dimensional analysis to find the general functional form of the expression for the time t in which the angular velocity falls to $\Omega/2$. Assume that this depends only on Ω, the radius of the disc a, the density of the disc σ, the density ρ and the viscosity μ of the fluid.

Could you designate any of the parameters involved as a Reynolds number?

In an experiment of this type using a fluid having $\rho = 1.0 \times 10^3\,\mathrm{kg\,m^{-3}}$ and $\mu = 1.0 \times 10^{-3}\,\mathrm{kg\,m^{-1}\,s^{-1}}$, t was observed to be $100\,\mathrm{s}$ when Ω was $20\,\mathrm{rad\,s^{-1}}$. What can you conclude from this about the motion of the same disc (a) in a fluid with $\rho = 1.0 \times 10^3\,\mathrm{kg\,m^{-3}}$, $\mu = 2.0 \times 10^{-3}\,\mathrm{kg\,m^{-1}\,s^{-1}}$, and (b) in a fluid with $\rho = 0.75 \times 10^3\,\mathrm{kg\,m^{-3}}$, $\mu = 3.0 \times 10^{-3}\,\mathrm{kg\,m^{-1}\,s^{-1}}$?

Suppose now that one argues on physical grounds that t must be directly proportional to σ (because this enters the problem only through the moment of inertia of the disc and not through the resisting fluid motion). How can the general expression for t be simplified?

What can you now conclude in cases (a) and (b) above?

18. A string of density σ, diameter d, length l is held in tension across a channel through which fluid of kinematic viscosity v flows. The string vibrates when its fundamental natural frequency coincides with the Kármán vortex street frequency. The following observations were made of the tension F at which resonance occurred for various flow speeds U of a fluid with $v = 1.0 \times 10^{-5}\,\mathrm{m^2\,s^{-1}}$:

$U(\mathrm{m\,s^{-1}})$	0.2	0.4	0.6	0.8	1.0	1.2	1.4
$F(\mathrm{N})$		0.096	0.43	1.03	1.92	3.00	4.3	5.8

Suppose now that the fluid is changed to one with $v = 5.0 \times 10^{-5}\,\mathrm{m^2\,s^{-1}}$. At which of the following speeds can the resonance tension be predicted and then what is it?

$$U(\mathrm{m\,s^{-1}}) \quad 0.4 \quad 1.0 \quad 5.5$$

(You may assume that the only properties of the string that enter the problem are σ, d, and l, and that changes in the density of the fluid have no effect (except through the kinematic viscosity); and that no harmonics are generated.)

19. A plate bounding a region containing fluid of density ρ and kinematic viscosity v is subjected to a tangential oscillatory force, F per unit area, of frequency n so that

$$F = F_0 \sin 2\pi nt.$$

As a result the plate oscillates with frequency n and with amplitude A_0. Experiments show that, when the resulting motion of the fluid is laminar, A_0 is proportional to F_0 (all other quantities being held constant) and, when it is fully turbulent, $A_0 \propto F_0^{1/2}$. How would you expect A_0 to change (a) for laminar motion, (b) for fully turbulent motion, and (c) in the transition region between the two, for the following cases?

(1) when the frequency is doubled (at constant F_0, ρ, and v)
(2) when the frequency is doubled and the fluid is changed to one of half the original viscosity (at constant F_0 and ρ).

You may suppose that the other boundaries of the fluid region are distant (so that no length characteristic of the region enters the problem) and that the inertia of the plate is negligible (so that F_0 is entirely balanced by the fluid resistance to the motion).

Note: it is not necessarily possible to make a prediction in every case.

20. An effectively two-dimensional jet of fresh water is emitted horizontally from a slit of width a with an average speed U close to the top of a tank containing salt water. The density of the salt water exceeds the density, ρ_0, of the fresh water by a controllable amount, $\Delta\rho$. $\Delta\rho$ is sufficiently small that its only effect on

the dynamics of the jet is in changing the gravitational force on fluid particles (i.e. $\Delta\rho g$ is a parameter but not $\Delta\rho$ or g individually). In a series of experiments the lowest value of U at which the jet became turbulent was observed for various values of $\Delta\rho$. The results were:

$$a = 5.0 \text{ mm}; \quad \rho_0 = 1000 \text{ kg m}^{-3}; \quad \mu(\text{viscosity}) = 1.0 \times 10^{-3} \text{ kg m}^{-1}\text{s}^{-1}$$

$\Delta\rho(\text{kg m}^{-3})$	0	2	4	8	12	16
$U(\text{mm s}^{-1})$	10	25	33	43	51	58

Similar experiments can be done with changed slit width and with different fluids and contaminants. Will the flow be laminar or turbulent in each of the following cases?

	(1)	(2)	(3)
a (mm)	10	5.0	50
ρ_0 (kg m^{-3})	1000	1.0	2.0
μ (10^{-3} kg m^{-1} s^{-1})	1.0	0.010	0.020
$\Delta\rho$ (kg m^{-3})	2.0	0.020	0.0010
U (mm s^{-1})	25	400	20

(You may assume that the walls of the tank have no effect on the motion; and that the diffusivity of the contaminant may be ignored. g may be taken as 10 m s^{-2}.)

21. Determine the conditions for dynamical similarity of steady incompressible flow of an electrically conducting fluid in a magnetic field, governed by the equations

$$\nabla \cdot \boldsymbol{u} = 0$$

$$\nabla \cdot \boldsymbol{B} = 0$$

$$\boldsymbol{u} \cdot \nabla \boldsymbol{u} = -\frac{1}{\rho}\nabla p + \frac{1}{\rho\mu}(\nabla \times \boldsymbol{B}) \times \boldsymbol{B} + \nu\nabla^2\boldsymbol{u}$$

$$\boldsymbol{u} \cdot \nabla \boldsymbol{B} = \boldsymbol{B} \cdot \nabla \boldsymbol{u} + \frac{1}{\sigma\mu}\nabla^2\boldsymbol{B}.$$

Briefly make any comments on the results that you consider to be of interest. (Notation: $\boldsymbol{u}$ = velocity; $\boldsymbol{B}$ = magnetic field; p = pressure; ρ = density; ν = kinematic viscosity; μ = magnetic permeability; σ = electrical conductivity.)

22. (1) What is the largest size of water drop (assumed spherical) for which the rate of free fall in air can be calculated using Stokes's formula (9.17) and what is its rate of fall?

(2) How small must a water drop be for its fall under its own weight to be negligible when it is in a 1 m s^{-1} wind generating turbulence of intensity equal to 10 per cent of the mean velocity?

23. (1) Is the motion incompressible for the flows given by the following velocity potentials? If so, determine the corresponding stream functions.

(a) $\phi = C(x^2 + y^2)$

(b) $\phi = C(x^2 - y^2)$.

(2) Is the motion irrotational for the flows given by the following stream functions? If so, determine the corresponding velocity potentials.

(a) $\psi = C(x^2 + y^2)$

(b) $\psi = C(x^2 - y^2)$.

Sketch the streamlines for all cases and the lines of constant ϕ where possible.

24. Show that if the effect of the hydrostatic pressure is significant, Bernoulli's equation becomes

$$p + \tfrac{1}{2} \rho q^2 + \rho g z = \text{constant}$$

where p is the true pressure and z is the vertical coordinate (positive upwards).

Hence, show that the velocity of liquid emerging from a small hole in a tank with a free surface is the same as the velocity that would be acquired by free fall from the level of the surface to that of the hole.

25. Find how the constant in Bernoulli's equation varies with radius for each of the two velocity fields, eqn (6.17) and eqn (6.19).

26. Fluid flows out of a reservoir (so large that the fluid in it can be considered stationary) through a circular tube with a Venturi constriction in it. The radius of the tube is 10 mm, reducing to 5 mm in the constriction. If the pressure (above atmospheric) is 200 N m^{-2} in the reservoir and 190 N m^{-2} in the main part of the tube, what is it in the constriction? Suppose that the flow is incompressible and inviscid (low Mach number, thin boundary layers).

27. (1) Sensitive liquid level manometers can be read with an accuracy of about $\pm 10^{-2}$ mm of water. What is the lowest air velocity that can be measured to within 5 per cent with a Pitot tube?

(2) It is found empirically that buoyancy effects upset the calibration of a hot-wire anemometer when $Gr^{1/3} > Re/2$. (Why is this of a different form from eqn (14.20)?) What is the lowest air velocity that can be measured when the ratio of the absolute temperature of the wire to that of the air is 1.5?

28. Show that all the conditions for inviscid incompressible flow past a fixed circular cylinder of radius a (with uniform velocity at large distances) are satisfied by a velocity potential in cylindrical polar coordinates of the form

$$u_0\left(r + \frac{a^2}{r}\right)\cos \phi.$$

Show that the corresponding expression in spherical polar coordinates for flow past a sphere is

$$u_0\left(r + \frac{1}{2}\frac{a^3}{r^2}\right)\cos\theta.$$

What are the pressure distributions over the surfaces of the cylinder and sphere?

29. In spherical polar coordinates (r, θ, ϕ), the velocity potential

$$\Phi = -kr^2 P_2(\cos\theta) = -\tfrac{1}{2}kr^2(3\cos^2\theta - 1)$$

represents the axisymmetric flow produced by the confluence of two equal and opposite streams. Sketch the streamline pattern in any plane through the axis of symmetry.

If now a solid sphere with surface $r = a$ is placed in this flow, how is the velocity potential modified? Hence, determine the distribution of velocity and pressure over the surface of the sphere.

State briefly the principal ways in which a real flow would be expected to depart from this ideal one.

30. (1) For (a) inviscid flow past a sphere and (b) Stokes flow past a sphere, determine the distance to the side of the sphere (measured in sphere radii) at which the difference from the free-stream velocity falls to 1 per cent.

(2) For high Reynolds number flow past a cylinder, estimate the distance at which the difference from the free-stream velocity falls to 1 per cent (a) to the side of the cylinder and (b) directly downstream of the cylinder.

Note: refer to the solutions of Question 28 and to caption of Fig. 21.5.

31. Compare the two-dimensional boundary layers on plane and curved surfaces below the same inviscid velocity distribution $u_0(x)$, where x is the curvilinear coordinate in the surface. Find the order of magnitude of the pressure variation across the boundary layer for the latter case, by considering the balance between the pressure gradient and the centrifugal force associated with the curved flow. Hence, show that the boundary layer equations are the same for the two cases provided that $\delta/R \ll 1$, where δ is the boundary layer thickness and R is the radius of curvature of the surface.

32. In the two-dimensional flow away from a stagnation point on a flat wall, the inviscid velocity at the wall is

$$u_0 = ax$$

where x is the distance along the wall from the stagnation point (cf. the inviscid flow in Question 23 (1)(b)). Show that the boundary layer below this flow, in the region where the Reynolds number ax^2/v is large, has constant thickness. If the stream function in the boundary layer is written

$$\psi = kaxf(y/\delta)$$

show that it is appropriate to put

$$k = \delta = (v/a)^{1/2}$$

and that the governing differential equation is then

$$f''' + ff'' - f'^2 + 1 = 0.$$

Formulate the boundary conditions for this.

33. The growth of a boundary layer can be inhibited by sucking some of the fluid through a porous wall. In appropriate circumstances, this can give a boundary layer of which both the thickness and the velocity profile remain constant with distance downstream. Consider this situation for a two-dimensional flat plate boundary layer in zero external pressure gradient. Determine the velocity profile, by considering first the continuity equation and then the momentum equation with boundary conditions:

$$u = 0 \quad \text{and} \quad v = -v_0 \quad \text{at} \quad y = 0$$

$$u = u_\infty \quad \text{at large } y.$$

Whereabouts on a wall of finite extent would you expect this solution to apply?

How does it relate to the solution when $v_0 = 0$? Mention another situation in which the introduction of an extra parameter gives rise to a simpler solution in a similar way.

34. Show that the velocity profile (11.54) satisfies each of (11.64), (11.65), and (11.66). (Note: by use of the appropriate variable this can be done without detailed integrations.)

35. (1) Use a procedure similar to that in Section 11.7 to show that, in the wake behind a two-dimensional obstacle fixed in a stream of velocity u_0,

$$dM/dx = 0, \quad \text{where} \quad M = \rho \int_{-\infty}^{\infty} u(u_0 - u)\, dy.$$

Interpret M as the rate of transport of the momentum deficit produced by the drag on the obstacle.

(2) Correspondingly, the rate of transport of kinetic energy deficit is

$$E = \tfrac{1}{2}\rho \int_{-\infty}^{\infty} u(u_0^2 - u^2)\, dy.$$

Show that

$$E = u_0 M - \tfrac{1}{2}\rho \int_{-\infty}^{\infty} u(u_0 - u)^2\, dy.$$

Hence, for a self-propelled body (i.e. a body holding itself fixed against the flow) for which $M = 0$, show that the wake involves transport of kinetic energy away from the body.

36. In the Reynolds number range 10^2 to 3×10^5, a good approximation to the drag on a circular cylinder is given by the following model (see Figs. 3.15 and 12.9):

(a) the drag is entirely due to pressure variations over the cylinder surface;

(b) the value of $(p - p_0)/\frac{1}{2}\rho u_0^2$ (where p is the pressure on the cylinder and p_0 is the ambient pressure) is given by the inviscid flow solution (Question 28) for $-\pi/4 < \theta < \pi/4$ (where θ is as in Fig. 12.9) and is equal to -1 elsewhere.

Show that this model implies that the drag coefficient, $C_D = 0.94$.

37. An aeroplane has to reach a speed of $30 \mathrm{~m~s}^{-1}$ to gain sufficient lift for take-off at an airport at sea-level and $0°C$. What speed does the same aeroplane with the same load need to reach at an airport at an altitude of $1000 \mathrm{~m}$ and $30°C$?

38. Show that the velocity field (in two-dimensional polar coordinates (r, ϕ))

$$u_r = -u_0\left(1 - \frac{a^2}{r^2}\right)\cos \phi$$

$$u_\phi = u_0\left(1 + \frac{a^2}{r^2}\right)\sin \phi + \frac{\Gamma}{2\pi r}$$

represents a solution of the equations of inviscid irrotational motion around a circular cylinder of radius a, tending to a free-stream velocity u_0 far from the cylinder and with circulation Γ round the cylinder. Determine the pressure distribution over the surface of the cylinder, and hence show that the lift per unit length is given by the Kutta–Zhukovskii result, $L = -\rho u_0 \Gamma$.

39. From the information that an aircraft of mass $10^4 \mathrm{~kg}$ has wings of average chord $3 \mathrm{~m}$ and total span $30 \mathrm{~m}$, estimate the magnitude of the air speeds associated with circulation around the wings during level flight at a speed of $100 \mathrm{~m~s}^{-1}$.

40. Consider a smooth ball of diameter $40 \mathrm{~mm}$ and density $1.2 \times 10^3 \mathrm{~kg~m}^{-3}$ travelling (instantaneously) horizontally at a speed of $20 \mathrm{~m~s}^{-1}$. Check that the experimental data in Fig. 13.9 are applicable and use them to determine (a) the radius of curvature in a horizontal plane of its trajectory when it is spinning about a vertical axis at an angular velocity of $600 \mathrm{~rad~s}^{-1}$, and (b) how rapidly it would need to spin about a horizontal axis in order to counteract 20 per cent of the gravitational force.

41. (1) Consider forced convection in a circular pipe of radius a with the Poiseuille velocity profile, eqn (2.17). Suppose that the temperature of the pipe wall increases linearly with distance along the pipe, and that, in consequence, the same constant axial temperature gradient $\partial T/\partial x$ exists everywhere in the pipe. Determine the radial temperature distribution. Determine also the heat transfer from the pipe wall. Express the latter result in the non-dimensional form, eqn (14.28) (defining the temperature difference scale as the temperature change in a length of the pipe equal to its diameter, $\Theta = 2a\, \partial T/\partial x$).

(2) Show how the above solution can be extended to apply to Poiseuille flow occurring with uniform internal heat generation in a pipe with a thermally insulating wall. Explain why, for this flow to establish a uniform axial temperature gradient, there must be a temperature difference between the fluid at the wall and that at the centre, and derive an expression for this difference.

42. Free convection in a layer between two vertical walls at different uniform temperatures (as described in Section 4.3) of large enough aspect ratio (h/d) can have a region in the middle vertically in which the velocity and the temperature depend only on the horizontal coordinate and not on the vertical. Show that, in this region, the temperature profile is unaffected by the flow and determine the velocity profile. Express the latter result in terms of the dependence of the Reynolds number, based on the maximum velocity and the layer width, on the Grashof number.

43. (1) Movements of the Earth's mantle are inferred with speeds around $10^{-9}\,\mathrm{m\,s^{-1}}$ (see Section 26.5). Supposing these are produced by free convection in a homogeneous fluid of kinematic viscosity $10^{17}\,\mathrm{m^2\,s^{-1}}$ and coefficient of expansion $10^{-5}\,\mathrm{K^{-1}}$ and that the length scale of both the velocity and temperature fields is $10^6\,\mathrm{m}$, what are the temperature differences associated with the flow?

(2) If the thermal diffusivity of the mantle is $10^{-7}\,\mathrm{m^2\,s^{-1}}$, what is the thickness of the thermal boundary layers associated with the above flow? If the temperature variations are confined to such boundary layers, how is the answer to (1) modified?

44. In Section 15.1, it is shown that when the Reynolds and Péclet numbers are both large, the criterion for stratification to affect a flow strongly is that Fr^2 should be small. Show that the corresponding criteria in the cases of (a) small Reynolds number but large Péclet number and (b) small Reynolds number and small Péclet number are respectively that $\mathrm{Fr}^2/\mathrm{Re}$ and $\mathrm{Fr}^2/\mathrm{RePe}$ should be small.

45. It is desired to investigate the lee waves associated with the flow of a wind of $10\,\mathrm{m\,s^{-1}}$ past a hill in an isothermal (and therefore subadiabatic) atmosphere at $300\,\mathrm{K}$, by towing a $1:10^4$ scale model of the hill at $50\,\mathrm{mm\,s^{-1}}$ in a channel in which a uniform vertical salt gradient has been established. If the channel depth is $200\,\mathrm{mm}$ and the water at the top is fresh, what should be the salt concentration (expressed as weight of salt per unit weight of water) at the bottom, assuming that the water volume does not change when salt is dissolved in it? (Assume that viscosity and diffusion may be neglected.)

46. A horizontal cylindrical body is oscillated periodically but *non-*sinusoidally in a stratified fluid at a frequency equal to 0.3 times the Brunt–Väisälä frequency. For each of the internal wave modes generated, determine the direction of the group velocity, the direction of the phase velocity, and the ratio of the group velocity to the phase velocity.

47. A tank of fluid rotates in rigid body motion. Superimposed on this is a 'jet-stream' type of motion in which some of the fluid is moving azimuthally with

a different angular velocity. Show that an observer at rest in the laboratory regards the modification to the force field by this stream as a change in the centrifugal force; that an observer rotating with the tank regards it as either a Coriolis force or a Coriolis force plus a contribution to $u \cdot \nabla u$ (depending on whether the Rossby number is low or not); but that the two are agreed on the radial pressure gradient associated with the motion.

48. A demonstration of the action of the Coriolis force in fluid dynamics can be made by introducing a capillary tube into a uniform flow in a rotating channel. The basic flow, the tube, and the axis of rotation are all mutually perpendicular. Explain briefly why there is a flow through the tube, and calculate the ratio of the mean speed of this flow to the speed of the basic flow when the angular velocity of rotation is $0.5 \, \text{rad s}^{-1}$, the radius of the capillary tube $0.2 \, \text{mm}$, and the kinematic viscosity of the fluid $1 \, \text{mm}^2 \, \text{s}^{-1}$.

To an order of magnitude, how much higher is the water level on the French coast than on the English coast when there is an eastward tidal flow through the English Channel at a typical speed of $1 \, \text{m s}^{-1}$?

49. Apply the laminar theory of the Ekman spiral to the case of a semi-infinite expanse of fluid with a boundary moving in its own plane at a constant velocity U with respect to a frame of reference rotating with angular velocity Ω about an axis perpendicular to the boundary. The fluid far from the boundary is at rest in this frame. How does the flow vary with distance from the boundary? What is the smallest distance at which the flow direction is exactly opposite to that of the boundary?

This theory may be applied to the motion of the upper layers of the ocean under the action of the wind stress, with the difference that the viscous stress at the surface, not the velocity, is prescribed. Show that the surface moves in a direction at 45° to the direction of the stress, and the net mass flux over the whole flow is at 90° to the direction of the stress. (It may be assumed that only the vertical component of the Earth's angular velocity need be considered.)

50. An Ekman layer becomes unstable at a Reynolds number (based on the geostrophic velocity and the layer thickness) of about 55. Experiments are performed in water in a rotating annulus of inner and outer radii 0.45 and 0.55 m. Ekman layers are observed on boundaries normal to the rotation axis (a) by moving the boundary relative to the annulus at a speed of $30 \, \text{mm s}^{-1}$; (b) by producing a transient geostrophic flow by suddenly changing the rotation rate by 10 per cent; and (c) by producing a geostrophic flow by maintaining a pressure difference between the inner and outer walls 10 per cent greater than that needed to balance the centrifugal force. In each case, what is the maximum or minimum rotation rate at which the instability could be observed? Also, at this rotation rate, is the assumption that the Ekman layer is thin compared with other relevant dimensions fulfilled?

51. For Rossby waves of the type analysed in Section 16.7, show that the criterion for non-linear inertial effects to be negligible is that the Rossby number based on the velocity in the direction of the depth gradient and on the wavelength should be small.

52. Derive the principal properties of inertial waves in a rotating fluid (as shown in Fig. 16.11) by the procedure applied to internal waves in a stratified fluid in Section 15.4. (Start with the linearized, inviscid, but time-dependent equations for motion in a rotating fluid and substitute velocity and pressure fields of the form of eqns (15.28) and (15.29).) In particular, show that

$$\omega = 2\Omega \, |\cos \theta|$$

where θ is the angle between k and Ω, and that the group velocity is perpendicular to k in the plane of k and Ω. Consider the significance of the limiting cases $\omega = 0$ and $\omega = 2\Omega$.

53. (1) A cyclone has pressure differences of about 10 mbar ($10^3 \, \text{N m}^{-2}$) over horizontal distances of about 10^3 km. What is a characteristic wind speed?

(2) A tornado has pressure differences of about 50 mbar over horizontal distances of about 10^2 m. What is a characteristic wind speed?

(Briefly justify your choice of procedure in each case.)

54. (1) For free convection in a rotating fluid, determine the non-dimensional parameter that indicates the relative importance of Coriolis and inertia forces (ignoring viscous forces).

(2) For free convection in a fluid rotating about a vertical axis sufficiently rapidly that the Coriolis force is dominant, show that the Taylor–Proudman theorem still applies to the vertical component of the velocity, but is modified for the horizontal components. Illustrate the physical significance of the new form by discussing briefly the vorticity balance associated with the increase in speed with height of westerly winds at mid-latitudes.

55. The capillary instability of a stationary cylindrical column of liquid may be investigated by introducing a disturbance of a controlled wavelength. The table below gives the amplification rate (reciprocal of e-folding time), in units of s^{-1}, as a function of the radius of the column and of the wavelength, for a liquid with density $1.5 \times 10^3 \, \text{kg m}^{-3}$, surface tension $6 \times 10^{-2} \, \text{N m}^{-1}$ and kinematic viscosity $1.0 \times 10^{-4} \, \text{m}^2 \, \text{s}^{-1}$. For a 1 mm radius column of another liquid with density $1.0 \times 10^3 \, \text{kg m}^{-3}$, surface tension $2 \times 10^{-2} \, \text{N m}^{-1}$, and kinematic viscosity $0.5 \times 10^{-4} \, \text{m}^2 \, \text{s}^{-1}$, what would be the amplification rate of a 10 mm wavelength disturbance and what wavelength would give the maximum amplification rate?

	Radius (mm)		
	0.5	1	2
5	660	0	0
10	750	310	0
15	580	340	85
Wavelength (mm) 20	480	300	135
30	340	230	135
40	290	180	120
60	180	140	90

56. Suppose that the behaviour of a buoyant flow with an exothermic chemical reaction may be modelled by the equations

$$dX/dt = (AX + B)Y - CX$$

$$dY/dt = D + EX - FY.$$

(Temperature X is increased by the reaction at a rate depending on the temperature and, through the supply of reagents, on the flow rate; it is decreased by heat loss. The flow rate Y is increased by a constant pressure gradient and a buoyancy force and is decreased viscously.) Suppose also that an appropriate non-dimensionalization gives

$$A = 1, \quad B = 2, \quad C = 3, \quad D = 2, \quad E = 1, \quad F = 3.$$

Show that there are two steady state solutions of the equations and carry out a linear stability analysis to find out whether both, one, or neither may actually occur.

57. (A highly simplified ecological model that illustrates the methods of stability analysis.)

The populations of two animal species S_x and S_y are proportional to X and Y which are governed by the equations

$$dX/dt = AX - BX^2 - CXY$$

$$dY/dt = -DY + EXY$$

where A, B, C, D, and E are positive constants. (When X is small it tends naturally to increase, but this growth is limited both by competition for food at larger X and by S_x being preyed upon by S_y; S_y is sustained by preying on S_x and would otherwise die out.) Consider the case in which (with t appropriately non-dimensionalized) one may put

$$B = 3, \quad C = E = 1.$$

Find steady state solutions in which (1) Y is zero but X is non-zero, (2) X and Y are both non–zero. Show that the latter exists only when A and D are appropriately related.

Investigate the stability of these solutions and show that (1) is stable when the alternative does not exist and unstable when it does, and that (2) is stable whenever it exists. (Note: it is meaningful to perturb a solution for which $Y = 0$ by a perturbation with $y \neq 0$; it corresponds, for example, to the effect of introducing one pair of S_y into a population of S_x.)

An interesting extension: suppose the governing equations are modified to

$$dX/dt = AX - 3X^2 - XY - F$$

$$dY/dt = -DY + XY$$

where F is a positive constant (S_x is culled at a rate that pays no regard to the current population). Show that, provided that F is not too large, there is still a steady solution with both X and Y non-zero, but that the effect of introducing F is to change Y, not X. (Comment!) However, show also that this solution may now be stable or unstable and that the instability may take the form of either growing oscillations or monotonic growth, depending on values of the parameters. (Hints: (1) you can readily show that the introduction of F modifies the stability analysis only through the changed value of the steady-state Y to be used; (2) the different types of behaviour of a perturbation are readily seen by considering examples of the parameter values: compare, for example,

(a) $A = 12D$, $F = 2D^2$; (b) $A = 12D$, $F = 4D^2$; (c) $A = 12D$, $F = 8D^2$.)

58. Write down the velocity distribution in rotating Couette flow for the two special cases of (a) two cylinders rotating in the same sense with the same angular velocity, and (b) a single cylinder rotating in an infinite expanse of fluid. Hence, show that according to the inviscid stability criterion (the Rayleigh criterion) the former is stable and the latter is neutrally stable.

Show that the Rayleigh criterion may be reformulated as indicating that the motion is unstable if the vorticity has the opposite sign to the fluid angular velocity.

59. The tendency for buses on a frequent service to form bunches can be understood as follows: if one bus is slightly delayed, more passengers will accumulate at the next stop and the delay will be increased; another bus that gets slightly early will similarly become earlier still. A simple quantitative model of this can be formulated by considering an infinite homogeneous bus route. If the nth bus is a time t_n late, its additional delay at a stop may be taken proportional to the excess time since the last bus; that is to $(t_n - t_{n-1})$. After ΔN stops

$$\Delta t_n = A(t_n - t_{n-1})\Delta N.$$

If both the buses and the stops are sufficiently close together, t_n can be treated as a continuous function of n and N. Then one may write

$$\partial t_n / \partial N = A(t_n - t_{n-1}).$$

Or, approximating by the first two terms of a Taylor expansion,

$$\partial t_n / \partial N = A \partial t_n / \partial n - \tfrac{1}{2} A \partial^2 t_n / \partial n^2.$$

Consider the evolution in time (i.e. with respect to N) of a disturbance that is periodic in space (i.e. with respect to n).

Discuss the extent to which this treatment is analogous to the methods used for studying hydrodynamic instability.

60. Consider whether any significant part of the boundary layers might be laminar in the cases of (a) a ship of length 100 m sailing at a speed of $10\ \mathrm{m\,s}^{-1}$, (b) a fish of length 0.5 m swimming at $2\ \mathrm{m\,s}^{-1}$, and (c) the aircraft of Question 39.

61. It is planned to carry out a vibrating ribbon experiment to verify the boundary layer stability curve shown in Fig. 17.17 for Reynolds numbers up to 5000, using a wind-tunnel with speeds ranging from 1 to $50\ \mathrm{m\,s}^{-1}$. Owing to

limitations in the electrical and mechanical system, the ribbon can be vibrated only in the range 15–300 Hz. Also, so that the ribbon does not significantly block the flow, it is desired that the boundary layer thickness should be at least 5 mm. Show that the experiment can be performed within these constraints, but only by varying the distance of the ribbon from the leading edge. Determine the range of variation needed.

62. The flow in a pipe is in the regime described in Section 18.3 in which turbulent slugs are produced periodically. The pressure difference is such that, if the flow were entirely laminar, the Reynolds number would be 9000. A slug is generated close to the pipe entry whenever the Reynolds number rises above 6000. Use the information in Figs 2.11 and 18.7 to estimate the fraction of the flow that is turbulent (a) when a slug is just being generated, (b) when the rear of a slug is passing out of the pipe, and (c) when the front of a slug is passing out of the pipe. Estimate also the range over which the Reynolds number oscillates during the cycle. (Ignore entry length effects on laminar and turbulent flow properties.)

63. Formulate an equation for the energy of the mean motion in a turbulent flow (by multiplying eqn (19.13) by U_i) and rearrange the term representing the interaction with the fluctuations to exhibit (a) a loss term equivalent to the production term in the turbulence energy equation, and (b) a term representing energy transport by the turbulence (integrating to zero over the whole flow).

64. Show that eqn (20.1) for the correlation functions in isotropic turbulence implies that

$$\int_0^\infty gr \, dr = 0$$

provided that $f \to 0$ more rapidly than r^{-2} as $r \to \infty$. Indicate why one would expect this result from continuity considerations.

65. (1) The energy dissipation in isotropic turbulence (denoted by ϵ in Section 20.3) depends only on the r.m.s. velocity fluctuation $(\overline{u^2})^{1/2}$ and a length scale l of the large eddies. Use dimensional analysis to infer the form of this dependence.

(2) The length scale η of the dissipating eddies (the Kolmogorov length scale) depends only on the energy dissipation and the kinematic viscosity. Use dimensional analysis to infer the form of this dependence.

(3) Supposing that one requires $\eta/l < 10^{-3}$ for the inertial sub-range to be observable (so that there is a decade in wave number between $10/l$ and $1/10\,\eta$), express this as a condition on the Reynolds number based on $(\overline{u^2})^{1/2}$ and l. (Assume that the proportionality constants involved in parts (1) and (2) are ~ 1.)

(4) With the observations that $(\overline{u^2})^{1/2} \sim U/30$ and $l \sim M/3$, where U is the mean velocity and M is the grid mesh length, determine the minimum UM/ν for an experiment to observe the inertial sub-range. Show that, with the condition that the flow must be incompressible, the experiment requires a very large wind-tunnel.

66. Determine the pattern of two-dimensional, irrotational, incompressible motion produced when a spatially sinusoidal normal velocity is imposed on the boundary of a semi-infinite region of fluid; that is the boundary conditions are

$$v = v_0 \cos kx \quad \text{at} \quad y = 0,$$

$$u, v \to 0 \quad \text{as} \quad y \to \infty.$$

This analysis is the starting point for a theory of the irrotational velocity fluctuations in fluid adjacent to a turbulent region. Explain why analysis of the response to a sinusoidal disturbance is relevant to this, and show that the resulting motion does not generate a Reynolds stress.

67. (1) If the velocity and length scales of an axisymmetric jet are of the form

$$u_{max} \propto x^m; \quad \Delta \propto x^n$$

what are m and n for (a) laminar flow (b) turbulent flow?

(2) If the velocity, length and temperature difference scales of an axisymmetric plume are of the form

$$u_{max} \propto x^m; \quad (T_{max} - T_0) \propto x^p; \quad \Delta \propto x^n$$

what are m, p, and n for (a) laminar flow (b) turbulent flow?

68. The ratio u_τ/U_0 of the wall stress velocity to the free-stream velocity for a turbulent boundary layer on a smooth wall varies with Reynolds number sufficiently slowly that it may be taken as typically 1/30. Estimate the total skin friction (i.e. that part of the drag due directly to viscous forces on the surface) acting on the wings of the aircraft in Question 39.

If the engines have a power of 2000 kW, compare the above drag with the total drag on the aircraft.

For the assumption that the wings are smooth to be valid, any roughnesses must be within the viscous sub-layer. How small does this require them to be?

69. Evaluate the Monin–Obukhov length L for the atmospheric boundary layer over smooth flat ground for conditions in which the wall stress velocity $u_\tau = 0.3 \text{ m s}^{-1}$ and the vertical heat flux $H = 200 \text{ W m}^{-2}$. What are the gradients of the mean velocity and the mean temperature at a height of $L/10$? (Check that the adiabatic gradient may be neglected in these considerations.)

70. (1) Formulate the equation for the mean temperature in a turbulent boundary layer type flow (i.e. apply to eqn (14.12), with $J = 0$, the procedure leading from the Navier–Stokes equation to eqn (19.14)). Show that the turbulence produces a heat transfer across the flow proportional to $\overline{w\theta}$, in the notation of Section 21.7. Consider the physical significance of non-zero $\overline{w\theta}$.

(2) Rederive eqn (19.22) for a flow in which buoyancy forces are significant; i.e. starting with eqn (14.9). Show that for a horizontal mean flow, there is an additional term, proportional to the vertical heat transfer, representing a

production or removal of turbulence energy depending on whether the flow is unstably or stably stratified.

(3) Show that if the Reynolds stress is assumed proportional to the mean velocity gradient and the heat transfer to the mean temperature gradient, then the ratio of the additional term in (2) to the production term, $-\overline{uw}\,\partial U/\partial z$, is essentially the Richardson number.

71. Show that the non-dimensional parameter governing the onset of the variable surface tension instability described in Section 4.5 is $\chi\Delta Td/\rho\nu\kappa$ (the Marangoni number) where χ is the negative rate of change of the surface tension with temperature, ΔT is the horizontally averaged temperature difference across the layer, and other symbols are as in the text.

The critical Marangoni number in the absence of buoyancy effects is 80. The critical Rayleigh number (for free upper surface with no horizontal heat transfer) in the absence of surface tension effects is 670. For a layer of oil with $\rho = 1.0 \times 10^3\,\mathrm{kg\,m^{-3}}$, $\nu = 1.0 \times 10^{-4}\,\mathrm{m^2\,s^{-1}}$, $\kappa = 1.1 \times 10^{-7}\,\mathrm{m^2\,s^{-1}}$, $\alpha = 1.0 \times 10^{-3}\,\mathrm{K^{-1}}$, and $\chi = 6 \times 10^{-5}\,\mathrm{N\,m^{-1}\,K^{-1}}$, estimate the depth at which the predominant destabilizing mechanism changes. What is the minimum temperature difference necessary to produce motion at depths of each of $1/10$ and 10 times the above depth?

72. A variant of Bénard convection is given by having a layer of fluid in which heat J per unit volume is generated uniformly, the top boundary being at a fixed temperature and the bottom one being thermally insulating. Show that the relevant form of the Rayleigh number is $g\alpha Jd^5/\nu\kappa^2\rho C_p$.

The solar radiation falling on the planet Venus has an intensity of $2.6 \times 10^3\,\mathrm{W\,m^{-2}}$. About 70 per cent of this is reflected and about 1 per cent reaches the planet's surface. The remainder is absorbed in a cloud layer, 15–30 km thick. Cellular convection is observed in the sub-solar region of this cloud layer. Suppose that such convection can occur only if the Rayleigh number is in the range 10^3–10^6, and that it is brought into this range by small-scale turbulence producing an effective kinematic viscosity and thermal diffusivity (equal to one another). In what range must this eddy viscosity lie? (For order-of-magnitude purposes, suppose that the atmosphere has the properties of CO_2 at 300 K and $10^5\,\mathrm{N\,m^{-2}}$, and that g is the same as for the Earth.)

73. Estimate the Brunt–Väisälä period for a region close to the ocean surface in which the temperature decreases from 25°C to 15°C and the salinity increases from 3.50 per cent to 3.55 per cent by weight over 100 m increase in depth.

If the salinity change were a decrease instead of an increase, would salt fingering be likely to occur?

If, alternatively, the temperature change were an increase instead of a decrease, would double diffusive layering be likely to occur?

HINTS AND ANSWERS TO PROBLEMS

1. cf. Section 2.2 with G replaced by $\rho g \sin \theta$
   Answer. $\rho g \sin \theta (hy - \frac{1}{2} y^2)/\mu$; $2/3$; $\rho^2 g h^3 \sin \theta /3\mu$.

2. The applied pressure difference must exceed the hydrostatic pressure; the difference between these then produces Poiseuille flow (Section 2.3)
   Answer. 1.02 m; 3.8×10^{-6} m^3 s^{-1}.

3. cf. Section 2.2 with modified boundary conditions.
   Answer. $G > 2\mu U/d^2$; stress is zero.

4. Critical value of $\rho M/\mu^2$ (independent of diameter)

5. If the criteria are taken as being respectively $L_1 < 10^{-3} d = 10^{-5}$ m (with volume L_1^3 containing 100 molecules) and $\lambda < 10^{-3} d$, requirements are $p > \sim 10^{-3}$ N m^{-2} (10^{-8} atmospheres) and $p > \sim 10^3$ N m^{-2} (10^{-2} atmospheres).

6. Geometrically $DV/Dt = \nabla \cdot \boldsymbol{u}$.

7. (1) Follow derivation of (5.59) with (5.52) replaced by $\Delta P/L \sim \mu U/L^2$.
   (2) Procedure similar to that in Section 5.8 gives $\Delta \rho/\rho \sim UL/a^2 \Psi$. For $\partial \rho/\partial t \ll \rho \nabla \cdot \boldsymbol{u}$, one needs $\Delta \rho/\Psi \ll \rho U/L$. Hence requirement is $L^2/a^2 \Psi^2 \ll 1$.

8. Mass flux Q is given by (2.19) with $(p_1 - p_2) = \rho gh$. Also $\rho A\, dh/dt = -Q$ where A is the cross-sectional area of the reservoir. This leads to $h = h_0 \exp(-t/\tau)$ with $\tau = 8vlA/g\pi a^4$. The requirements are (a) $\frac{1}{2} \rho u^2 \ll \rho gh$, and (b) $\partial u/\partial t \ll v\partial^2 u/\partial y^2$; i.e. $u/\tau \ll vu/a^2$.
   Answer. $\tau \ln 2$ (with τ defined above); $gha^4/v^2 l^2 \ll 1$; $ga^6/v^2 lA \ll 1$; 3×10^5 s; yes; yes.

9. The first equation in (6.1) integrates directly; the second integrates after substitution of the first result.

10. (a) Eliminate t; (b) differentiate (x, y, z) to give (u, v, w) and then use original expressions to eliminate (x_0, y_0, z_0); (c) the flow is steady if the previous stage has also eliminated t; (d) is div $\boldsymbol{u} = 0$? (e) integrate (6.1).
   Answer. (1) $(x/x_0) = (y_0/y)^2 = (z_0/z)^2$; $(u, v, w) = (-2x/s, y/s, z/s)$; yes; yes; $x = Ay^{-2} = Bz^{-2}$ (cf. particle path: steady flow). The flow is converging from $\pm x$ directions (e.g. two jets blowing at one another) with axisymmetric divergence.
   (2) $xyz = x_0 y_0 z_0$ and $y = y_0[1 - \frac{1}{2} \ln(x/x_0)]^2$ (or equivalent expressions); $(u, v, w) = (-2x/s, 2y/(t+s), 2zt/s(t+s))$; no; yes; $y = Ax^{-s/(t+s)}$ and $z = Bx^{-t/(t+s)}$ (contrast particle path: unsteady flow). The flow is similar to case (1)

except that the divergence is entirely in the $\pm y$ directions at $t = 0$ and gradually changes to the $\pm z$ directions as $t \to \infty$.

11. Streamline: $y = v_0 \sin(kx)/ku_0$.
Particle path is given by integrating $dx/dt = u$, $dy/dt = v$ with the result of the former being used to put the integrand of the latter in terms of t alone. The result is: $y = v_0 \sin[(k - \alpha/u_0)x]/(ku_0 - \alpha)$.
When $\alpha = 0$ the flow is steady and the two are the same. When $k = 0$ the streamline is a straight line but the particle path is a sinusoid.

13. The equation becomes $\partial u/\partial t = v\partial^2 u/\partial y^2$; cf. Ref. [116].
Answer. (2) Amplitude $= U_0 \exp[-(\omega/2v)^{1/2}y]$; phase angle $= (\omega/2v)^{1/2}y$.
(3) $\rho U_0(v/\pi t)^{1/2}$; $2\rho U_0^2(vt/\pi)^{1/2}$ per unit area; $(1 - 1/\sqrt{2})$ as kinetic energy, the rest dissipated.

14. The non-linear terms in the vorticity equation (6.26) are zero for this geometry. Hence $\partial \zeta/\partial t = v\nabla^2 \zeta$.
Answer.

$$u_\phi(r) = \frac{1}{r} \int_0^r \zeta r' \, dr' = \Phi[1 - \exp(-r^2/4vt)]/2\pi r.$$

15 Dimensional analysis gives $Ga^5/\rho Q^2 = f(\rho Q/a\mu)$.
Answer. $G \propto \rho^{3/4} \mu^{1/4}$.

16. Dimensional analysis gives $P/\rho U^3 L^2 = f(UL/v)$ and the two cases agree in the overlapping range of UL/v. For the larger object $P/\rho U^3 L^2$ is independent of UL/v; this corresponds to the drag coefficient being independent of Reynolds number, as typical of high Re. The results for the smaller object extend to smaller Re, where there are variations with Re.

17. *Answer.* $\mu t/\sigma a^2 = f(\sigma/\rho, \rho\Omega a^2/\mu)$; $\rho\Omega a^2/\mu$ is Reynolds number based on velocity Ωa and length a; (a) if $\Omega = 40 \, \text{rad s}^{-1}$, then $t = 50 \, \text{s}$; (b) no inference is possible. $\mu t/\rho a^2 = f(\rho\Omega a^2/\mu)$; (a) no additional inference; (b) if $\Omega = 80 \, \text{rad s}^{-1}$, then $t = 33.3 \, \text{s}$.

18. $F/\sigma U^2 d^2 = f(l/d, Ud/v)$.
Answer. No prediction; 2.4 N; $\simeq 90$ N.

19. Dimensional analysis gives $A_0^2 n/v = f(F_0/\rho n v)$. Thus for laminar motion $A_0^2 n/v \propto (F_0/\rho n v)^2$; for turbulent motion $A_0^2 n/v \propto F_0/\rho n v$.
Answer. (1) (a) A_0 is decreased by a factor $2\sqrt{2}$; (b) A_0 is decreased by a factor 2; (c) no prediction is possible.
(2) A_0 is decreased by a factor of 2 in all three cases.

20. There are two independent non-dimensional governing parameters, most conveniently taken as $\rho_0 \Delta\rho ga^3/\mu^2$ and $\rho_0 Ua/\mu$. Plot a graph of these (or any

other independent pair) for transition and see which side of the line the new cases lie.

Answer. (1) laminar; (2) turbulent; (3) laminar.

21. Non-dimensionalize the equations as in Section 7.2, with the addition of a magnetic field scale B_0. Three governing non-dimensional parameters emerge: UL/v, $B_0^2/\mu\rho U^2$, and $\sigma\mu UL$.

The second of these is the ratio of magnetic to kinetic energy per unit volume and thus a measure of the importance of magnetic effects. The ratio of the third (the magnetic Reynolds number) to the first (the Reynolds number) is a property of the fluid alone; cf. Pr, Section 14.4.

22. (1) Calculate the rate of fall assuming Stokes's law and then check when $\text{Re} \leqslant \frac{1}{2}$. *Answer.* $d \simeq 1/10$ mm; 100 mm s^{-1}.

(2) *Answer.* $\ll 1/10$ mm.

23. *Answer.* (1) (a) No; (b) yes; $2Cxy$. (2) (a) No; (b) yes; $-2Cxy$.

25. *Answer.* constant $+ \rho\Omega^2 r^2$; constant.

26. Use Bernoulli's equation and mass conservation.
Answer. 40 N m^{-2}.

27. *Answer.* (1) 1.3 m s^{-1}; (2) 0.12 m s^{-1}; (14.20) does not apply when Re is low.

28. Conditions to be fulfilled are: continuity equation; impermeability condition at body surface; velocity tends to uniform u_0 far from body. Pressure distributions are given by Bernoulli's equation: $p - p_0 = \frac{1}{2}\rho u_0^2(1 - 4\sin^2\phi)$ for cylinder; $p - p_0 = \frac{1}{2}\rho u_0^2(1 - \frac{9}{4}\sin^2\theta)$ for sphere.

29. This question assumes knowledge of general solutions of Laplace's equation in spherical polar coordinates. Radial variation associated with P_l is of the form $(Ar^l + B/r^{l+1})$. Hence, in the present case, add a $1/r^3$ term and match to the impermeability condition that $(\partial\Phi/\partial r)_{r=a} = 0$, giving $\Phi = -\frac{1}{2}k(r^2 + 2a^5/3r^3)(3\cos^2\theta - 1)$. Hence,

$$u = \frac{1}{r}\frac{\partial\Phi}{\partial\theta} = 5ka^2\sin\theta\cos\theta \quad \text{over} \quad r = a.$$

Bernoulli's equation then gives

$$p = \text{constant} - \frac{25}{2}\rho k^2 a^4 \sin^2\theta\cos^2\theta.$$

A real flow would involve separation in regions where p is increasing in the flow direction over the surface of the sphere.

30. *Answer.* (1) (a) 3.7; (b) 75. (2) (a) ~ 10 (using inviscid flow solution); (b) $\sim 10^4$.

32. One looks for a solution of the form $u/u_0 = g(y/\delta)$ to

$$u\frac{\partial u}{\partial x} + v\frac{\partial u}{\partial y} = a^2 x + v\frac{\partial^2 u}{\partial y^2}$$

(cf. Section 11.4). This leads to

$$\frac{k^2 a^2 x}{\delta^2} f'^2 - \frac{k^2 a^2 x}{\delta^2} ff'' = a^2 x + \frac{kvx}{\delta^3} f'''$$

and so to the results stated. Boundary conditions: $f = f' = 0$ at $y/\delta = 0$; $f' \to 1$ as $y/\delta \to \infty$.

33. *Answer.* $u = u_\infty[1 - \exp(-v_0 y/v)]$. This applies far enough downstream for the boundary layer to have thickened to $\sim v/v_0$. When $v_0 = 0$, this gives $\delta = \infty$; the boundary layer is thickening throughout the flow (cf. Section 11.4). Another example is the Ekman layer (Section 16.5).

34. Writing $u = ax^{-1/3}\text{sech}^2\eta$ where $\eta = bx^{-2/3}y$,

$$\int u \, dy = \frac{ax^{1/3}}{b}\int \text{sech}^2\eta \, d\eta \propto x^{1/3}$$

$$\int u^2 \, dy = \frac{a^2}{b}\int \text{sech}^4\eta \, d\eta, \quad \text{independent of } x$$

$$\int u^3 \, dy = \frac{a^3}{bx^{1/3}}\int \text{sech}^6\eta \, d\eta \propto x^{-1/3}.$$

35. For $M = 0$, $E < 0$; transport of energy deficit is negative; i.e. more energy is transported downstream than in the absence of the body.

36.
$$D = a\int_{-\pi/2}^{\pi/2}(p_F - p_R)\cos\theta \, d\theta \quad (F = \text{front}, R = \text{rear})$$

$$= 2a(\tfrac{1}{2}\rho u_0^2)\int_0^{\pi/4}[(1 - 4\sin^2\theta) + 1]\cos\theta \, d\theta$$

$$C_D = 2\sqrt{2}/3.$$

37. ρU^2 must be the same.
Answer. 33 m s^{-1}.

38. The solution given is the same as that in Question 28 with an additional term that does not alter the match to any of the conditions. The pressure distribution is given by Bernoulli's equation:

$$p - p_0 = \tfrac{1}{2}\rho u_0^2(1 - 4\sin^2\phi - 2\Gamma\sin\phi/\pi a u_0 - \Gamma^2/4\pi^2 a^2 u_0^2).$$

The pressure difference between corresponding points at the top and bottom, $\Delta p = -2\rho\Gamma u_0 \sin\phi/\pi a$. Integrate the appropriate component of the force due to this to determine the lift.

39. *Answer.* $10\,\mathrm{m\,s^{-1}}$.

40. *Answer.* (a) 400 m; (b) $800\,\mathrm{rad\,s^{-1}}$.

41. (1) Integrate

$$u\partial T/\partial x = \frac{\kappa}{r}\frac{\partial}{\partial r}\left(r\frac{\partial T}{\partial r}\right)$$

with u given by (2.17).
Answer.

$$T - T_{\text{wall}} = \frac{u_{\text{max}}\,\partial T/\partial x}{\kappa}\left[\frac{r^2}{4} - \frac{r^4}{16a^2} - \frac{3a^2}{16}\right]$$

$$H = \frac{kau_{\text{max}}\,\partial T/\partial x}{4\kappa}; \quad \mathrm{Nu} = \mathrm{RePr}/4.$$

(2) Add $J/\rho C_p$ to the right-hand side of the above differential equation. Heat is generated uniformly but advected more rapidly at the centre than at the sides; heat must therefore be conducted towards the centre.
Answer.

$$\Delta T = \frac{Ja^2}{4k} = \frac{a^2 u_{\text{max}}\,\partial T/\partial x}{8\kappa}.$$

42. Integrate $v\partial^2 w/\partial y^2 = -g\alpha[T - (T_1 + T_2)/2] = -g\alpha(T_2 - T_1)y/d$.
Answer. $w = g\alpha(T_2 - T_1)(yd^2/4 - y^3)/6vd$; $\mathrm{Re} = \mathrm{Gr}/72\sqrt{3}$.

43. *Answer.* (1) $\sim 1°$; (2) $\sim 10^4\,\mathrm{m}$; $\sim 100°$.

44. (a) The left-hand side of (15.5) is changed to $-\mu\nabla^2\omega$. (b) Additionally the first term of (15.3) is changed to $-\kappa\nabla^2\rho'$.

45. Fr (eqn (15.7)) must be the same in the two cases. In the atmospheric case, $d\rho_0/dz$ to be used in Fr is that associated with the potential temperature gradient, which is equal and opposite to the adiabatic gradient (eqn (14.55)).
Answer. 2 parts per 100.

46. Harmonics generating waves have angular frequencies $0.3N$, $0.6N$, and $0.9N$. The group velocity is at angle θ to the horizontal and the phase velocity at angle θ to the vertical, where $\theta = 17.5°$, $37°$, and $64°$ respectively. $c_g/c_p = 3.2$, 1.33, and 0.48.

48. A pressure difference between the sides of the channel is needed to balance the Coriolis force; this produces Poiseuille flow through the tube.
Answer. 1/200.

The pressure difference associated with the Coriolis force is balanced by the hydrostatic pressure difference.

Answer. 1 m.

49. The theory of Section 16.5 applies with $\partial p_0/\partial y = 0$ and changed boundary conditions. The result can be shown as a polar diagram like Fig. 16.7 with the origin at the other end of the spiral.

Answer. (smallest distance) $z = \pi(v/\Omega)^{1/2}$.

Components of surface stress: $\mu(\partial u/\partial z)_{z=0}$; $\mu(\partial v/\partial z)_{z=0}$. Components of mass flux: $\int_0^\infty \rho u\, dz$; $\int_0^\infty \rho v\, dz$.

50. *Answer.* (a) unstable when $\Omega < 0.30\,\text{rad s}^{-1}$; (b) unstable when $\Omega > 1.2\,\text{rad s}^{-1}$ $(u_0 = \Omega r/10)$; (c) unstable when $\Omega > 2.4\,\text{rad s}^{-1}$ $(2\Omega u_0 = \Omega^2 r/10)$. Yes in all cases.

51. One requires $u\, \partial u/\partial x \ll \partial u/\partial t$. Using $\partial u/\partial x \sim \gamma v/h$ (eqn (16.53)) and $\partial u/\partial t \sim \omega u$ with eqn (16.60) gives the stated result.

52. $\omega = 0$, corresponding to steady motion: k is perpendicular to $\boldsymbol{\Omega}$, implying no variation parallel to the axis; cf. the Taylor–Proudman theorem.

$\omega = 2\Omega$: k is parallel to $\boldsymbol{\Omega}$, and the motion is in planes perpendicular to it; this relates to the 'displaced particle' argument in Section 16.6.

53. *Answer.* (1) $10\,\text{m s}^{-1}$; (2) $70\,\text{m s}^{-1}$. (The Rossby number is respectively low and high—as has to be checked *a posteriori*—and so (1) involves a Coriolis force/pressure force balance and (2) a centrifugal force/pressure force balance.)

54. (1) Determine the Rossby number when U is given by each of an inertia/buoyancy balance and a Coriolis/buoyancy balance.

Answer. $g\alpha\Theta/\Omega^2 L$.

(2) Addition of a buoyancy force to (16.9) changes (16.13) to

$$\boldsymbol{\Omega} \cdot \nabla u = \nabla \times (g\alpha\Delta T) = \alpha g \times \nabla(\Delta T),$$

cf. discussion of (14.48). Westerly winds involve a balance between gravitational generation of vorticity by the north–south temperature gradient and the twisting of the background vorticity by vertical wind shear.

55. Dimensional analysis gives $\sigma v/r^2 = f(\lambda/r, Sr/\rho v^2)$. The third column has the same value of $Sr/\rho v^2$ as the new case.

Answers. $67.5\,\text{s}^{-1}$; 12.5 mm.

56. Follow the procedures applied to the Lorenz equations in Section 17.3. Note: it is readily seen that, when σ is governed by a quadratic equation, $\sigma^2 + b\sigma + c = 0$, then stability requires both $b > 0$ and $c > 0$.

Answer. $X = 1$, $Y = 1$, stable; $X = 4$, $Y = 2$, unstable.

57. Follow the procedures applied to the Lorenz equations in Section 17.3.
Answer. (1) $X = A/3$, $Y = 0$; (2) $X = D$, $Y = A - 3D$; criterion for existence of solution (2), instability of solution (1), and stability of solution (2): $A > 3D$. Extension: steady solution becomes $X = D$, $Y = A - 3D - F/D$; (a) stable, (b) unstable by overstability, (c) unstable by monotonic growth.

59. Substitute $t_n = t_{n0} \exp(ikn + \sigma N)$. The real part of σ is a function of k and positive.

60. Boundary layers remain laminar for Re_δ up to $\sim 10^3$ and $Re_x \sim Re_\delta^2$.
Answer. (a) No; (b) yes; (c) no.

61. Consider two values of Re close to the ends of the range of interest. (1) $Re = 2000$. The requirement on the boundary layer thickness implies $u_0 < 6\ \text{m s}^{-1}$. The requirement that $\sigma_i v / u_0^2$ goes down to 10^{-4} (Fig. 17.17) implies $u_0 > 4\ \text{m s}^{-1}$. (Note that 15–300 Hz corresponds approximately to $100 < \sigma_i < 2000\ \text{rad s}^{-1}$). $u_0 x / v = (Re/5)^2$ (eqn (11.31)). $u_0 = 5\ \text{m s}^{-1}$ implies $x = 0.5\ \text{m}$. (2) $Re = 5000$. Similar considerations (with $\sigma_i v / u_0^2$ going down to 2×10^{-5}) give $u_0 < 15\ \text{m s}^{-1}$ and $u_0 > 9\ \text{m s}^{-1}$. $u_0 = 12\ \text{m s}^{-1}$ implies $x = 1.25\ \text{m}$.

62. Denote $(p_1 - p_2) d^3 \rho / \mu^2 l$ by C, and let suffices S, L, and T indicate specified values and values in laminar and turbulent regions. Then $l_L + l_T = l_S$ and $C_L l_L + C_T l_T = C_S l_S$ $(=(p_1 - p_2) d^3 / \rho v^2$, which is fixed). From Fig. 2.11, $C_S = 3 \times 10^5$. When a new slug is just forming $C_L = 2 \times 10^5$ and $C_T = 6 \times 10^5$. The above equations then give $l_T / l_S = 0.25$ (*Answer* (a)). From Fig. 18.7, $U_F / u_{av} \simeq 1.5$ and $U_R / u_{av} \simeq 0.6$. After formation of the new slug, the rear of the old slug has to travel $0.25\ l_S$ before leaving the pipe. In this time the front of the new slug travels $0.625\ l_S$ and its rear $0.25\ l_S$. Hence, $l_T / l_S = 0.375$ (*Answer* (b)). When the front of the new slug reaches the end, its rear has traveled $0.4\ l_S$. $l_T / l_S = 0.6$ (*Answer.* (c)).

63. *Answer.*

$$\tfrac{1}{2} U_j \, \partial U_i^2 / \partial x_j = -\frac{1}{\rho} U_i \, \partial P / \partial x_i + v U_i \, \partial^2 U_i / \partial x_j^2 + \overline{u_i u_j} \, \partial U_i / \partial x_j - \partial(U_j \overline{u_i u_i}) / \partial x_j.$$

(a) fourth term; (b) fifth term.

65. *Answer.* (1) $\epsilon \propto u^3 / l$; (2) $\eta \propto (v^3 / \epsilon)^{1/4}$; (3) $ul/v > 10^4$; (4) $UM/v > 10^6$; with $U \ll a$, say $U < 60\ \text{m s}^{-1}$, this implies $M = 0.25\ \text{m}$ and the grid must have many mesh lengths for the tunnel walls not to affect the turbulence.

66. Find the appropriate solution of Laplace's equation for the velocity potential; i.e. $\phi = -v_0 \exp(-ky) \cos(kx)/k$. Turbulent flucutations can be considered as a Fourier superposition of sinusoidal ones. $u \propto \sin kx$ and $v \propto \cos kx$; since $\sin kx \cos kx = 0$, there is no Reynolds stress.

67. cf. two-dimensional flows in Sections 11.6, 14.8, 21.1.
Answer. (1) (a) $m = -1$, $n = 1$; (b) $m = -1$, $n = 1$.
(2) (a) $m = 0$, $p = -1$, $n = \frac{1}{2}$; (b) $m = -\frac{1}{3}$, $p = -\frac{5}{3}$, $n = 1$.

68. *Answer.* 2×10^3 N; 2×10^4 N; 0.05 mm (taking $u_\tau h / \nu < 10$).

69. *Answer.* 12 m; $\partial U / \partial z = u_\tau / Kz = 0.6 \, \text{s}^{-1}$; $\partial T / \partial z = H / \rho C_p u_\tau Kz =$
1.1 K m^{-1} (at $z = L/10$ stratification has little effect); cf. $dT_a / dz \simeq 10^{-2}$ K m^{-1}.

70. (1) $U \partial T / \partial x + W \partial T / \partial z = \kappa \partial^2 T / \partial z^2 - \partial (\overline{w\theta}) / \partial z$. $\rho C_p \overline{w\theta}$ is the heat transfer due to correlation between velocity and temperature fluctuations; e.g. positive $\overline{w\theta}$ implies that rising fluid is typically hotter than falling.
(2) Additional term $g \alpha \overline{w\theta}$ appears on the right-hand side of (19.22) (with z replacing y and w replacing v). This is positive when the heat transfer is upwards, i.e. for unstable stratification.

71. Estimate the depth at which the two critical conditions are reached for the same ΔT.
Answer. ~7 mm; ~20° (due to Marangoni instability); ~2×10^{-3} deg (due to Bénard instability).

72. Either proceed from first principles (cf. Sections 7.2, 14.5) or determine the temperature difference across the layer in the absence of convection and thus the Rayleigh number based on this.
Answer. 10^3 to 3×10^4 m^2 s^{-1}.

73. *Answer.* Period $= 2\pi / N \simeq 400$ s; yes and no respectively (because the density change due to temperature change is larger than that due to salinity change and double diffusive effects require overall stability).

BIBLIOGRAPHY AND REFERENCES

1. Introductory reading

This list includes books which give a good elementary account of fluid dynamical topics in the context of an application.
1. Cottrell, A. H. (1964). *The Mechanical Properties of Matter.* Wiley, London.
2. Sutton, O. G. (1965). *Mastery of the Air.* Hodder and Stoughton, London.
3. Maunder, L. (1986). *Machines in Motion.* Cambridge University Press.
4. Goody, R. M. and Walker, J. C. G. (1972). *Atmospheres.* Prentice-Hall, Englewood Cliffs, NJ.
5. Shapiro, A. H. (1961). *Shape and Flow.* Heinemann, London.

2. Classical texts

References [6] and [7] together still serve as sourcebooks for work done up to the 1930s, the former primarily for theory and the latter primarily for experimental work. The other books, of which Refs. [8] and [10] are also primarily experimental, are less comprehensive but full of useful insights.
6. Lamb, H. (1932). *Hydrodynamics,* 6th edn. Cambridge University Press: Dover, New York.
7. Goldstein, S. (ed.) (1938). *Modern Developments in Fluid Dynamics,* 2 vols. Oxford University Press: Dover, New York.
8. Prandtl, L. (1952). *Essentials of Fluid Dynamics.* Blackie, London.
9. Prandtl, L. and Tietjens, O. G. (1934). *Fundamentals of Hydro- and Aeromechanics.* McGraw-Hill, New York: Dover, New York.
10. Prandtl, L. and Tietjens, O. G. (1934). *Applied Hydro- and Aeromechanics.* McGraw-Hill, New York: Dover, New York.

3. Theoretical texts

Of the books on fluid dynamics written for applied mathematicians or theoretical physicists, the following may be the most useful to readers of the present book.
11. Batchelor, G. K. (1970). *An Introduction to Fluid Dynamics.* Cambridge University Press.
12. Landau, L. D. and Lifshitz, E. M. (1959). *Fluid Mechanics.* Pergamon, Oxford.
13. Lighthill, J. (1986). *An Informal Introduction to Theoretical Fluid Mechanics.* Oxford University Press.
14. Paterson, A. R. (1983). *A First Course in Fluid Dynamics.* Cambridge University Press.

4. Engineering texts

Many books on fluid dynamics have been written for engineers; the following are among those giving extensive treatment of basic fluid dynamics.

15. Daily, J. W. and Harleman, D. R. F. (1966). *Fluid Dynamics*. Addison-Wesley, Reading, Mass.
16. Duncan, W. J., Thom, A. S. and Young, A. D. (1970). *Mechanics of Fluids*, 2nd edn. Arnold, London.
17. Li, W.-H. and Lam, S.-H. (1964). *Principles of Fluid Mechanics*. Addison-Wesley, Reading, Mass.
18. Massey, B. S. (1983). *Mechanics of Fluids*, 5th edn. Van Nostrand Reinhold, Wokingham.

5. Other general texts

The books in this section are less readily classified but are all useful sources of information over a broad range of topics. (Refs. [22] and [23] are broader in their scope than is suggested by their titles.) Reference [21] is an updated version of the original text of Ref. [8]. Reference [25] is a beautiful collection of photographs of fluid flow.

19. Panton, R. L. (1984). *Incompressible Flow*. Wiley, New York.
20. Raudkivi, A. J. and Callander, R. A. (1975). *Advanced Fluid Mechanics, an Introduction*. Arnold, London.
21. Prandtl, L., Oswatitsch, K. and Wieghardt, K. (1984). *Führer durch die Strömungslehre*, 8th edn. Vieweg, Braunschweig.
22. Rosenhead, L. (ed.) (1963). *Laminar Boundary Layers*. Oxford University Press.
23. Schlichting, H. (1979). *Boundary Layer Theory*, 7th edn. McGraw-Hill, New York.
24. Lugt, H. J. (1983). *Vortex Flow in Nature and Technology*. Wiley, New York.
25. Van Dyke, M. (1982). *An Album of Fluid Motion*. Parabolic Press, Stanford.

6. Books on particular branches of fluid dynamics

The books in this section contain introductory material and could be read alongside the present book. (See also Refs. [22] and [23].)

26. Aris, R. (1962). *Vectors, Tensors, and the Basic Equations of Fluid Mechanics*. Prentice-Hall, Englewood Cliffs, N.J.
27. Isaacson, E. de St. Q. and Isaacson, M. de St. Q. (1975). *Dimensional Methods in Engineering and Physics*. Arnold, London.
28. Pankhurst, R. C. (1964). *Dimensional Analysis and Scale Factors*. Chapman and Hall, London.
29. Bradshaw, P. (1964). *Experimental Fluid Mechanics*. Pergamon, Oxford.
30. Atkinson, B. W. (ed.) (1981). *Dynamical Meteorology: an Introductory Selection*. Methuen, London.

31. Holton, J. R. (1979). *An Introduction to Dynamic Meteorology*, 2nd edn. Academic Press, New York.
32. Bradshaw, P. (1971). *Introduction to Turbulence and its Measurement.* Pergamon, Oxford.
33. Tennekes, H. and Lumley, J. L. (1972). *A First Course in Turbulence.* M.I.T. Press, Cambridge, Mass.
34. Monin, A. S. and Yaglom, A. M. (1971). *Statistical Fluid Mechanics: Mechanics of Turbulence,* Vol. I. M.I.T. Press, Cambridge, Mass.
35. Monin, A. S. and Yaglom, A. M. (1975). *Statistical Fluid Mechanics: Mechanics of Turbulence,* Vol. II. M.I.T. Press, Cambridge, Mass.
36. Merzkirch, W. (1987). *Flow Visualization*, 2nd edn. Academic Press, New York.

7. Further books on particular branches

The books here are more specialized than those above.

37. Turner, J. S. (1973). *Buoyancy Effects in Fluids.* Cambridge University Press.
38. Yih, C. S. (1980). *Stratified Flows.* Academic Press, New York.
39. Greenspan, H. P. (1968). *The Theory of Rotating Fluids.* Cambridge University Press.
40. Gill, A. E. (1982). *Atmosphere-Ocean dynamics.* Academic Press, New York.
41. Pedlosky, J. (1979). *Geophysical Fluid Dynamics.* Springer-Verlag, New York.
42. Chandrasekhar, S. (1961). *Hydrodynamic and Hydromagnetic Stability.* Oxford University Press.
43. Drazin, P. G. and Reid, W. H. (1981). *Hydrodynamic Stability.* Cambridge University Press.
44. Hinze, J. O. (1975). *Turbulence,* 2nd edn. McGraw-Hill, New York.
45. Batchelor, G. K. (1953). *The Theory of Homogeneous Turbulence.* Cambridge University Press.
46. Townsend, A. A. (1976). *The Structure of Turbulent Shear Flow,* 2nd edn. Cambridge University Press.

8. Multi-author books on particular branches

The books listed here have all been edited in a way that gives systematic coverage.

47. Zierep, J. and Oertel, H. (eds.) (1982). *Convective Transport and Instability Phenomena.* Braun, Karlsruhe.
48. Meyer, R. E. (ed.) (1981). *Transition and Turbulence.* Academic Press, New York.
49. Swinney, H. L. and Gollub, J. P. (eds.) (1981). *Hydrodynamic Instabilities and the Transition to Turbulence.* Springer-Verlag, Berlin.
50. Bradshaw, P. (ed.) (1976). *Turbulence.* Springer-Verlag, Berlin.

51. Goldstein, R. J. (ed.) (1983). *Fluid Mechanics Measurements*. Hemisphere: Springer-Verlag, Berlin.

9. Films

Many fluid dynamical phenomena are, of course, much better illustrated by cine-films than by still photographs. Of the valuable series of films produced by the National Committee for Fluid Mechanics Films, and distributed by Encyclopaedia Britannica Educational Corporation, the following are the most relevant to this book. Many extracts from these films are also available as short film loops.
52. Shapiro, A. H. *Vorticity*.
53. Taylor, G. I. *Low Reynolds Number Flows*.
54. Abernathy, F. H. *Fundamentals of Boundary Layers*.
55. Fultz, D. *Rotating Flows*.
56. Long, R. R. *Stratified Flow*.
57. Mollo-Christensen, E. L. *Flow Instabilities*.
58. Stewart, R. W. *Turbulence*.
59. Kline, S. J. *Flow Visualization*.

10. Specific references

60. Akiyama, M., Hwang, G. J. and Cheng, K. C. (1971). *J. Heat Transfer* **93,** 335.
61. Andereck, C. D., Liu, S. S. and Swinney, H. L. (1986). *J. Fluid Mech.* **164,** 155.
62. Anderson, A. B. C. (1956). *J. Acoust. Soc. Amer.* **28,** 914.
63. Antonia, R. A. (1981). *Ann. Rev. Fluid Mech.* **13,** 131.
64. Antonia, R. A., Satyaprakash, B. R. and Hussain, A. K. M. F. (1982). *J. Fluid Mech.* **119,** 55.
65. Antonia, R. A. et al. (1983). *J. Fluid Mech.* **134,** 49.
66. Arya, S. P. S. (1975). *J. Fluid Mech.* **68,** 321.
67. Arya, S. P. S. and Plate, E. J. (1975). *J. Atmos. Sci.* **26,** 656.
68. Asanuma, T. (ed.) (1979). *Flow Visualization*. Hemisphere; McGraw-Hill, New York.
69. Atkinson, B. W. (1981). *Meso-scale Atmospheric Circulations*. Academic Press, New York.
70. Badri Narayanan, M. A. (1968). *J. Fluid Mech.* **31,** 609.
71. Bagnold, R. A. (1941). *The Physics of Blown Sand and Desert Dunes*. Methuen, London.
72. Baines, P. G. (1979). *Tellus* **31,** 351.
73. Baines, P. G. and Gill, A. E. (1969). *J. Fluid Mech.* **37,** 289.
74. Baines, P. G. and Hoinka, K. P. (1985). *J. Atmos. Sci.* **42,** 1614.
75. Baker, D. J. (1966). *J. Fluid Mech.* **26,** 573.
76. Bandyopadhyay, P. R. (1986). *J. Fluid Mech.* **163,** 439.

77. Bardsley, W., Hurle, D. T. J. and Mullin, J. B. (eds.) (1979). *Crystal Growth, a Tutorial Approach.* North-Holland, Amsterdam.
78. Barkla, H. M. and Auchterlonie, L. J. (1971). *J. Fluid Mech.* **47**, 437.
79. Barnard, B. J. S. and Pritchard, W. G. (1975). *J. Fluid Mech.* **71**, 43.
80. Barrow, J. et al. (1984). *J. Fluid Mech.* **149**, 319.
81. Barry, M. D. J. and Ross, M. A. S. (1970). *J. Fluid Mech.* **43**, 813.
82. Barsoum, M. L., Kawall, J. G. and Keffer, J. F. (1978). *Phys. Fluids* **21**, 157.
83. Bearman, P. W. (1980). *J. Fluids Engng.* **102**, 265.
84. Becker, H. A. and Massaro, T. A. (1968). *J. Fluid Mech.* **31**, 435.
85. Benjamin, T. B. and Mullin, T. (1982). *J. Fluid Mech.* **121**, 219.
86. Bergé, P. et al. (1980). *J. Physique Lett.* **41**, L-341.
87. Bergé, P. and Pomeau, Y. (1980). *La Récherche* **11**, 422.
88. Bergé, P., Pomeau, Y. and Vidal, C. (1986). *Order within Chaos.* Wiley, New York: Hermann, Paris.
89. Berger, E. (1964). *Jahrbuch 1964 WGLR,* 164.
90. Berger, E. and Wille, R. (1972). *Ann. Rev. Fluid Mech.* **4**, 313.
91. Billing, B. F. (1964). *Thermocouples: their Instrumentation, Selection, and Use.* Inst. Engng. Inspection, London.
92. Binnie, A. M. and Fowler, J. S. (1947). *Proc. Roy. Soc.* A **192**, 32.
93. Blackman, R. B. and Tukey, J. W. (1959). *The Measurement of Power Spectra.* Dover, New York.
94. Blackwelder, R. F. (1983). *Phys. Fluids* **26**, 2807.
95. Bloor, M. S. (1964). *J. Fluid Mech.* **19**, 290.
96. Bourque, C. and Newman, B. G. (1960). *Aero. Quart.* **11**, 201.
97. Bradshaw, P. (1967). *J. Fluid Mech.* **30**, 241.
98. Bradshaw, P. (1981). *Sci. Prog.* **67**, 185.
99. Bray, R. J., Loughhead, R. E. and Durrant, C. J. (1984). *The Solar Granulation,* 2nd edn. Cambridge University Press.
100. Brennan, C. and Winet, H. (1977). *Ann. Rev. Fluid Mech.* **9**, 339.
101. Brorson, S. D., Davey, D. and Linsey, P. S. (1983). *Phys. Rev.* A **28**, 1201.
102. Browand, F. K. and Winant, C. D. (1972). *Geophys. Fluid Dynam.* **4**, 29.
103. Brown, R. A. (1980). *Rev. Geophys. Space Phys.* **18**, 683.
104. Browne, L. W. B., Antonia, R. A. and Bisset, D. K. (1986). *Phys. Fluids* **29**, 3612.
105. Busse, F. H. (1978). *Rep. Prog. Phys.* **41**, 1929.
106. Busse, F. H. (1981). In ref. [49], p. 97.
107. Busse, F. H. and Clever, R. M. (1979). *J. Fluid Mech.* **91**, 319.
108. Busse, F. H. and Whitehead, J. A. (1971). *J. Fluid Mech.* **47**, 305.
109. Busse, F. H. and Whitehead, J. A. (1974). *J. Fluid Mech.* **66**, 67.
110. Caldwell, D. R. and Van Atta, C. W. (1970). *J. Fluid Mech.* **44**, 79.
111. Caldwell, D. R., Van Atta, C. W. and Helland, K. N. (1972). *Geophys. Fluid Dynam.* **3**, 125.
112. Calkins, D. E. (1984). *Amer. Inst. Aero. Astro. J.* **22**, 1216.
113. Cantwell, B. J. (1981). *Ann. Rev. Fluid Mech.* **13**, 457.
114. Cantwell, B. J., Coles, D. and Dimotakis, P. (1978). *J. Fluid Mech.* **87**, 641.
115. Carr, M. H. (1981). *The Surface of Mars.* Yale University Press.

116. Carslaw, H. S. and Jaeger, J. C. (1959). *Conduction of Heat in Solids*. Oxford University Press.
117. Cascais, J., Dilão, R. and Noronha da Costa, A. (1983). *Phys. Lett.* A **93**, 213.
118. Central Electricity Generating Board, Report of the Committee of Inquiry into Collapse of Cooling Towers at Ferrybridge 1 November 1965.
119. Cermak, J. E. (1976). *Ann. Rev. Fluid Mech.* **8**, 75.
120. Chashechkin, Yu. D. and Popov, V. A. (1979). *Sov. Phys. Dokl.* **24**, 827.
121. Cheesewright, R. (1968). *J. Heat Transfer* **90**, 1.
122. Chen, M. M. and Whitehead, J. A. (1968). *J. Fluid Mech.* **31**, 1.
123. Chu, T. Y. and Goldstein, R. J. (1973). *J. Fluid Mech.* **60**, 141.
124. Clark, R. B. and Tritton, D. J. (1970). *J. Zool.* **161**, 257.
125. Clark, R. P. and Edholm, O. G. (1985). *Man and his Thermal Environment*. Arnold, London.
126. Cockayne, B., Lent, B. and Roslington, J. M. (1976). *J. Materials Sci.* **11**, 259.
127. Čolak-Antić, P. (1964). *Sitzungsberichte der Heidelberger Akademie der Wissenschaften, Mathem.-naturw. Klasse, Jahrg.* 1962/4, 6.Abhandlung. Springer-Verlag, Berlin, 313.
128. Coles, D. (1965). *J. Fluid Mech.* **21**, 385.
129. Corrsin, S. (1949). *J. Aero. Sci.* **16**, 757.
130. Corrsin, S. (1959). *J. Geophys. Res.* **64**, 2134.
131. Coutanceau, M. and Bouard, R. (1977). *J. Fluid Mech.* **79**, 231.
132. Cox, R. N. and Clark, R. P. (1973). *Rev. Gén. Therm.* **12**, 11.
133. Creveling, H. F. et al. (1975). *J. Fluid Mech.* **67**, 65.
134. Cvitanovic, P. (ed.) (1984). *Universality in Chaos*. Adam Hilger, Bristol.
135. Davies, J. M. (1949). *J. Appl. Phys.* **20**, 821.
136. Davies, P. A. (1972). *J. Fluid Mech.* **54**, 691.
137. Debler, W. and Fitzgerald, P. (1971). Univ. Michigan, Dept. Engng. Mech., Tech. Rep. EM-71-3.
138. Debler, W. R. and Vest, C. M. (1977). *Proc. Roy. Soc.* A **358**, 1.
139. Delany, N. K. and Sorenson, N. E. (1953). Nat. Adv. Comm. Aero., Tech. Note 3038.
140. Dennis, S. C. R. and Chang, G.-Z. (1970). *J. Fluid Mech.* **42**, 471.
141. Dhawan, S. (1952). Nat. Adv. Comm. Aero., Tech. Note 2567.
142. Dhawan, S. and Narasimha, R. (1958). *J. Fluid Mech.* **3**, 418.
143. Diaz, F. et al. (1983). *Phys. Fluids* **26**, 3454.
144. DiPrima, R. C. and Stuart, J. T. (1983). *J. Appl. Mech.* **50**, 983.
145. DiPrima, R. C. and Swinney, H. L. (1981). In Ref. [49], p. 139.
146. Donnelly, R. J. (1958). *Proc. Roy. Soc.* A **246**, 312.
147. Donnelly, R. J. and Glaberson, W. (1966). *Proc. Roy. Soc.* A **290**, 547.
148. Douglas, H. A., Hide, R. and Mason, P. J. (1972). *Quart. J. Roy. Met. Soc.* **98**, 247.
149. Drain, L. E. (1980). *The Laser-Doppler Technique*. Wiley, New York.
150. Dubois, M., Rubio, M. A. and Bergé, P. (1983). *Phys. Rev. Lett.* **51**, 1446.
151. Durrani, T. S. and Greated, C. A. (1977). *Laser Systems in Flow Measurement*. Plenum, New York.

152. Durst, F., Melling, A. and Whitelaw, J. H. (1981). *Principles and Practice of Laser-Doppler Anemometry,* 2nd edn. Academic Press, New York.
153. Elder, J. W. (1965). *J. Fluid. Mech.* **23,** 77.
154. Elder, J. W. (1965). *J. Fluid. Mech.* **23,** 99.
155. Fabris, G. (1979). *J. Fluid. Mech.* **94,** 673.
156. Faller, A. J. and Kaylor, R. (1967). *Phys. Fluids* **10,** S-212.
157. Farell, C. and Blessman, J. (1983). *J. Fluid. Mech.* **136,** 375.
158. Fenstermacher, P. R., Swinney, H. L. and Gollub, J. P. (1979). *J. Fluid. Mech.* **94,** 103.
159. Finn, R. K. (1953). *J. Appl. Phys.* **24,** 771.
160. Fischer, H. B. et al. (1979). *Mixing in Inland and Coastal Waters.* Academic Press, New York.
161. Fitzjarrald, D. E. (1982). *J. Phys. E., Sci. Instrum.* **15,** 911.
162. Foster, T. D. and Waller, S. (1985). *Phys. Fluids* **28,** 455.
163. Fowler, A. C. and McGuinness, M. J. (1982). *Physica D* **5,** 149.
164. Freymuth, P. (1966). *J. Fluid. Mech.* **25,** 683.
165. Friehe, C. A. (1980). *J. Fluid. Mech.* **100,** 237.
166. Frisch, U., Sulem, P. and Nelkin, M. (1978). *J. Fluid. Mech.* **87,** 719.
167. Fukui, K., Nakajima, M. and Ueda, H. (1983). *Quart. J. Roy. Met. Soc.* **109,** 661.
168. Fultz, D. and Nakagawa, Y. (1955). *Proc. Roy. Soc. A* **231,** 211.
169. Gad-el-Hak, M., Blackwelder, R. F. and Riley, J. J. (1981). *J. Fluid Mech.* **110,** 73.
170. Garbini, J. L., Forster, F. K. and Jorgensen, J. E. (1982). *J. Fluid. Mech.* **118,** 445 & 471.
171. Garg, V. K. (1981). *J. Fluid. Mech.* **110,** 209.
172. Gass, I. G., Smith, P. J. and Wilson, R. C. (eds.) (1971). *Understanding the Earth.* Artemis, Horsham.
173. Gawthorpe, R. G. (1978). *Railway Engr. Internat. (Inst. Mech. Engrs. Lond.)* **3,** part 3 p. 7 and part 4 pp. 38, 41.
174. Gawthorpe, R. G. (1982). *Internat. J. Vehicle Design* **3,** 263.
175. Gerrard, J. H. (1966). *J. Fluid Mech.* **25,** 401.
176. Gerrard, J. H. (1978). *Phil. Trans. Roy. Soc. A* **288,** 351.
177. Giglio, M., Muzatti, S. and Perini, U. (1981). *Phys. Rev. Lett.* **47,** 243.
178. Giovanelli, R. (1984). *Secrets of the Sun.* Cambridge University Press.
179. Glickman, M. E., Coriell, S. R. and McFadden, G. B. (1986). *Ann. Rev. Fluid Mech.* **18,** 307.
180. Goedde, E. F. and Yuen, M. C. (1970). *J. Fluid. Mech.* **40,** 495.
181. Gollub, J. P. and Benson, S. V. (1980). *J. Fluid. Mech.* **100,** 449.
182. Gollub, J. P., Benson, S. V. and Steinman, J. (1980). *Ann. N. Y. Acad. Sci.* **357,** 22.
183. Gollub, J. P., McCarrier, A. R. and Steinman, J. F. (1982). *J. Fluid. Mech.* **125,** 259.
184. Gorman, M., Widmann, P. J. and Robbins, K. A. (1986). *Physica D* **19,** 255.
185. Graebel, W. P. (1969). *Quart. J. Mech. Appl. Math.* **22,** 39.
186. Grant, H. L. (1958). *J. Fluid. Mech.* **4,** 149.

187. Grant, H. L., Stewart, R. W. and Moilliet, A. (1962). *J. Fluid. Mech.* **12**, 241.
188. Grant, R. P. and Middleman, S. (1966). *Amer. Inst. Chem. Engrs. J.* **12**, 669.
189. Grass, A. J. (1971). *J. Fluid. Mech.* **50**, 233.
190. Gray, D. D. and Giorgini, A. (1976). *Internat. J. Heat Mass Transfer* **19**, 545.
191. Greeley, R. and Iversen, J. D. (1985). *Wind as a Geological Process on Earth, Mars, Venus and Titan.* Cambridge University Press.
192. Happel, J. and Brenner, H. (1965). *Low Reynolds Number Hydrodynamics.* Prentice-Hall, Englewood Cliffs, N.J.
193. Head, M. R. and Bandyopadhyay, P. (1981). *J. Fluid Mech.* **107**, 297.
194. Head, M. R. and Surrey, N. B. (1965). *J. Sci. Instrum.* **42**, 349.
195. Hess, F. (1968). *Sci. Amer.* **219**(5), 124.
196. Hide, R. and Ibbetson, A. (1966). *Icarus* **5**, 279.
197. Hide, R., Ibbetson, A. and Lighthill, M. J. (1968). *J. Fluid Mech.* **32**, 251.
198. Hide, R. and Mason, P. J. (1975). *Adv. Phys.* **24**, 47.
199. Hoffmann, P. H., Muck, K. C. and Bradshaw, P. (1985). *J. Fluid Mech.* **161**, 371.
200. Holder, D. W. and North, R. J. (1963). *Nat. Phys. Lab., Notes Appl. Sci.* 31.
201. Howard, L. N. (1963). *J. Appl. Mech.* **30**, 481.
202. Howroyd, G. C. and Slawson, P. R. (1975). *Boundary Layer Met.* **8**, 201.
203. Humphreys, J. S. (1960). *J. Fluid Mech.* **9**, 603.
204. Huner, B. and Hussey, R. G. (1977). *Phys. Fluids* **20**, 1211.
205. Huppert, H. E. and Linden, P. F. (1979). *J. Fluid Mech.* **95**, 431.
206. Huppert, H. E. and Turner, J. S. (1981). *J. Fluid Mech.* **106**, 299.
207. Hurle, D. T. J. (1983). *J. Crystal Growth* **65**, 124.
208. Hussain, A. K. M. F. (1983). *Phys. Fluids* **26**, 2816.
209. Ibbetson, A. and Phillips, N. (1967). *Tellus* **19**, 1.
210. Isataev, S. I. (1985). In Kozlov, V. V. (ed.) *Laminar-turbulent Transition,* Springer-Verlag, Berlin, p. 411.
211. Jayaweera, K. O. L. F. and Mason, B. J. (1965). *J. Fluid Mech.* **22**, 709.
212. Jeffreys, H. (1961). *Cartesian Tensors.* Cambridge University Press.
213. Jeffries, C. and Perez, J. (1983). *Phys. Rev. A* **27**, 601.
214. Johnston, J. P., Halleen, R. M. and Lezius, D. K. (1972). *J. Fluid Mech.* **56**, 533.
215. Jonas, P. R. and Kent, P. M. (1979). *J. Phys. E., Sci. Instrum.* **12**, 604.
216. Jones, A. D. W. (1984). *J. Crystal Growth* **69**, 165.
217. Jones, C. A. (1981). *J. Fluid Mech.* **102**, 249.
218. Joseph, D. D. (1976). *Stability of Fluid Motions,* 2 vols. Springer-Verlag, Berlin.
219. Kachanov, Yu. S. and Levchenko, V. Ya. (1984). *J. Fluid Mech.* **138**, 209.
220. Kim, H. T., Kline, S. J. and Reynolds, W. C. (1971). *J. Fluid Mech.* **50**, 133.
221. Kim, J. and Moin, P. (1986). *J. Fluid Mech.* **162**, 339.
222. Kiya, M. and Matsumura, M. (1985). *Bull. Japan Soc. Mech. Engnrs.* **28**, 2617.

223. Klebanoff, P. S. (1955). Nat. Adv. Comm. Aero., Rep. 1247.
224. Klebanoff, P. S. and Tidstrom, K. D. (1959). Nat. Adv. Comm. Aero., Tech. Note D-195.
225. Klebanoff, P. S., Tidstrom, K. D. and Sargent, L. M. (1962). *J. Fluid Mech.* **12**, 1.
226. Kline, S. J. and Shapiro, A. H. (1965). Film-loop FM-48: *Pathlines, Streaklines and Streamlines in Unsteady Flow* (extract from Ref. [59]).
227. Klinker, T., Meyer-Ilse, W. and Lauterborn, W. (1984). *Phys. Lett.* A **101**, 371.
228. Koh, R. C. Y. and Brooks, N. H. (1975). *Ann. Rev. Fluid Mech.* **7**, 187.
229. Komoda, H. (1967). *Phys. Fluids* **10**, S-87.
230. Kourta, A. et al (1985). *PhysicoChem. Hydrodynam.* **6**, 703.
231. Kovasznay, L. S. G., Kibens, V. and Blackwelder, R. F. (1970). *J. Fluid Mech.* **41**, 283.
232. Kovasznay, L. S. G., Komoda, H. and Vasudeva, B. R. (1962). *Proc. 1962 Heat Transfer and Fluid Mech. Inst.*, Stanford, 1.
233. Kraus, W. (1955). *Messungen des Temperatur- und Geschwindigkeitsfeldes bie freier Konvektion.* Braun, Karlsruhe.
234. Krishnamurti, R. (1970). *J. Fluid Mech.* **42**, 309.
235. Kuo, A. Y. and Corrsin, S. (1971). *J. Fluid Mech.* **50**, 285.
236. Kurosaka, M. and Sundaram, P. (1986). *Phys. Fluids* **29**, 3474.
237. Kurzweg, H. (1933). *Ann. Phys.* (5) **18**, 193.
238. Lachmann, G. V. (ed.) (1961). *Boundary Layer and Flow Control*, 2 Vols. Pergamon, Oxford.
239. Langlois, W. E. (1985). *Ann. Rev. Fluid Mech.* **17**, 191.
240. Laufer, J. (1950). *J. Aero. Sci.* **17**, 277.
241. Lauterborn, W. and Vogel, A. (1984). *Ann. Rev. Fluid Mech.* **16**, 223.
242. Leconte, J. (1858). *Phil. Mag.* (4) **15**, 235.
243. Lee, Y. and Karpala, S. A. (1983). *J. Fluid Mech.* **126**, 91.
244. Lesieur, M. (1982). *La Récherche* **13**, 1412.
245. Lewis, H. E. et al. (1969). *Lancet* **1**, 1273.
246. Libchaber, A. (1985). *Proc. Internat. School Phys. 'Enrico Fermi', Course LXXXVIII* (Soc. Ital. Phys.), p. 18.
247. Libchaber, A., Fauve, S. and Laroche, C. (1983). *Physica* D **7**, 73.
248. Liebster, H. (1927). *Ann. Phys.* (4) **82**, 541.
249. Liepmann, H. W. (1943). Nat. Adv. Comm. Aero., Wartime Rep. ACR 3H30.
250. Lighthill, J. (1969). *Ann. Rev. Fluid Mech.* **1**, 413.
251. Lighthill, J. (1978). *SIAM Rev.* **18**, 161.
252. Linden, P. F. (1978). *J. Geophys. Res.* **83**, 2902.
253. Lindgren, E. R. (1957). *Ark. Fys.* **12**, 1.
254. Lindgren, E. R. (1959). *Ark. Fys.* **15**, 97.
255. Lindgren, E. R. (1959). *Ark. Fys.* **15**, 503.
256. Lindgren, E. R. (1969). *Phys. Fluids* **12**, 418.
257. Lomas, C. G. (1986). *Fundamentals of Hot-wire Anemometry.* Cambridge University Press.
258. Long, R. R. (1972). *Ann. Rev. Fluid Mech.* **4**, 69.
259. Lorenz, E. N. (1963). *J. Atmos. Sci.* **20**, 130.

260. Lumley, J. L. and Panofsky, H. A. (1964). *The Structure of Atmospheric Turbulence.* Interscience, New York.
261. Maccoll, J. W. (1928). *J. Roy. Aero. Soc.* **32,** 777.
262. Madden, R. A. (1979). *Rev. Geophys. Space Phys.* **17,** 1935.
263. Martien, P. et al. (1985). *Phys. Lett.* A **110,** 399.
264. Mason, B. J. (1971). *Nature* **233,** 382.
265. Matsui, T. (1980). in Eppler, R. and Fasel, H. (eds.) *Laminar-turbulent Transition,* Springer-Verlag, Berlin, p. 288.
266. Matsui, T. and Okude, M. (1985). in Kozlov, V. V. (ed.) *Laminar-turbulent Transition,* Springer-Verlag, Berlin, p. 625.
267. Mattingley, G. E. and Criminale, W. O. (1972). *J. Fluid Mech.* **51,** 233.
268. Maurer, J. and Libchaber, A. (1979). *J. Physique Lett.* **40,** L-419.
269. Maurer, J. and Libchaber, A. (1980). *J. Physique Lett.* **41,** L-515.
270. Maxworthy, T. (1970). *J. Fluid Mech.* **40,** 453.
271. Merk, H. J. (1958). *Appl. Sci. Res.* A **8,** 100.
272. Miles, C. A. and Holwill, M. E. J. (1969). *J. Exp. Biol.* **50,** 683.
273. Miles, J. (1984). *Physica* D **11,** 309.
274. Miles, J. (1984). *Adv. Appl. Mech.* **24,** 189.
275. Miles, J. (1984). *J. Fluid Mech.* **149,** 15.
276. Milne-Thompson, L. M. (1962). *Theoretical Hydrodynamics,* 4th edn. Macmillan, London.
277. Ministry of Technology (1968). *Hydraulics Research* 1967, Rep. Hyd. Res. Stat. HMSO, London, p. 24.
278. Ministry of Technology (1970). *Hydraulics Research* 1969, Rep. Hyd. Res. Stat. HMSO, London, p. 11.
279. Moin, P. and Kim, J. (1985). *J. Fluid Mech.* **155,** 441.
280. Moore, D. W. and Saffman, P. G. (1969). *Phil. Trans. Roy. Soc.* A **264,** 597.
281. Mori, Y. and Uchida, Y. (1966). *Internat. J. Heat Mass Transfer* **9,** 803.
282. Mowbray, D. E. and Rarity, B. S. H. (1967). *J. Fluid Mech.* **28,** 1.
283. Muck, K. C., Hoffmann, P. H. and Bradshaw, P. (1985). *J. Fluid Mech.* **161,** 347.
284. Mueller, T. J. et al. (1981). *Amer. Inst. Aero. Astro. J.* **19,** 1607.
285. Müller, D. et al. (1985). *Quart. J. Roy. Met. Soc.* **111,** 761.
286. Mumford, J. C. (1983). *J. Fluid Mech.* **137,** 447.
287. Narasimha, R. and Sreenivasan, K. R. (1979). *Adv. Appl. Mech.* **19,** 221.
288. Newland, D. E. (1975). *An Introduction to Random Vibrations and Spectral Analysis.* Longman, London.
289. Nicholl, C. I. H. (1970). *J. Fluid Mech.* **40,** 361.
290. Nishioka, M. and Asai, M. (1984). in Tatsumi, T. (ed.) *Turbulence and Chaotic Phenomena in Fluids,* North-Holland Amsterdam, p. 87.
291. Nishioka, M., Asai, M. and Iida, S. (1981). in Ref. [48] p. 113.
292. Normand, C., Pomeau, Y. and Velarde, M. G. (1977). *Rev. Mod. Phys.* **49,** 581.
293. Oler, J. W. and Goldschmidt, V. W. (1984). *J. Fluids Engng.* **106,** 187.
294. Ostrach, S. (1953). Nat. Adv. Comm. Aero., Rep. 1111.
295. Owen, P. R. (1964). *J. Fluid Mech.* **20,** 225.

296. Pantulu, P. V. (1962). M.Sc. Thesis, Dept. Aero. Engng., Indian Inst. Sci., Bangalore.
297. Papailiou, D. D. and Lykoudis, P. S. (1974). *J. Fluid Mech.* **62,** 11.
298. Peltier, W. R. (1985). *Ann. Rev. Fluid Mech.* **17,** 561.
299. Pera, L. and Gebhart, B. (1975). *J. Fluid Mech.* **68,** 259.
300. Perry, A. E. (1982). *Hot-wire Anemometry.* Clarendon Press, Oxford.
301. Perry, A. E., Henbist, S. and Chong, M. S. (1986). *J. Fluid Mech.* **165,** 163.
302. Pfenninger, W. (1961). in Ref. [238], p. 970.
303. Phillips, O. M. (1955). *Proc. Camb. Phil. Soc.* **51,** 220.
304. Phillips, O. M. (1963). *Phys. Fluids* **6,** 513.
305. Pilsbury, R. K. (1969). *Clouds and Weather.* Batsford, London.
306. Pitty, A. (1984). *Geomorphology.* Blackwell, Oxford.
307. Platzman, G. W. (1968). *Quart. J. Roy. Met. Soc.* **94,** 225.
308. Pocheau, A. and Croquette, V. (1984). *J. Physique* **45,** 35.
309. Polymeropoulos, C. E. and Gebhart, B. (1967). *J. Fluid Mech.* **30,** 225.
310. Press, F. and Siever, R. (1986). *Earth,* 4th edn. Freeman, San Francisco.
311. Preston, J. H. (1972). *J. Phys. E, Sci. Instrum.* **5,** 277.
312. Priestley, C. H. B. (1959). *Turbulent Transfer in the Lower Atmosphere.* University of Chicago Press.
313. Pritchard, W. G. (1986). *J. Fluid Mech.* **165,** 1.
314. Purcell, E. M. (1977). *Amer. J. Phys.* **45,** 3.
315. Purtell, L. P., Klebanoff, P. S. and Buckley, F. T. (1981). *Phys. Fluids* **24,** 802.
316. Quinn, T. J. (1983). *Temperature.* Academic Press, New York.
317. Ramdas, L. A. and Malurkar, S. L. (1932). *Ind. J. Phys.* **7,** 1.
318. Reynolds, O. (1883). *Phil. Trans. Roy. Soc.* **174,** 935.
319. Richter, F. M. (1978). *J. Fluid Mech.* **89,** 553.
320. Richter, F. M. and Parsons, B. (1975). *J. Geophys. Res.* **80,** 2529.
321. Riley, J. J. and Gad-el-Hak, M. (1985). in Davis, S. H. and Lumley, J. L. (eds.) *Frontiers in Fluid Mechanics,* Springer-Verlag, Berlin, p. 123.
322. Rogallo, R. S. and Moin, P. (1984). *Ann. Rev. Fluid Mech.* **16,** 99.
323. Rollins, R. W. and Hunt, E. R. (1982). *Phys. Rev. Lett.* **49,** 1295.
324. Rosenhead, L. (1954). *Proc. Roy. Soc.* A **226,** 1.
325. Roshko, A. (1961). *J. Fluid Mech.* **10,** 345.
326. Rossby, H. T. (1969). *J. Fluid Mech.* **36,** 309.
327. Rotta, J. (1956). *Ing. Arch.* **24,** 258.
328. Rotta, J. C. (1962). *Prog. Aero. Sci.* **2,** 1.
329. Sachs, P. (1978). *Wind Forces in Engineering,* 2nd edn. Pergamon, Oxford.
330. Saric, W. S. and Thomas, A. S. W. (1984). in Tatsumi, T. (ed.) *Turbulence and Chaotic Phenomena in Fluids,* North-Holland, Amsterdam, p. 117.
331. Sarpkaya, T. (1975). *J. Fluid Mech.* **68,** 345.
332. Sato, H. and Kuriki, K. (1961). *J. Fluid Mech.* **11,** 321.
333. Savaş, Ö. (1985). *J. Fluid Mech.* **152,** 235.
334. Sawyer, R. A. (1960), *J. Fluid Mech.* **9,** 543.
335. Schmidt, E. (1932). *Forsch. Geb. IngWes.* **3,** 181.
336. Schmiedel, J. (1928). *Phys. Z.* **29,** 593.
337. Schubauer, G. B. (1954). *J. Appl. Phys.* **25,** 188.

338. Schubauer, G. B. and Skramstad, H. K. (1947). *J. Res. Nat. Bur. Stand.* **38**, 251.
339. Schubert, G. (1979). *Ann Rev. Earth Planet. Sci.* **7**, 289.
340. Schuster, H. G. (1984). *Deterministic Chaos.* Physik Verlag, Weinheim.
341. Scorer, R. S. (1972). *Clouds of the World.* Lothian, David & Charles, Melbourne.
342. Scorer, R. S. (1986). *Cloud Investigation by Satellite.* Ellis Horwood, Chichester.
343. Seki, N., Fukusako, S. and Inaba, H. (1978). *J. Fluid Mech.* **84**, 695.
344. Shirtcliffe, T. G. L. (1967). *Nature* **213**, 489.
345. Shirtcliffe, T. G. L. and Turner, J. S. (1970). *J. Fluid Mech.* **41**, 707.
346. Skinner, G. T., Dunn, M. G. and Hiemenz, R. J. (1982). *Rev. Sci. Instrum.* **53**, 342.
347. Smith, A. M. O. (1960). *J. Fluid Mech.* **7**, 565.
348. Smith, C. V. (1959). Met. Office, Met. Rep. 21.
349. Smith, G. P. and Townsend, A. A. (1982). *J. Fluid Mech.* **123**, 187.
350. Smith, R. B. (1979). *Adv. Geophys.* **21**, 87.
351. Snyder, H. A. (1970). *Internat. J. Non-linear Mech.* **5**, 659.
352. Soloukhin, R. I., Curtis, C. W. and Emrich, R. J. (1981). in Emrich, R. J. (ed.) *Methods of Experimental Physics* 18B, *Fluid Dynamics*, Academic Press New York, p. 499.
353. Sparrow, C. (1982). *The Lorenz Equations: Bifurcations, Chaos, and Stange Attractors.* Springer-Verlag, New York.
354. Sparrow, E. M., Husar, R. B. and Goldstein, R. J. (1970). *J. Fluid Mech.* **41**, 793.
355. Sreenivasan, K. R. (1985). in Davis, S. H. and Lumley, J. L. (eds.) *Frontiers in Fluid Mechanics.* Springer-Verlag, Berlin. p. 41.
356. Stevenson, T. N. (1968). *J. Fluid Mech.* **33**, 715.
357. Stuart, J. T. (1964). *J. Fluid Mech.* **18**, 481.
358. Stuart, J. T. (1981). in Ref. [48] p. 77.
359. Sun, Y. C. (1984). *Arch. Mech.* **36**, 365.
360. Sutton, O. G. (1953). *Micrometeorology.* McGraw-Hill, New York.
361. Swinney, H. L. (1983). *Physica* D **7**, 3.
362. Szewczyk, A. A. (1962). *Internat. J. Heat Mass Transfer* **5**, 903.
363. Tamai, N. and Asaeda, T. (1984). *J. Geophys. Res.* **89**, 727.
364. Taneda, S. (1956). *J. Phys. Soc. Japan* **11**, 302.
365. Taneda, S. (1965). *J. Phys. Soc. Japan* **20**, 1714.
366. Taneda, S. (1979). *J. Phys. Soc. Japan* **46**, 1935.
367. Taneda, S., Honji, H. and Tatsuno, M. (1974). *J. Phys. Soc. Japan* **37**, 784.
368. Tani, I. (1969). *Ann. Rev. Fluid Mech.* **1**, 169.
369. Tani, I. (1980). in Eppler, R. and Fasel, H. (eds.) *Laminar-turbulent Transition.* Springer-Verlag, Berlin, p. 263.
370. Tanida, Y., Okajima, A. and Watanabe, Y. (1973). *J. Fluid Mech.* **61**, 796.
371. Tanner, L. H. (1966). *J. Sci. Instrum.* **43**, 878.
372. Taylor, G. I. (1923). *Proc. Roy. Soc.* A **104**, 213.
373. Taylor, G. I. (1923). *Phil. Trans. Roy. Soc.* A **223**, 289.
374. Taylor, G. I. (1937). *J. Aero. Sci.* **4**, 311.

375. Taylor, G. I. (1938). *Proc. Roy. Soc.* A **164,** 15.
376. Teitgen, R. (1979). in Eppler, R. and Fasel, H. (eds.) *Laminar-turbulent Transition,* Springer-Verlag, Berlin, p. 27.
377. Thomas, F. O. and Brehob, E. G. (1986). *Phys. Fluids* **29,** 1788.
378. Thompson, J. M. T. and Stewart, H. B. (1986). *Nonlinear Dynamics and Chaos.* Wiley, Chichester.
379. Thorpe, S. A. (1971). *J. Fluid Mech.* **46,** 299.
380. Timme, A. (1957). *Ing. Arch.* **25,** 205.
381. Townsend, A. A. (1949). *Aust. J. Sci. Res.* A **2,** 451.
382. Townsend, A. A. (1959). *J. Fluid Mech.* **5,** 209.
383. Townsend, A. A. (1970). *J. Fluid Mech.* **41,** 13.
384. Townsend, A. A. (1972). *J. Fluid Mech.* **55,** 209.
385. Townsend, A. A. (1979). *J. Fluid Mech.* **95,** 515.
386. Tritton, D. J. (1959). *J. Fluid Mech.* **6,** 547.
387. Tritton, D. J. (1967). *J. Fluid Mech.* **28,** 439.
388. Tritton, D. J. (1983). *Bull. Inst. Math. Appl.* **19,** 19.
389. Tritton, D. J. (1985). *Proc. Internat. School Phys. 'Enrico Fermi', Course* LXXXVIII (Soc. Ital. Phys.), p. 172.
390. Tritton, D. J. (1986). *Eur. J. Phys.* **7,** 162.
391. Tritton, D. J., Rayburn, D. M. and Forrest, M. A. (1980). in Davies, P. A. and Runcorn, S. K. (eds.) *Mechanisms of Continental Drift and Plate Tectonics,* Academic Press, London, p. 267.
392. Tsinober, A. B., Yaholem, Y. and Shlien, D. J. (1983). *J. Fluid Mech.* **135,** 199.
393. Tsoar, H. (1983). *Sedimentology* **30,** 567.
394. Tsuji, Y., Morikawa, Y. and Mizuno, O. (1985). *J. Fluids Engng.* **107,** 484.
395. Turner, J. S. (1985). *Ann. Rev. Fluid Mech.* **17,** 11.
396. Uberoi, M. S. (1963). *Phys. Fluids* **6,** 1048.
397. Unal, M. F. and Rockwell, D. (1984). *Phys. Fluids* **27,** 2598.
398. Van Atta, C. (1966). *J. Fluid Mech.* **25,** 495.
399. Vaziri, A. and Boyer, D. L. (1971). *J. Fluid Mech.* **50,** 79.
400. Veronis, G. (1970). *Ann. Rev. Fluid Mech.* **2,** 37.
401. Vest, C. M. and Arpaci, V. S. (1969). *J. Fluid Mech.* **36,** 1.
402. Walden, R. W. and Donnelly, R. J. (1979). *Phys. Rev. Lett.* **42,** 301.
403. Walshe, D. E. and Isles, D. C. (1969) Nat. Phys. Lab., Aero. Special Rep. 023.
404. Warner, C. Y. and Arpaci, V. S. (1968). *Internat. J. Heat Mass Transfer* **11,** 397.
405. Watson, A. (1985) *Quart. J. Engng. Geol.* **18,** 237.
406. Webb, J. E. (1973) *J. Zool.* **170,** 325.
407. Webster, C. A. G. (1964) *J. Fluid Mech.* **19,** 221.
408. Wei, T. and Smith, C. R. (1986) *J. Fluid Mech.* **169,** 513.
409. Werlé, H. (1973) *Ann. Rev. Fluid Mech.* **5,** 361.
410. Wieselsberger, C. (1921) *Phys. Z.* **22,** 321.
411. Willis, G. E. and Deardorff, J. W. (1965) *Phys. Fluids* **8,** 2225.
412. Willmarth, W. W. (1975) *Adv. Appl. Mech.* **15,** 159.
413. Wortmann, F. X. (1953) *Z. angew. Phys.* **5,** 201.

414. Wygnanski, I. J. and Champagne, F. H. (1973) *J. Fluid Mech.* **59,** 281.
415. Wygnanski, I. J., Sokolov, M. and Friedman, D. (1976). *J. Fluid Mech.* **78,** 785.
416. Yorke, J. A. and Yorke, E. D. (1981) in Ref. [49], p. 77.
417. Zaman, K. M. B. Q. and Hussain, A. K. M. F. (1980) *J. Fluid Mech.* **101,** 449.
418. Zdravkovich, M. M. (1969) *J. Fluid Mech.* **37,** 491.

INDEX